T0141352

About Island Press

Since 1984, the nonprofit organization Island Press has been stimulating, shaping, and communicating ideas that are essential for solving environmental problems worldwide. With more than 1,000 titles in print and some 30 new releases each year, we are the nation's leading publisher on environmental issues. We identify innovative thinkers and emerging trends in the environmental field. We work with world-renowned experts and authors to develop cross-disciplinary solutions to environmental challenges.

Island Press designs and executes educational campaigns in conjunction with our authors to communicate their critical messages in print, in person, and online using the latest technologies, innovative programs, and the media. Our goal is to reach targeted audiences—scientists, policymakers, environmental advocates, urban planners, the media, and concerned citizens—with information that can be used to create the framework for long-term ecological health and human well-being.

Island Press gratefully acknowledges major support of our work by The Agua Fund, The Andrew W. Mellon Foundation, The Bobolink Foundation, The Curtis and Edith Munson Foundation, Forrest C. and Frances H. Lattner Foundation, The JPB Foundation, The Kresge Foundation, The Oram Foundation, Inc., The Overbrook Foundation, The S.D. Bechtel, Jr. Foundation, The Summit Charitable Foundation, Inc., and many other generous supporters.

PEOPLE, FORESTS, AND CHANGE

People, Forests, and Change

Lessons from the Pacific Northwest

Edited by

Deanna H. Olson and Beatrice Van Horne

Library of Congress Control Number: 2016962434

Printed on recycled, acid-free paper ♲

Manufactured in the United States of America

10 9 8 7 6 5 4 3 2 1

Keywords: human-forest ecosystems, dynamic, adaptive management, forest management, ecosystem services, rural communities, socioeconomic, collaborative groups, ecosystem resilience, biodiversity, riparian management, watershed, climate change, wood products, trust

CONTENTS

PREFACE

This book germinated several years ago in recognition of the need to synthesize and reconnect social and ecological perspectives that have emerged over the last few decades in Pacific Northwest moist coniferous forests in light of potential new directions for management. This is clearly a human-forest ecosystem, as people cannot be separated from the ecological components, functions, or processes of the forest. A wealth of new research has followed a regional cross-ownership planning and management effort initiated in 1994. The outcomes of this work provide lessons in successes, failures, and adaptive trajectories, and what we hope is the beginning of a more cohesive vision to meet future management challenges. As we press onward through the twenty-first century, we hope our knowledge compilation for this case study of forest management will provide important context for future forest planning efforts and policy and management changes in the Pacific Northwest and elsewhere.

In navigating the chapters of this book, please note that we envisioned a relatively short synthesis of key selected topics. Hence we restricted chapter length and referencing. The book has four sections: (1) background; (2) the dynamic socioeconomic and ecological context; (3) science advances; and (4) future directions. We use common names of species in the text and provide a species list at the end the book with both common and scientific names.

We thank several key visionaries who have helped us to frame this book's content, including Paul Anderson, Bernard Bormann, Bob Deal, Jerry Franklin, Cheryl Friesen, Norm Johnson, Gordon Reeves, and Tom Spies. We thank all contributing authors for generously donating their time to this effort, in many cases providing comments on others' chapters. We sincerely thank the anonymous peer reviewers of our chapters; the editorial wizardry of Kathryn Ronnenberg and Rachel White; graphic design and organizational assistance by Kathryn Ronnenberg; and assistance with map design by Kelly Christiansen—our product is greatly improved by their keen insights. Funding was provided by the US Forest Service, Pacific

Northwest Research Station. Any mention of trade names does not imply endorsement by the US government. Opinions expressed in this book are those of the authors and do not necessarily reflect the views of the US government.

SECTION 1

Framework for Moist Temperate Forest Management

The last century of forest management has been driven by the increasing momentum to sustain both socioeconomic and ecological systems. Recognition of complex dynamics in social, economic, and forest ecological systems is leading us to novel forest management approaches that address not only sustainability and restoration, but also ecological resilience and resistance. In that context, management of moist coniferous forests has developed into a complex interaction of sociopolitical and science-based decision making. Our goals and objectives have changed over time, differing in space across landownerships with significant interactions that cross ownership boundaries. This section introduces several of these subjects and lays the conceptual foundation for subsequent chapters.

Chapter 1 provides an introduction to the contemporary moist coniferous forest that is managed for combined societal and ecological productivity: the human-forest ecosystem. With a focus on the US Pacific Northwest, we introduce the ongoing debate between commodity production and ecological priorities and how it plays out at a landscape scale. Chapter 2 characterizes the moist coniferous forests of the Northwest region and their dynamics. The productive capacity of these forests, as measured by capacity for biomass production and carbon storage, is among the greatest for temperate forests worldwide. In chapter 3, traditional knowledge and approaches to forest resource management by Native peoples of the Pacific Northwest are framed to provide a historical foundation for an integrated socioecological forest-management system. Chapter 4 addresses the subsequent historical development of forest-based economies in the US Northwest during the twentieth century through the early twenty-first century—development that has produced both regional prosperity and discord. Rural communities have experienced persistent poverty, and there is now a greater political focus on creating the conditions for socioeconomic sustainability and resilience. Chapter 5 describes the rapidly developing value system of forest ecosystem services, spanning wood products, water, biodiversity, and recreation opportunities, and outlines a framework for an ecosystem services–based approach to forest planning.

Chapter 1

Introduction: The Human-Forest Ecosystem

Deanna H. Olson, Beatrice Van Horne, Bernard T. Bormann, Paul D. Anderson, and Richard W. Haynes

Close your eyes while standing in a mature forest along the North Pacific coast of North America in spring and you will smell the wet moss and feel its softness beneath your feet. You might detect the scent of a nearby cedar or hear the long and trembling song of a wren. You would sense the presence of tall and stately conifers nearby, and perhaps the wind lifting and bouncing their rain-heavy branches. Coniferous forests of the North Pacific coast of North America have an air of permanence where they remain intact, although this book will explore their heterogeneity and the uncertainties in their future. Aesthetically they are unparalleled in grandeur, being graced with long days during the growing season, mild winter and summer temperatures, and abundant rainfall. An active geology has been favorable as well: as the Juan de Fuca plate dives under the North American plate, the products of mountain uplift and volcanoes often mix with rocks ground by glaciers to provide young, nutrient-rich substrates. This setting supports forests of extraordinary productivity and biodiversity from northwestern California to Southeast Alaska (fig. 1.1; plate 1; chap. 2) that are greatly valued by their human inhabitants. Abundant moist temperate plants and animals have nurtured some of the wealthiest nonagricultural Native cultures ever known from North America. Visions of this productive landscape drove many of the Europeans migrating along the Oregon Trail early in the nineteenth century, and the forests provided a strong economic engine for the growing region through the twentieth century.

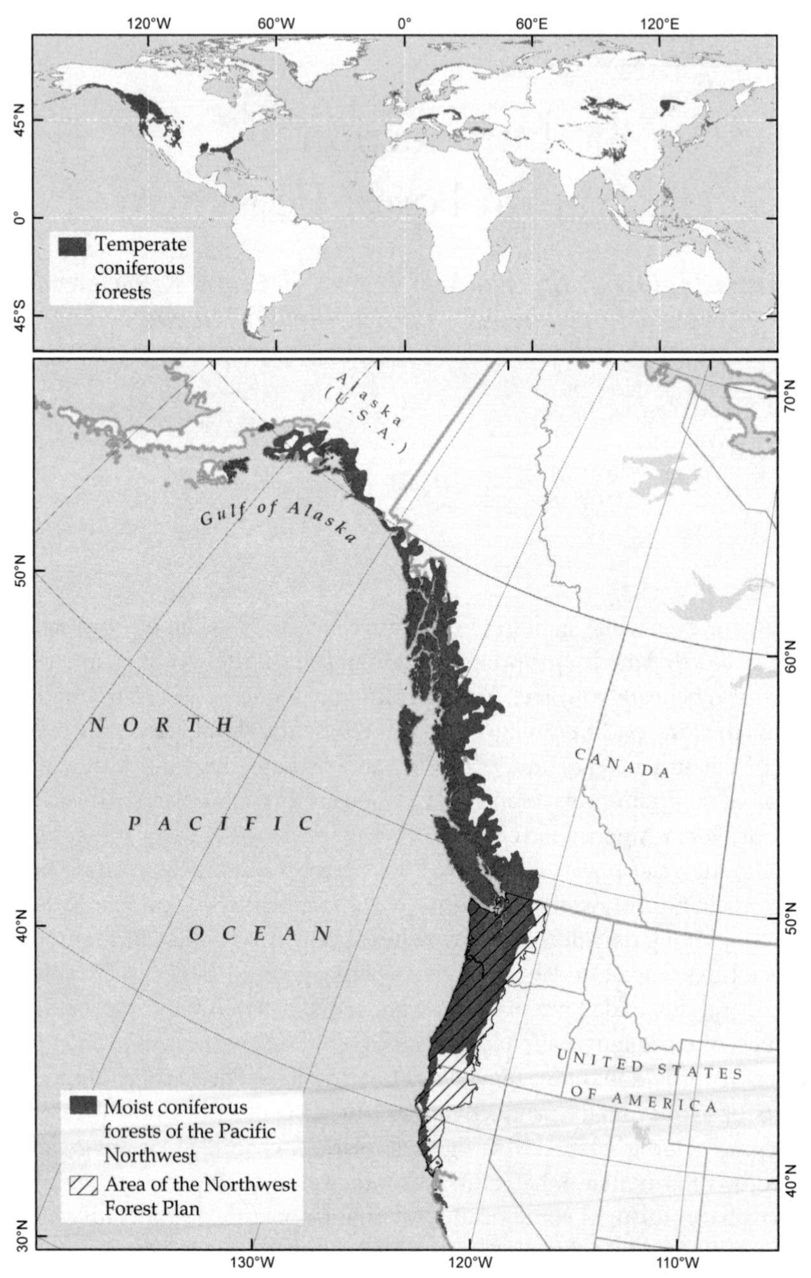

FIGURE I.1. Temperate coniferous forests around the globe (upper figure; data source: Nature Conservancy). Moist coniferous forests of the Pacific Northwest in North America (lower figure; adapted from Little 2013), with hatching indicating the area of the US federal Northwest Forest Plan (USDA and USDI 1994).

These forests not only exert a strong influence on the local culture and sense of place in the Pacific Northwest, but continue to provide wood, water, and other forest products that support local communities and broader economies. People intimately tied to these forests include Native tribes, those employed in wood industries, anglers, recreationists, and anyone who values Northwest forests as a public commons. As old-growth forests diminish globally, their value for ecosystem services is increasingly recognized. Water, wood, rare species, and carbon are dominant flash points in debates about forest management for commodities and services, human stewardship of natural communities, and contributions to stabilizing the greenhouse gases that are affecting world climates. These topics are not restricted to forests and cultures in the Northwest; worldwide there is a new focus on providing forest resources for social and economic sustainability while maintaining ecological integrity.

Debates about forest management are made more challenging by the constant change forests undergo as a result of human activities and natural processes. Active forest management affects the mosaic of forest conditions and its trajectory through time and is itself changing rapidly in response to heightened socioeconomic considerations, altered forest conditions, scientific advances in understanding forest complexity and dynamics, and integration of adaptive management into decision making. Recognition of an accelerated pace of change from human factors as well as altered climate, fire, and pest and pathogen outbreaks triggers new understanding of the importance of human stewardship for retaining a balance among forest types, developmental stages, and resources.

The more we learn about the dynamics of these forests, the more complex and nuanced management guidance has become. Indeed, we are experiencing a paradigm shift in our understanding of the interrelationship of forests and people—people and their values and objectives are now recognized as an intimate component of these systems, the *human-forest ecosystem*. It is time to assess what new perspectives have emerged over the last few decades and how they can contribute to the long-term value of these forests. This book builds on past and current forest ecological, social, and economic contexts and their interplay to derive insights into the development of a new future for moist coniferous forests and their component human communities. Forest sustainability concepts are changing as landscape conditions and socioeconomic conditions change. Lessons learned from the Pacific Northwest of North America bear on processes unfolding in other ecosystems, especially where conflicts among human values and priorities are acute.

Retrospective to Prospective Insights

Northwest forests have a relatively long history of forest uses that mirror cultural phases of development in North America.

How Did the Pacific Northwest Forest Landscape Reach Its Current State?

Several phases of human interactions with Northwest forests can be identified. Native Americans managed forests for multiple resources (chap. 3). A more singular focus on commodity-driven forest management emerged along with the industrialization of softwood lumber manufacturing in the latter half of the nineteenth century, however, along with the development of a railroad-based distribution system for sawn wood. By the early 1900s, forest ownership and management patterns were well established in the Northwest. Private lands consolidated by forest industries have been among the most productive in the world. These timberlands have provided the wood-fiber resource base for a number of forest industrial firms that produce a wide variety of products. Forest industries are dynamic, with changing ownerships, harvest practices, and products. Industrial forest management reflects a heightened responsiveness to the economics of wood markets and to changing social priorities.

Federal lands in the US Northwest have followed a different trajectory. Federal land policy initially revolved around the transfer of land from the public domain into private ownership, and in some cases, from private to public domains. In the late 1800s, much of the land remaining in the public domain was placed in the National Forest System to be managed on behalf of the US public. Nearly half of US Northwest timberland is federally managed. Harvest from federal lands is influenced by sociopolitical factors, which vary over time. For example, during World War I, a special division of the US Army, the Spruce Production Division, worked to supply Sitka spruce to support the war effort. Between World War I and World War II, federal timber harvest was relatively small owing to efforts by the forest industry to maintain stumpage prices high enough to support active management of private timberlands as well as to the relative inaccessibility of the national forests. After World War II, affordable housing for a growing middle class became a priority in a country recovering from war (chap. 4). Federal timber harvest was accelerated to keep wood prices low in support of that goal. With growing prosperity and leisure in the last decades of the twentieth century, however, came a growing awareness of nontimber forest values.

In Search of Sustainability for US Pacific Northwest Forests

Throughout this history there have been advocates for sustainable forest management across multiowner forested landscapes. Regional attempts to promote sustainable forest-management practices and limit harvest on public timberlands started in the 1930s, with the three goals of (1) stabilizing stumpage prices to provide an incentive for forest-management practices on both public and private timberlands; (2) developing sustained-yield calculations as an alternative to earlier cut-and-run practices; and (3) for public timberlands, establishing trust arrangements in which timber revenues substituted for property tax and other revenues. This was also the period in which state regulation of forest practices on private land first began, along with the establishment of the first state forests.

More recently, advocates for sustainability have focused on regulations to balance competing interests for timberlands. With conflicting natural resource priorities at the forefront of debates on forest sustainability goals, the importance of information from scientific research has become elevated as a foundation and rationale for management decisions (box 1.1, 1.2). Researchers, managers, policy makers, and the public have long debated how to integrate science into forest management, and a breakthrough in bridging science and management occurred in the development of the US federal Northwest Forest Plan (FEMAT 1993; USDA and USDI 1994). The Plan was adopted in 1994 as a reaction to judicial injunctions on the sale of federal timber due to threatened species concerns and was seen as a way to manage federal forestlands to meet joint ecological and socioeconomic goals. The approach was implemented by allocating management responsibility to federal agencies (US Forest Service, Bureau of Land Management, National Park Service, Fish and Wildlife Service, and Department of Defense), with different regulations and guidelines depending on the agency and what was on the land, such as streams and adjacent riparian zones, old-growth forest, or logged forest stands. In this context, conservation biology principles guided Plan development for ecosystem management through the designation of large protected areas to anchor ecological system components, functions, and processes. The Plan became a series of policies and guidelines governing federal land use across a vast landscape of ~10 million ha (24 million ac) from northwestern California to northwestern Washington. It was originally conceived with the intent of protecting critical habitat for the northern spotted owl, yet as it developed, the Plan came to include many more species and much broader habitat-protection goals. Because much of this landscape was previously clearcut, and new goals were established for recovery of species listed under the US Endangered Species Act, a new focus on forest restoration (box 1.1) developed as part of this Plan. The Plan amounted to a reset of the strategy for achieving socioeconomic sustainability from the federal forest landscape. Several interacting factors were also at play, however, affecting US Northwest timber socioeconomics. The reduction in wood production from federal lands that had begun before the Plan's implementation from earlier court rulings continued. These reductions targeted larger logs from older stands, which had been a dominant segment of pre-Plan markets. Reductions in federal wood sales were exacerbated by a steep decline in the log-export market (a result of economic struggles in Japan and other Asian countries). These joint factors led to mill closures and job losses in timber-dependent communities, although industrial forests partially compensated by ramping up harvest schedules on private lands. Other major factors driving job losses included mechanized harvest techniques, increased mill efficiency, and other changes in timber markets (chap. 4).

Today's moist coniferous forest landscape, after a couple of decades of Plan implementation, has been profoundly influenced by decisions of diverse owners with different management objectives and concerns. The fixed boundaries within federal ownership and among all landowners, developed or acknowledged in the Plan, were an attempt to simplify management by designating portions of the landscape for different uses and thus simultaneously meeting diverse objectives. But if forest landscapes are dynamic systems with a shifting mosaic of stand conditions created by frequent and irregularly bounded patterns of disturbance, then fixed boundaries become an obstacle to wise management. Such fixed boundaries may produce conflicts among landowners as resource patterns change with time, or as the valuation of critical forest resources changes. Public and private mature forests may be too small and disconnected to manage effectively for a wide array of "old-growth" benefits on each parcel or ownership. Our thinking on boundaries and how to manage resources across them is undergoing rapid evolution. In this light, we explore the emergence of collaborative groups in this volume (chaps. 9, 18). Some of our insights into regional forest management result from large-scale studies (e.g., Poage and Anderson 2007), in addition to the numerous studies at Experimental Forests and Research Natural Areas.

BOX 1.1. SUSTAINABILITY, RESILIENCE, RESISTANCE, AND RESTORATION

Long-term and large-scale environmental changes (e.g., climate change, invasive species, dams) can produce long-lasting fundamental alterations in ecosystem structure and function. The terms *sustainability, resilience, resistance,* and *restoration* are used to describe management objectives and ecological responses to such changing ecosystems, but can be vague in meaning and interpretation.

Sustainability means that an ecosystem will continue to produce the same types and levels of services, such as clean water, biodiversity, carbon sequestration, and timber. Sustainability is often a goal of restoration.

Resilience refers to the ability of an ecosystem to recover its range of components, processes, and functions in the face of disturbance or permanent change, and to the speed with which these return following disturbance. Resilience is a function of the levels of water, nutrients, and energy input to the ecosystem or how many different organisms occur within the system and can substitute for one another in maintaining ecosystem function. Resilience provides buffering that allows managers to learn from the results of their actions when ecosystem processes and functions are monitored. For any ecosystem, there is a threshold to the type or magnitude of disturbance beyond which it cannot maintain or recover its range of states. Restoration can sometimes be used to intervene when a threshold is crossed, or to enhance resilience.

Resistance is the property of an ecological component, function, or process to remain essentially unchanged when subject to a disturbance. In contrast, *sensitive* ecological components (e.g., species) are prone to change when subject to disturbance.

Restoration refers to practices that counter the effects of past human actions or natural disturbances that have degraded or destroyed ecosystem function, and that promote or create resilient and sustainable landscapes in an era of global environmental change.

Forest management has been simplified by establishing uniform regulations attached to each type of ownership. Such an approach fails to acknowledge how much forest type, conditions, and context vary across a landscape, with differences in soil type, moisture availability, latitude, and topography. For example, the same forest treatment at two different places in a similar forest type may produce very different results, but broad regulations are not likely to acknowledge such differences. Perhaps one of the most gradually and painfully learned lessons from the practice of forestry is

BOX 1.2. THE SCIENCE UNDERLYING SUSTAINABILITY

Success in achieving sustainability through time depends on the science used to determine what is sustainable. Traditionally, ecosystem science has considered humans as extrinsic to the system, perhaps acting as a source of disturbance or nutrient or pollution inputs. For example, the classical ecological approach to understanding how forests and stream systems are affected by disturbances such as logging or fire is to apply treatments to a series of plots and to then document the response of the forest over time. The first half century of terrestrial ecosystem research focused on ecological function with and without human influence. Additionally, considerable past forest science research addressed promoting efficiency of harvest systems. Initially this work used stands of uniform age or treatment or plots within stands; it then progressed to larger, more variable areas such as small catchments and watersheds. Beginning in 1908, the US Forest Service established a network of experimental forests with the initial intent of examining effects of tree cutting on hydrology. The value of having specific areas where multiple stand- and watershed-scale experiments could be designed and conducted was obvious, and studies expanded to include the effects of management on nutrient cycling and many other ecological functions. Insights gained from such experiments can inform policies developed to enhance sustainability across larger landscapes. But these policies also respond to other human concerns and interests that are outside the bounds of traditional ecological science. Current approaches to understanding and predicting landscape change share a goal of integrating social and ecological elements of the ecosystem to consider a far greater range of human involvements in the design of experiments. Managers, scientists, and others now recognize that the full ramifications of management alternatives need to be considered beyond the bounds of experimental forests and across the regional landscape.

the requirement that management be tailored to site characteristics. Geographically broad, unchanging standards and guides, or golden rules, adopted as a compromise with the argument that we can only track simple things such as a fixed maximum age or the diameter of Douglas-fir that can be thinned from a stand, surely are a violation of this lesson. Equally problematic are the dramatic differences in stream and riparian management objectives as streams cross ownerships.

Is there a more effective way to manage these forests to meet multiple objectives while adapting to, rather than working against, the dynamic nature of the forests through time and space? To inspire new conversations to address that question, we offer this book as a compendium of what the

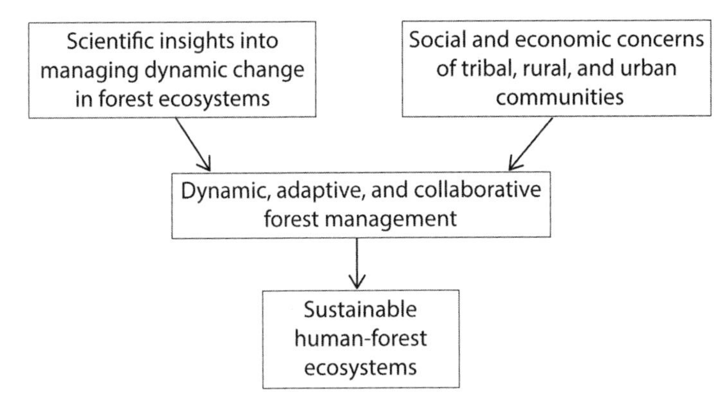

FIGURE 1.2. A current view of management in moist coniferous forests. New social, economic, and scientific understandings and priorities drive a more collaborative and adaptive approach that improves outcomes for human and ecological communities.

authors view as key lessons learned about landscape dynamics in moist coniferous forests, focusing on the two-plus decades since the Plan took effect in 1994. Of course, there is much that we still do not know about the complexity of interactions that influence people and their landscapes. It is important that we admit what we do not know and acknowledge that our knowledge base will continue to develop. Rather than assuming the results of management activities or regulations, scientists can work with managers to learn about what happens before and after management activities take place or regulatory policies are adopted. In this way, both can learn and adapt. Adaptive management is a baseline premise of dynamic system approaches, and insights from a variety of field trials will contribute to learning (chap. 8).

The challenge then is to redefine measures of dynamic sustainability for this landscape and its people—the human-forest ecosystem. We know that forests and linked rural and urban communities are in flux, with constantly changing economic and ecological values and products, as well as overlying natural disturbances. In recognition of the system's dynamism, an adaptive-management framework that more fully integrates changing knowledge and conditions is warranted (fig. 1.2). We envision this changing landscape supporting a diverse biota and a variety of human economic, recreational, and natural-resource uses through time, if it is managed holistically—that is, with an understanding of how decisions today will affect the mosaic of ecosystem services tomorrow. As parts of the landscape change through ecological succession or various disturbances, the whole landscape will continue to provide the services we value and expect. Ecosystem-management goals are retained, yet a multistate system across the landscape develops.

Central to this vision is a new perspective on what sustainability means, which has emerged in part from work in developing countries, where it is apparent that wildlife conservation must benefit local people in order to be sustainable. With local people benefiting by participating in collective ownership processes of forest-associated materials, uses, and values, their stewardship to mediate disturbances (e.g., overexploitation, risk of fire) becomes more effective. People are part of the ecosystem, and their well-being must be considered as ecological patterns and processes are shaped to achieve management goals. This human-forest ecosystem is composed of (1) rural ecosystems, including forested areas and people in local communities dependent upon them; (2) urban ecosystems, including the lands and people in multiple local urban-suburban communities who work together to create a city; and (3) regional ecosystems, which combine the above and contribute to regional sustainability.

A broadly inclusive philosophy has the potential to get people to think first about what they want to get from the forest before debating details of individual silvicultural systems or piecemeal objectives. They can then use scientific information to project the effects of the proposed resource management and use. This order of debate has the real potential to allow a host of creative and individualized management strategies to emerge and contribute to better solutions. Recognition of the divergent priorities among landowners and the autonomy afforded US private forest landowners is part of such inclusive collaborative stewardship planning. A foundational question is, What are desired distributions among an array of ecological, societal, and local community well-being benefits, and are such desired distributions sustainable?

Overview of This Volume

In this book, we aim to synthesize the relevant societal dynamics and scientific advances that beg for a new management and conservation paradigm—one that fully acknowledges the dynamic nature of the landscape and the need to consider both ecological and community well-being at multiple geographic and temporal scales. For the sake of brevity and cohesion, we focus primarily on moist forests of Oregon and Washington. However, in various chapters we provide selected examples from the moist forests of Alaska, Canada, and Northern California; in a few instances, due to common socioeconomic and ecological issues and challenges, we reference relevant examples from drier forests of Oregon and Washington.

In a broad sense, our more than two-decade "experiment" with the Northwest Forest Plan on federal forests in the Pacific Northwest moist coniferous zone has yielded success, failure, and surprises. Lessons learned continually feed into the emerging human-forest integrated system, with incremental and pulsed changes for management and policy. After several decades it is time to pause and reflect on how to refine approaches to human-forest sustainability.

In section I, we present the background for understanding these Northwest moist forests, their historical management contexts, and the outcomes for the landscape, its ecosystems, and its stakeholders. The history of forest management is partly a history of the changing human societies of the Northwest and of how people have variously depended on the forests in the development of their economies and cultures. We also frame the developing concept of ecosystem services, which is used for contemporary prioritization of natural resource management relative to the goods and services valued by today's society.

In section II, we explore the dynamic nature of forest management, including the diverse and changing objectives of forest owners and managers, socioeconomic processes affecting natural resource management decisions and governance, and how these interact with both deliberate and uncontrolled disturbances across the landscape. We can then more fully explore how management can be designed and implemented to better incorporate multiple societal aims and scientific uncertainties.

In section III, we highlight research advances on forest ecosystem structures, functions, and processes and their management. Science has accelerated at an extraordinary pace over the last two decades. In particular, the sciences of silviculture, vegetation ecology, long-term productivity and belowground processes, carbon, biodiversity, aquatic-riparian ecosystems, and watershed and landscape ecology have burgeoned. New knowledge documenting the multistate system of the forest landscape and consideration of multiple disturbances to forest ecosystems occurring at unprecedented levels have contributed to the rapid development of new forest management models. These go beyond classical concepts of sustainability and restoration to include the resilience and resistance (box 1.1) of forest components to change. In this section, we describe the development of these concepts from empirical data and models, and how they are now shaping our view of a dynamic ecology and management of these landscapes.

In section IV, we consider emerging topics framing the future of forest landscape management, in which reconnecting people and forest habitats are of paramount importance. Rapidly developing arenas for forest management now include knitting back into a more seamless ecosystem a landscape that has diverged into a bimodal (frequent timber harvest and no timber harvest) pattern, managing trust issues among stakeholders with diverse forest objectives and priorities to reconcile conflicts, and creating climate-smart forests. Our consideration of alternative futures includes the rapid development of novel forest products and incentives for managing for carbon, water, and biodiversity that are changing our valuation of forest systems and the markets to which they contribute. We explore the need to incorporate these developments, along with diverse objectives, scientific uncertainty, adaptive management, and long-term sustainability, into forest planning processes.

We hope that this book will contribute to a new and better future for these and other forests and landscapes, and all of the people dependent on them, wherever they may live. We believe that science and management trajectories in Northwest moist coniferous forests resonate across ecosystem boundaries, with ripple effects for new socioecological approaches to broader natural resource management.

Literature Cited

FEMAT (Forest Ecosystem Management Assessment Team). 1993. *Forest ecosystem management: An ecological, economic, and social assessment. Report of the Forest Ecosystem Management Assessment Team*. Portland, OR: US Department of Agriculture; US Department of the Interior [and others].

Little, E. L. Jr. 2013. *Digital representations of tree species range maps from "Atlas of United States Trees" by Elbert L. Little, Jr. (and other publications)*. United States Department of Interior, US Geological Survey, Geosciences and Environmental Change Science Center. http://esp.cr.usgs.gov/data/little/.

Poage, N. J., and P. D. Anderson. 2007. *Large-scale silviculture experiments of western Oregon and Washington*. General Technical Report PNW-GTR-713. Portland, OR: USDA Forest Service, Pacific Northwest Research Station.

USDA and USDI (US Department of Agriculture and US Department of the Interior). 1994. *Record of decision for amendments to Forest Service and Bureau of Land Management planning documents within the range of the northern spotted owl* [plus Attachment A: Standards and Guides]. [Place of publication unknown]:

US Department of Agriculture and US Department of the Interior. http://www.reo.gov/library/reports/newroda.pdf.

Wilson, B. T., C. W. Woodall, and D. M. Griffith. 2013. Imputing forest carbon stock estimates from inventory plots to a nationally continuous coverage. *Carbon Balance and Management* 8:1. doi:10.1186/1750-0680-8-1.

Chapter 2

Setting the Stage: Vegetation Ecology and Dynamics

Jerry F. Franklin, Thomas A. Spies, and Frederick J. Swanson

The moist coniferous forests of the Pacific Northwest are notable for the dominance of long-lived evergreen conifers, productivity, and the massiveness of the older forests (Franklin and Dyrness 1988; Franklin and Halpern 2000) (plate 2A). The environment of this region is extremely favorable to forest growth, with its moderate temperatures, high precipitation, and relatively fertile soils. The dominance of evergreen conifers is unusual, however, as most moist temperate forest regions of the world are dominated by hardwoods (angiosperms) (e.g., Kuchler 1946; Askins 2014). The climate allows photosynthesis to occur on essentially a year-round basis, which has been identified as one factor favoring evergreen conifers (Waring and Franklin 1979). Although the high productivity of the forests is important, the massiveness of the older forests is as much the consequence of the dominance of tree species present in these forests: for example, Douglas-fir is capable of continuous growth over many centuries and also produces decay-resistant heartwood, which persists for several more centuries. Hence, older forests have accumulations of organic matter (biologically sequestered carbon) that are among the greatest in the world (plate 1B).

Douglas-fir is often identified with this moist forest region because of its overwhelming importance in both natural and managed forests. Western hemlock is the most constant associate. Significant tree diversity is present, however, and much of it is associated with subregional climate variability. For example, common low-elevation associates of Douglas-fir and western hemlock include western redcedar, grand fir, Pacific yew, red alder, big-leaf maple, black cottonwood, and Pacific dogwood. Oregon white oak and Pacific madrone occur in drier lowland areas. Sitka spruce is found primarily near the Pacific Ocean and in river valleys with strong marine fog influences, and the coastal subspecies of lodgepole pine is encountered on dunes along the coast. Noble fir (in the Cascade Range) and Pacific silver fir (in both the Cascade Range and Olympic Mountains) are common additions in midelevation forests. Western white pine is a less common but widespread associate occurring in both lowland (e.g., Puget Trough) and montane environments. Douglas-fir largely drops out in the subalpine forests, with their heavy and persistent winter snowpack. In these forests mountain hemlock and Alaska yellow-cedar join Pacific silver fir as a dominant species; subalpine fir, Engelmann spruce, and lodgepole pine may also be present. Some trees more characteristic of warmer and drier regions appear in southerly parts of the region, including sugar pine, incense cedar, ponderosa pine, tanoak, canyon live oak, and California black oak.

The moist coniferous forests we see today are a consequence of natural disturbance regimes and forest developmental patterns as well as human activities since European settlement of the region. We begin by providing a generalized natural history of the forests, including disturbance regimes and the legacies that they leave. We follow with consideration of modern human influences on the forest. We finish with speculations about the potential vulnerabilities of these extraordinary forests, including those associated with climate change and invasive species.

Disturbance Regimes

Prior to European settlement, the primary large-scale disturbance agents in Northwest moist forests were wildfires and marine-generated windstorms. They created a diversity of forest age classes and conditions by disrupting existing forests sufficiently to allow for successful natural regeneration of Douglas-fir and associated tree species. Other less common disturbances included volcanic eruptions (which also provided massive aerial deposition of soil parent materials), floods, and landslides. Insect-generated disturbances were (and are) surprisingly rare; the Douglas-fir bark beetle is not known to be a stand killer, and large-scale outbreaks of defoliating insects are very uncommon. Fungi are important in the mortality of individual trees and creation of gaps or forest openings.

Large, infrequent wildfires (100- to 400-year intervals) are the dominant natural disturbance agent creating large forest patches. These wildfires vary greatly in their severity, burning over large areas and often over long time periods (weeks or even months), during which burning conditions vary greatly. These events always include a significant high-severity stand-replacement component (killing >90% of the trees) as a consequence of the immense accumulations of fine fuels (e.g., duff, litter, branches, shrubs, and small trees) and extreme burning conditions that are associated with these large fires. However, extensive areas of low- and moderate-intensity burn are also created during these events, which create partial stand-replacement conditions. It is also important to note that even in areas of high-severity fire, some individuals and clusters of live trees usually survive along with unburned patches.

Fire return intervals have also varied across a broad gradient in the moist coniferous forest region. There is evidence that intervals were broadly similar in western Washington and the northern Oregon Cascade and Coast Ranges (200 to 400 years). However, there is also evidence for a shift to shorter fire return intervals south of about 44.5° north latitude (south of the Santiam River in the Western Cascades of Oregon), which may be related to both a greater incidence of lightning and a drier climate. There was also significant localized variability in fire occurrence associated with Native American activities (such as those associated with travel routes and foraging grounds) (Agee 1993) and, since the early nineteenth century, with European settlement.

Areas growing back following wildfires are hypothesized to be more vulnerable to additional wildfire for several decades afterward, when dead fuels (shrubs, branches, small trees) accumulate during natural thinning, the heights of live tree crowns are low, closing canopies create more homogeneous canopy fuels, and snags (dead standing trees) created by the first fire are still present; snags are major contributors to fire spread. Extensive areas burned in large historic wildfires, such as the Yacolt Burn of 1902 and Tillamook Burn of 1933, suffered repeated wildfire (reburns) in the decades following the original fires. Young forests are more vulnerable to burning than mature and old forests, which have trees with thicker bark, higher canopy-base heights, and more compact and moist surface fuels than younger forests. Repeated burns can create challenging environments for natural tree regeneration owing to competition from aggressive fire-resilient communities of shrubs and herbs, which survive fires and reproduce vegetatively (e.g., vine maple and beargrass); some of the outstanding, culturally important huckleberry fields of the region are products of repeated fire (Richards and Alexander 2006).

Old-growth Douglas-fir–western hemlock forests are highly resistant to fire, as it is difficult for wildfires to initiate and spread in the cool, moist, and calm interior conditions characteristic of these dense forests. Certainly it can and does happen. However, early foresters and lumbermen referred to them as "asbestos forests," because of the difficulty of fires successfully igniting and spreading in old growth. Essentially all of the major post-settlement fires that burned old growth in northern Oregon and western Washington are believed to have started outside of these forests and burned into them, generally during the weather pattern in which hot and very dry east winds blow across the Cascade Range into western Washington and Oregon.

Ocean-generated windstorms are the other major natural stand-disrupting agent in the moist coniferous forests of the Northwest. These storms generally move onto the continent from the southwest and track north and east into the interior of the region. The Columbus Day Windstorm of 1962 was the largest single disturbance event in recorded history in this forest region. It affected essentially all of the coastal mountain systems (Oregon Coast Range, Willapa Hills, and Olympic Mountains), as well as much of the western slope of the Cascade Range in northern Oregon and western Washington. The 1921 Blow on the west side of the Olympic Peninsula was another historic windstorm. There is a strong west-east gradient in the importance of windthrow (trees felled by high winds), with wind having its greatest impact in the near-coastal forests, where it may have been the primary natural disturbance agent. Henderson et al. (1989) suggested that stand-opening windstorms returned at about 200-year intervals on the Olympic Peninsula. East winds produce more localized tree mortality in the Cascade Range, occasionally resulting in stand-replacement levels of damage, such as in the vicinity of the Columbia Gorge (Sinton et al. 2000).

Wildfire and windstorm contrast significantly in their effects on subsequent forest conditions. Douglas-fir is the species most likely to survive wildfire and remain as an in situ source of tree seed. Postfire conditions—well-lighted, extensive mineral soil seedbeds—also favor Douglas-fir over its major tree competitors. Hence subsequent stands, especially in the drier parts of the region, have a high probability of being dominated by Douglas-fir relative to other species: for example, western hemlock seedlings are much smaller than those of Douglas-fir, which makes them more sensitive to the summer heat and drought characteristic of much of the region. On partially burned areas, where a significant overstory (upper tree canopy) remains, western hemlock may dominate the new cohort. Many old-growth forests throughout the region have experienced this developmental pattern.

Windstorms, on the other hand, uproot and break overstory trees, but still leave behind significant areas of undamaged understory. This understory often includes abundant seedlings and saplings of western hemlock and other shade-tolerant species, which grow rapidly in response to increased light and moisture. Hence, windstorms favor postdisturbance forests dominated by western hemlock and other shade-tolerant tree species and provide only limited opportunities for Douglas-fir regeneration.

In addition to these major forest disturbances, small-scale within-stand disturbances kill individual trees and small tree groups, producing openings in the overstory. These disturbances and the resulting canopy gaps are very important in creating and maintaining the complex structural conditions characteristic of older moist coniferous forests. The disturbance agents include insects (e.g., bark beetles), pathogens (e.g., bole, butt, and root rots), snow and ice, and wind (uprooting and breaking individual or small groups of trees). These will be discussed in the following section on forest stand development.

Forest Stand Development

Development of a moist Douglas-fir–western hemlock forest following a major partial or complete stand-replacement disturbance is a long and complex process, which culminates in an old-growth forest, provided another major disturbance doesn't intercede first. We call the developmental sequence a *sere* (fig 2.1). Under historical conditions in the moist forests, the old-growth stage was relatively common on the landscape (>40%–60% of the region at any one time). This dominance of older forests suggests that over millennia, stand-replacing fires were infrequent enough that forests were able to develop for centuries following large disturbances. These forests were resilient to natural disturbance agents. That resilience was expressed at both local scales (forests with tall canopies and Douglas-fir trees with thick bark), but especially at large spatial scales (landscapes and watersheds) and over long time frames (centuries). The development of these

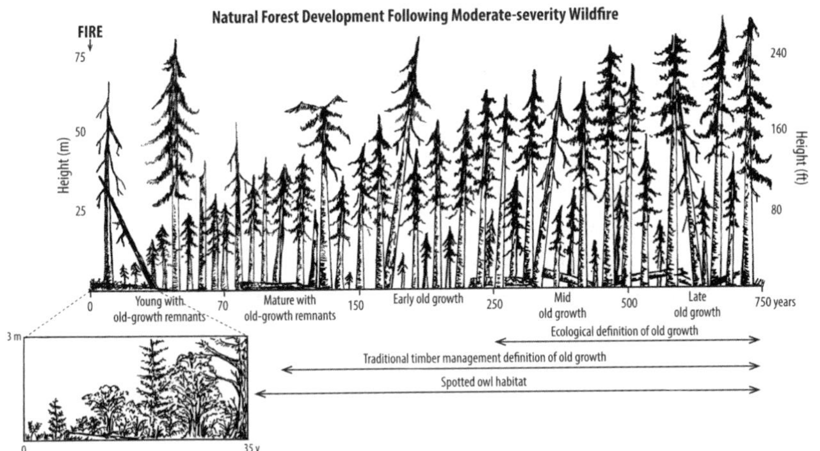

FIGURE 2.1. Moist forest seres—the forest developmental sequence following a disturbance—may go on for many centuries. These seres undergo gradual change through pre-forest, mature forest, and old-forest stages. Old-growth conditions typically take 200 to 250 years to develop, depending upon site productivity. Inset shows the pre-forest stage, also known as the early-successional stage, which is the most biologically diverse in terms of species richness. Art by Laura Hauck, USFS.

forests in the aftermath of intense disturbances follows multiple pathways toward increasing biomass and structural complexity. The pathways, which are gradual and continuous, differ in terms of rate, structure, and composition. No two old-growth forests are exactly the same, just as no postdisturbance conditions are the same.

For purposes of understanding the developmental process, we examine a simple linear pathway that we partition into four stages following the initiating disturbance: pre-forest, young forest, mature forest, and old forest. We will assume that the initiating event is a complete stand-replacement event (it killed most or all of the existing live trees), and that partial disturbances do not occur during the centuries that it takes the sere to develop; there are many alternative initiating disturbances and developmental patterns that we could show, but this is an important region-wide example. Finally, we assume the disturbance event was a wildfire, but we will also make some comparisons with a wind-generated sere.

We have included considerable detail on the sere and its developmental stages because it is important for forest managers, researchers, and stakeholders to understand and appreciate the differences between these stages and the unique roles that each stage plays in sustaining regional forest functions, including provision of habitat for wildlife and other biodiversity. This also helps emphasize that in order to sustain regional forest values, all the stages need to be represented—"we need the whole sere!" Historically, nature provided for disturbance regimes that produced all the stages—the whole sere. A forest landscape composed solely of plantations and old-growth forests would not be complete in the sense of providing habitat for all of the native biota, nor in terms of the important structural conditions and processes.

Disturbance and Legacy-Creation Event

The wildfire that initiates this sere kills most of the dominant trees as well as consuming the litter layers and ground vegetation. Although most of the trees are killed, most of the organic matter remains as standing dead trees (snags) and down logs and as soil organic matter; usually less than 5% of the organic matter is actually consumed in a wildfire. Similarly, much of the burned understory vegetation—shrubs, herbs, and non-conifer hardwood trees—has the ability to reproduce vegetatively (e.g., by sprouting from roots and stumps) and from viable seeds buried in the soil. These residual living and dead organic materials have been called ***biological legacies*** (Franklin et al. 2007). The live and dead structural legacies (e.g., live trees, snags, and logs) serve many important ecological roles in the post-disturbance ecosystem, including providing critical habitat for many animal species, such as birds, mammals, and amphibians (Swanson et al. 2011).

Hence, the post-disturbance ecosystem inherits an immense legacy of living and dead biological materials from the pre-disturbance forest, which is immediately and profoundly apparent as a sea of snags and down logs. These structures enrich the post-disturbance ecosystem and decay slowly, some of more decay-resistant tree species lasting and serving as important habitat for several centuries. The legacies of a windstorm contrast with post-fire conditions in consisting mostly of uprooted and broken fallen trees and leaving essentially intact understories, which often include an abundance of tree seedlings and saplings; essentially all of the organic matter remains. The root wads and soil depressions created by the uprooting process are additional important structural legacies and habitats.

Pre-forest Stage

The most important feature of this stage is the lack of dominance by the tree life-form (plate 2B), which allows for the development of ecosystems dominated by a diversity of plants. Opportunistic annual and perennial herbaceous species, such as groundsels, fireweeds, and thistles, dominate for the first two or three years, but shrubs and sometimes perennial herbs soon become dominant, reproducing by sprouting and growing from dormant seeds buried in the soil or transported to the site by animals. The pre-forest stage also is known as the early-successional stage, but we use the pre-forest label to emphasize its freedom from tree dominance.

The pre-forest stage is, hands down, the most biologically diverse of the four developmental stages (Swanson et al. 2011), including the old-forest stage, in terms of total species richness. This richness is measured in terms of the diversity of food chains as well as species numbers. The diverse and numerous food chains are based on the abundance of different kinds of foods, such as the diversity of green foliage, fruits, seeds, and nectar sources, and on an abundance of habitat structures, including snags and down wood. The majority of the pre-forest biological diversity is composed of habitat specialists—species that require the open and food-rich pre-forest conditions—and not opportunistic weedy species. Examples include a large array of animal species, such as neotropical migrant songbirds and nectar-dependent butterflies and moths. Of course, the pre-forest stage is also critical foraging habitat for elk and deer populations, which are important prey for top-tier predators, including humans. In addition, the pre-forest stage is critical for other important ecosystem functions, such as additions to stocks of biologically available nitrogen. Numerous plants that host nitrogen-fixing organisms in their root systems, such as members of the pea or legume family (e.g., lupines), alder trees, and ceanothus shrubs, are abundant during this developmental stage.

Following natural wildfires, the pre-forest stage often persisted for many decades—*it was not ephemeral*. This persistence was important because it allowed shrub and perennial herbaceous species to become fully mature and fruitful. Although tree canopy closure begins in two to three decades, complete reestablishment of closed forest canopies often took much longer (Freund et al. 2014). A similar extended period (45 years) was documented in forests that were only partially burned (Tepley et al. 2014). The pre-forest stage generally lasts a much shorter period of time following windstorm blowdown events because of the abundance of tree seedlings and saplings that survive and grow quickly in response to the open conditions.

Young-Forest Stage

The young-forest stage begins when trees grow up with sufficient density and height to develop a closed forest canopy and assume dominance over the site. This stage is one of rapid biomass accumulation and intense competitive interactions within the population of young and dense coniferous trees. However, this stage is also low in biodiversity. Structural complexity is high in the natural young forest developed following a wildfire because of the large legacy of snags and logs from the pre-disturbance forest, which will decay gradually.

The young-forest stage is most consistently and profoundly characterized by significant annual biomass accumulations, as young trees capitalize on the light, moisture, and nutrient resources of the site. This is the rapidly growing young forest on which plantations are modeled. A high proportion of the gross productivity of the site is being devoted to rapid height and diameter growth of trees and significant annual increments of wood, once conifer leaf areas are established.

Competition is the other major dynamic of the young-forest stage, the intensity of which depends upon the density and spatial pattern of trees in a stand. The competition is most obviously for light, and trees struggle to retain a position at the top of the rapidly rising tree canopy. Trees that fall behind and are overtopped by taller associates decline and quickly die, resulting in a pattern of tree death that is predictable among smaller trees. Hence, competition-based mortality is greatest where the stand is the densest and tends to result in development of a relatively uniform or homogeneous forest in terms of both tree density and size. Of course, competition from the tree life-form decimates the other plant life-forms (herbs and shrubs) that dominated the pre-forest stage. There may be other agents of mortality, such as damage from snow or ice storms or from falling snags, but competition is the most characteristic.

The young-forest stage has the lowest biodiversity of any of the four developmental stages in the sere. One reason is the large reduction in the abundance and variety of herbs and shrubs in the heavily shaded forest interior, which results in many fewer and less complex food webs; herbivory of the coniferous foliage is very low. Most food webs during this stage are based on dead organic materials (detritus) and composed of organisms that consume detritus as their energy source (detritivores), eat detritivores or their products (e.g., fungal fruiting bodies), or eat tree seeds.

Typically, the young-forest stage gradually transitions to the mature-forest stage at around 80 to 120 years. This involves a slow relaxation in the intensity of forest dominance as stands thin and canopies rise, along with the redevelopment of a more luxuriant understory, including seedlings and saplings of shade-tolerant tree species.

Mature-Forest Stage

The mature-forest stage is a period in which the pioneer forest cohort matures, major causes of tree mortality change, significant shifts in canopy architecture occur, and decadence begins to develop, all of which are associated with increases in biological diversity. Productivity remains very high in the mature forest, even though the rate of wood accumulation in live trees has slowed. In fact, at 100 years Douglas-fir trees have achieved only about two-thirds of their eventual height. They will achieve most of their remaining height and crown spread during their second century of growth. In traditional forestry terminology, forests that have grown beyond the young-forest stage are described as "overmature," but from an ecological perspective they are actually just reaching a mature stage of development.

Important structural changes related to canopy architecture and decadence (including abundance of snags and logs) occur during the mature-forest stage, much of it as a result of changes in spatial patterns and causes of tree mortality. Agents, such as insects, diseases, and wind, largely replace competition as the primary cause of tree death. Unlike competition, agent-based mortality often kills large dominant trees (rather than the smallest trees), is often spatially aggregated (multiple trees are killed), and is difficult to predict. Canopy gaps result, which contribute to structural complexity and stimulate growth of shade-tolerant tree species into the mid and upper canopy. This shift in mortality is a natural process and not an indication of poor forest health. More-open stand conditions also stimulate emergence of new (epicormic) branch systems in midbole positions on Douglas-fir trees. Both the growth of western hemlock and other shade-tolerant tree species and epicormic branch development contribute to establishment of a canopy that is largely continuous from the ground to the top of the largest trees. Mortality of larger trees also starts rebuilding the stocks of dead wood, which have been declining as much of the dead-wood legacy created by the stand-initiating disturbance has decayed away. Aspects of decadence, in addition to those resulting from tree mortality, also become evident as the pioneer tree cohort ages; these include cavity development and other decay in live trees, broken tops, and dense branch accumulations (brooms) stimulated by dwarf mistletoe infections.

The mature-forest stage commonly begins at about 80 to 120 years and lasts about 100 years, after which the forest gradually takes on more and more of the attributes of an old forest. In many ways it can be viewed as completion or maturation of the pioneer cohort and also a period of profound structural change between the relatively simple young forest and the complex old forest.

Old-Forest Stage

The old-forest stage encompasses the awesome old-growth forests for which the Pacific Northwest is known. This is the most structurally complex and biodiverse of the three forest-dominated developmental stages. Large to very large old trees are a conspicuous element of this forest, as are several iconic wildlife species, including the northern spotted owl and one of its primary prey species, the northern flying squirrel. Significant areas of this forest stage (>2.4 million ha, or >6 million ac) remain on federal forest lands, although the proportion of regional forests in the old-growth stage is probably less than half of its presettlement extent.

The old-forest stage has a complex structure. The types of individual structures—sizes, conditions, and species of live trees and decay states of snags and logs on the forest floor—and the spatial arrangement of these structures vary across the landscape. Live trees vary in size (the wide range in tree sizes is a key characteristic), age, species, and condition. Large old trees are an important and distinctive element, providing unique habitats such as large branches (including the epicormic branch systems), cavities, and other decadent features. Large accumulations of snags and logs are present in states of decay that vary from solid to brown mush. The canopy of green foliage is vertically continuous from ground to the top of the tallest trees, but it is bottom loaded rather than top loaded (as was the case in the young-forest stage), with the majority of the foliage in the lower third of the vertical dimension. As a consequence, you usually cannot see very far through the old forest. Much of the canopy is provided by shade-tolerant species, such as western hemlock, and this foliage is the dominant factor controlling the light environment near the ground.

Biodiversity is high and includes many habitat specialists. Some rare species may be found in older forests because they are poor dispersers and require long periods to spread across the landscape following disturbance, while others may be present because of the stable, milder microclimates (cooler and moister in summer and warmer and calmer in winter) of the old forest. Other old-growth-related species are strongly tied to decadence, including the abundance of large, well-decayed snags and logs, as well as other woody detritus in varying states of decay. The plant community in the understory is moderately well developed, particularly in gaps, but low in diversity, particularly in comparison with the pre-forest stage.

The old-forest stage is relatively stable in its structure and composition, with most of the dynamics associated with chronic individual tree- and gap-based mortality due to root, butt, and stem rots, insects, and wind. Douglas-firs are slowly declining in abundance and being replaced by hemlock, cedar, and other shade-tolerant species, since canopy gaps are usually not sufficient to allow for successful Douglas-fir regeneration. Mechanical failure due to a butt rot (velvet-top fungus) is responsible for about 60% of the old Douglas-fir mortality, with Douglas-fir bark beetle and wind (uprooting) killing most of the remainder (Franklin et al. 2002).

Old forests continue to evolve and gradually change in composition and structure over many centuries but do so relatively slowly. Rarely, forests may continue for 1,000 years or more before being subjected to the next major partial or complete stand-replacement disturbance. The majority of the existing old-growth Douglas-fir forests originated 400 to 600 years ago, when wildfires were widespread in the Northwest (Weisberg and Swanson 2003).

We have used an idealized and simplified picture of stand development to communicate the major successional and ecosystem processes that can operate during a multicentury sequence of forest succession following a single stand-replacing wildfire. Many forests have experienced a much more complex history in which live old-forest legacies survive partial disturbances and occur mixed with younger cohorts. Complexity also results from partial disturbances (fire, wind, disease) in mature- and old-forest stages that add patches of younger forests within areas of older forest. The frequency of partial stand-replacement fire and other disturbances also varies regionally (more partial disturbances to the south and east) and topographically (increased frequency on dry, southerly aspects and at lower elevations).

We have focused on "stand" scales (areas of tens to hundreds of hectares [acres]) of roughly similar age and structural conditions. At a landscape scale (thousands to millions of hectares [acres]), the net effect of disturbance and successional processes created a slowly shifting mosaic of successional stages dominated by mature- and old-forest structural conditions. Infrequently, climatic fluctuations would produce conditions conducive to stand-replacing disturbances, creating very large patches of pre-forest conditions. The mix of successional stages and habitats underpinned the biodiversity of the moist forests and contributed to their productivity and resilience in the face of natural disturbance agents.

Recent Human Impacts on the Forests

The moist coniferous forest landscapes and stands have been dramatically altered since European settlement of the region. Vast acreages of the most productive lowland areas have been converted to agriculture and to provide for towns, cities, transportation infrastructure, and the like. Exploitation of the forests for wood products began late in the nineteenth century, with little or no attention initially to reforestation or protection. Large areas were burned and reburned by extensive fires of European human origin beginning in the mid-nineteenth century, continuing well into the twentieth century to the Tillamook Burn of 1933.

Serious efforts at sustained protection and management of the forests began in the twentieth century as a result of efforts by both the federal government (e.g., establishment of the national forests) and state and private organizations (e.g., establishment of cooperative fire protection programs and the tree-farm movement). In some parts of the region, where wildfires were more common, fire suppression has reduced the occurrence of fire and thereby reduced the structural and compositional diversity of the forests at both stand and landscape scales.

Following World War II, management of forests on private, state, and federal lands intensified, with a very similar focus on the efficient production of wood. Although details varied, essentially all forest management organizations converged on even-aged management systems that consisted of development of road systems, clearcutting of existing natural forests, site preparation and slash disposal (often by broadcast burning), and establishment of plantations by artificial regeneration.

Plantations in this region are highly simplified versions of the young natural forests described earlier. They are created and managed to capture the high level of biomass (wood) accumulation characteristic of the young forest, using agronomic approaches such as weeding with herbicides to eliminate competing vegetation and thinning trees in order to concentrate growth on uniformly distributed crop trees. Early-successional vegetation is effectively eliminated or severely curtailed by multiple herbicide applications and rapid closure of tree canopies in the densely planted areas.

Management of federal and industrial forest lands diverged dramatically late in the twentieth century (chap. 4). Even-aged, plantation-based management of federal forest lands marginalized many important forest values and ultimately ran afoul of several US federal laws, including the National Environmental Policy Act, the Endangered Species Act, and the National Forest Management Act (Skillen 2015). Timber harvest levels, which had approached 5 billion board feet (~12 million cubic meters) of timber annually, declined to about 15% of that level as a result of the US federal Northwest Forest Plan adopted in 1994 (Skillen 2015).

Globalization and altered tax laws resulted in significant changes on what had historically been industrial forest lands. These included the vast majority of the highest-productivity forest lands in the moist forest region. The majority of these lands were held by vertically integrated forest-products corporations that managed them to provide wood supplies for manufacturing facilities. Late in the twentieth century, ownership of these lands was shifted to real estate investment trusts (REITs) and timber investment management organizations (TIMOs), where they are now managed to maximize return to investors; timber produced is sold to the highest bidder, which may mean that logs are shipped overseas to markets in Asia. The greater emphasis on return on investment has made economic models a much larger factor in decisions about forest management, with state and federal laws providing the only environmental constraints.

The Future of Moist Coniferous Forests

There is a general impression that the moist coniferous forests of the Northwest are invulnerable to climate change because of their high productivity and the mild climate of the region, which is expected to continue to have high precipitation. In fact, the moist forests in this region are at significant risk from both climate change and invasive species; arguably this is the most vulnerable major forest region in North America. This vulnerability is the result of three factors, the first two related to the profound ecological significance of the region's dry summers to forest composition and productivity, and to the region's major disturbance agent—wildfire (Franklin et al. 1991). The third risk factor is the overwhelming importance of a single species—Douglas-fir—which dominates the natural, seminatural, and plantation forests of the region. We will briefly review risks associated directly and indirectly with climate change, and those associated with biotic agents, both native and invasive. Climate futures of these forests also are addressed in chapter 16.

Climate projections for the Douglas-fir region include annual temperature increases of 3°C (5.4°F) to almost 10°C (18°F) by 2099, varying with the emissions scenario assumed and the global climate model used. Projections for change in annual precipitation are more uncertain, ranging from −11% to +18% (Mote and Salathé 2010; Mote et al. 2014). Summer precipitation is not expected to increase and may decline. These changes will almost certainly increase summer moisture stress in the Douglas-fir region and lengthen the intensity and duration of the region's dry season. This is likely to reduce growth and increase mortality on many sites, although there is evidence that there will be considerable variability in climate change associated with topography. In some higher-elevation (subalpine) forests where growth is limited by short growing seasons, tree-growth rates could increase with climate warming, although the geographical extent of the snowy habitats will decline. Also, some modeling results suggest that some high-elevation species may be increasingly under stress (Coops and Waring 2011).

Indirect effects of climate change may be as or more important than the direct physiological effects (Spies et al. 2010). For example, one study projects an eightfold increase in the area of fire in western Washington (Littell et al. 2010). Extreme weather and periods of drought could result in much larger fires than have occurred in most of the last 70 years. Large fires, coupled with years of postfire drought, could slow forest regeneration and succession following fires and shift tree species toward more fire- and drought-tolerant genotypes of Douglas-fir or other conifer species (e.g., grand fir, incense cedar, and ponderosa pine). Fires force established forests to undergo a regeneration cycle, and this is the period in which a forest is most sensitive to environmental conditions, which could result in tree-species shifts or even a transition to nonforest vegetation on dry sites.

Additional scenarios of forest disruption would be the introduction of a new invasive insect or disease, or development of novel (previously unexperienced) behaviors by native or naturalized insects or diseases of trees. The source of a new invasive most likely would be Eurasia, through import of live plant material (e.g., nursery stock) or raw wood (wood chips, logs, or unprocessed wood products). A recent example of such an introduction is sudden oak death, which has repeatedly been introduced on imported nursery stock. Novel behaviors of endemic pests and pathogens would result from warming climates that affect either the population dynamics of the pest organisms or the susceptibility of the host species. Swiss needle cast in plantations and natural forests in the Oregon Coast Range is an example of how an endemic disease can become a widespread epidemic as a result of warming climate and changes in land use—such as the spread of Douglas-fir plantations on sites better suited to western hemlock and Sitka spruce.

Given that all aspects of tree community and forest cover are so dominated by a single species, Douglas-fir, a virulent new pest or pathogen on this species could have an extraordinary impact on moist forest ecosystems. An agent that attacked Douglas-fir on the scale that white pine blister rust, chestnut blight, and the hemlock woolly adelgid have affected foundational species in other regions of North America would be disastrous for this region because of the unique prominence of Douglas-fir, from both ecological and commercial perspectives.

Literature Cited

Agee, J. K. 1993. *Fire ecology of Pacific Northwest forests*. Washington, DC: Island Press.

Askins, R. A. 2014. *Saving the world's deciduous forests*. New Haven, CT: Yale University Press.

Coops, N. C., and R. H. Waring. 2011. Estimating the vulnerability of fifteen tree species under changing climates in Northwest North America. *Ecological Modeling* 222:2119–2129.

Franklin, J. F., and C. T. Dyrness. 1988. *Natural vegetation of Oregon and Washington*. Corvallis: Oregon State University Press.

Franklin, J. F., and C. B. Halpern. 2000. Pacific Northwest forests. Pp. 123–159 in *North American terrestrial vegetation*. 2nd ed. Edited by M. G. Barbour and W. D. Billings. Cambridge, United Kingdom: Cambridge University Press.

Franklin, J. F., R. J. Mitchell, and B. J. Palik. 2007. *Natural disturbance and stand development principles for ecological forestry*. General Technical Report NRS-19. Newtown Square, PA: USDA Forest Service, Northern Research Station.

Franklin, J. F., T. A. Spies, R. Van Pelt, and 9 coauthors. 2002. Disturbances and structural development of natural forest ecosystems with silvicultural implications, using Douglas-fir forests as an example. *Forest Ecology and Management* 155:399–423.

Franklin, J. F., F. J. Swanson, M. E. Harmon, and 10 coauthors. 1991. Effects of global climatic change on forests in northwestern North America. *Northwest Environmental Journal* 7:233–254.

Freund, J. A., J. F. Franklin, A. J. Larson, and J. A. Lutz. 2014. Multi-decadal establishment for single-cohort Douglas-fir forests. *Canadian Journal of Forest Research* 44:1068–1078.

Henderson, J. A., D. H. Peter, R. D. Lesher, and D. C. Shaw. 1989. *Forested plant associations of the Olympic National Forest.* R6 Technical Paper 001-88. Portland, OR: USDA Forest Service, Pacific Northwest Region.

Kuchler, A. W. 1946. The broadleaf deciduous forests of the Pacific Northwest. *Annals of the Association of American Geographers* 36:122–147.

Littell, J. S., E. E. Oneil, D. McKenzie, J. A. Hicke, J. A. Lutz, R. A. Norheim, and M. M. Elsner. 2010. Forest ecosystems, disturbance, and climate change in Washington state, USA. *Climatic Change* 102:129–158.

Mote, P. W., and E. P. Salathé Jr. 2010. Future climate in the Pacific Northwest. *Climatic Change* 102:29–50.

Mote, P., A. K. Snover, S. Capalbo, S. D. Eigenbrode, P. Glick, J. Littell, R. Raymondi, and S. Reeder. 2014. Chapter 21: Northwest. Pp. 487–513 in *Climate change impacts in the United States: The third national climate assessment*. Edited by J. M. Melillo, T. C. Richmond, and G. W. Yohe. US Global Change Research Program. http://nca2014.globalchange.gov/report/regions/northwest.

Richards, R. T., and S. J. Alexander. 2006. *A social history of wild huckleberry harvesting in the Pacific Northwest*. General Technical Report PNW-GTR-657. Portland, OR: USDA Forest Service, Pacific Northwest Research Station.

Sinton, D. S., J. A. Jones, J. L. Ohmann, and F. J. Swanson. 2000. Windthrow disturbance, forest composition, and structure in the Bull Run basin, Oregon. *Ecology* 81:2539–2556.

Skillen, J. R. 2015. *Federal ecosystem management: Its rise, fall, and afterlife*. Lawrence: University Press of Kansas.

Spies, T. A., T. W. Giesen, F. J. Swanson, J. F. Franklin, and K. N. Johnson. 2010. Climate change adaptation strategies for federal forests of the Pacific Northwest, USA: Ecological, policy, and socio-economic perspectives. *Landscape Ecology* 25:1185–1199.

Swanson, M. E., J. F. Franklin, R. L. Beschta, C. M. Crisafulli, D. A. DellaSala, R. L. Hutto, D. B. Lindenmayer, and F. J. Swanson. 2011. The forgotten stage of forest succession: Early-successional ecosystems on forest sites. *Frontiers in Ecology and the Environment* 9:117–125.

Tepley, A. J., F. J. Swanson, and T. A. Spies. 2014. Post-fire tree establishment and early cohort development in conifer forests of the western Cascades of Oregon, USA. *Ecosphere* 5(7):80. http://dx.doi.org/10.1890/ES-14-00112.1.

Waring, R. H., and J. F. Franklin. 1979. Evergreen coniferous forests of the Pacific Northwest. *Science* 204:1380–1386.

Weisberg, P. J., and F. J. Swanson. 2003. Regional synchroneity in fire regimes of western Oregon and Washington. *Forest Ecology and Management* 172:17–28.

Chapter 3

People and Forest Plants

Susan Stevens Hummel and Jane E. Smith

Newcomers to the northwestern coast of North America can easily overlook the continuing legacy of settlement wars on forests. Battles were fought in homelands of indigenous people because settlers sought arable land and United States government policies encouraged westward expansion. Within two generations, profound changes have altered regional landscapes, cultures, and systems of learning. Women and men born in the early twentieth century share their knowledge of earlier times with children and grandchildren. Yet, as this elderly generation dwindles and immigration increases, fewer and fewer residents know directly of the traditional practices and historic events that influence regional waters, forests, people, and plants. Tribally reserved and ceded lands cover thousands of hectares along the Pacific Northwest coast, but estimates of their extent and use vary among state and federal agencies, treaty and nontreaty tribes (OTR 2015). Reserved lands are for exclusive tribal use, whereas ceded lands are off-reservation lands where tribal people retain some rights. Old lessons about people and forest plants can inform modern questions about the entire region. Relearning and adapting these old lessons will need new generations of leaders who understand the interrelationship of all life.

Human cultures that persist over centuries of environmental variation attest to the existence of practices that can be adapted when populations of plants and animals fluctuate or are scarce, and to the ability of people to react to change. Such adaptive and reactive practices rely on knowledge that is acquired when people interact with organisms in specific places and then is bequeathed to subsequent generations and nurtured by cultural traditions. If there is a goal to sustain the Northwest moist forest ecosystems of North America and all the people who now call the region home, then there is a role for management that includes traditional knowledge. This topic is central to contemporary land management, because indigenous systems for tending plants and animals have been influencing forests and sustaining humans for centuries.

Traditional Knowledge of Forest Plants

The sustained interactions of people and forests have reciprocal effects on populations and communities and the patterns that emerge. Human geography has forever been influenced by resource availability. Now, however, with more residents in the region and more urban than rural jobs, fewer people interact daily with plants for survival. Knowledge shrinks, and patterns of land use change accordingly. Yet there are still people who strive to maintain or renew traditional knowledge and uses of forest plants. A contemporary term, *traditional resource and environmental management* (TREM), refers to applying traditional ecological knowledge to maintain or enhance the productivity, diversity, availability, or other desired qualities of natural resources or ecosystems (Lepofsky 2009). TREM shows that many forest resources are obtainable by intentionally altering vegetation in different seasons, using various methods, and targeting different parts of a plant.

There is a debate in the anthropological and ethnobotanical literature over whether plants were cultivated by indigenous people of the Pacific Northwest coast or merely gathered. Blukis Onat (2002) asserted that actions rather than species distinguish cultivation, specifically interventions in life histories to promote desired harvest qualities. With this premise, indigenous cultivation of plants clearly occurred in moist coniferous forests. Traditional practices of cultivating wild plants have been referred to collectively as *tending* (Anderson 2013) and include rituals and activities that strengthen ties with kinfolk and with land. *Tending* means to apply oneself to the essential care of something, and the term conveys emotional links absent from *managing*. Tending affects plants but encompasses more than plants and connotes reciprocal seen and unseen effects. In this chapter, we illustrate the concept of tending with a short list of forest plants that are cultural keystone species on the Pacific Northwest coast. Keystone species in this sense are organisms that form the essential foundation of a culture and without which the associated societies would be very different (Garibaldi and Turner 2004). Despite evident omissions, the list hints at both the interactions supporting floristic and cultural diversity in the region and the value in delving further into source materials and consulting knowledgeable practitioners.

One hallmark of Northwest coastal forests is their layered canopy; another is the influence they exert. The stature and abundance of vegetation in these forests have been described, catalogued, studied, exclaimed over, and argued about for centuries (chap. 2). Trees reaching over 70 m (230 ft) in height and 5 m (16 ft) in girth have been measured from British Columbia south to California (Van Pelt 2001). Lichens, a remarkable fusion of fungi and algae, live in their crowns (plate 3A). Small trees and tall shrubs create intermediate layers, and at still lower levels are smaller shrubs, herbs, grasses, mosses, and fungi (plate 3). The overall assembly is architecturally varied in many dimensions, looking up, out, or down, and the associated biological diversity and biomass are globally renowned.

Early anthropologists mapped cultural patterns onto the regional geography of forests, woodlands, and high desert (Kroeber 1939). For some residents, the terms *natural resources* and *cultural resources* are synonymous because of the pervasiveness of nature in everything. Nature is culture, and vice versa, which means that some species are icons: they are emblematic of a place. Long-lived and massive trees such as western redcedar, Sitka spruce, coast redwood, and Douglas-fir are prominent examples from the Pacific coast, but iconic plants occupy every forest layer (plate 3), including shrubs like huckleberry and salmonberry.

During the twentieth century, the vertical and horizontal layers of Northwest forests were simplified wherever the goal was wood production (Hummel 2003). Brushy sites make for arduous work, and shrubs can outcompete valued tree seedlings (Zasada et al. 1994). Reducing intermediate layers can seem sensible if forests are valued for wood alone. Alternative management approaches are required when additional flora and fauna are desired. Some approaches for managing forest complexity and diversity are recent (chap. 10). Other approaches, like traditional resource and environmental management, predate the enabling legislation of national forests and other federal lands in the United States and have sustained people and biodiversity for centuries. Therefore it makes sense to get acquainted with TREM and some traditional practices.

TREM Practices

TREM practices influence the growth of plants by applying knowledge from people who observed population and community responses to actions undertaken in different seasons and conditions, adapted actions as needed, and shared the accumulating information with descendants. Actions include burning, cutting (e.g., pruning branches), transplanting, and gardening (e.g., aerating soil, weeding). The last two are discussed separately because transplanting physically changes the location of a plant and thus can influence its range, whereas gardening focuses on cultivation in place. The distinction underscores the obvious. Plants are sessile organisms, unable to pack up and move away from conditions of distress. In contrast, people and other animals are vagile organisms that are able to move. Plants have therefore developed mechanisms to sense and respond to abiotic factors, including light, wind, and moisture.

Plants respond to environmental signals with tropisms (from the Greek *trope*, meaning "turn"), allowing them to grow toward light and to send roots down and shoots up. How a plant senses and responds physiologically depends on hormones that trigger internal signals to bring about cellular changes influencing growth and development. (The word *hormone* comes from the Greek *horman*, meaning "to set in motion.") These chemical messengers, key to the development of flowers and fruit, may either stimulate or inhibit growth. The ability to cease growth, known as dormancy, allows plants to wait out hard times. Windstorms and pruning remove branches, suppressing a hormone in the terminal bud and stimulating other hormones that result in vigorous growth of lateral branches and a rounder and bushier appearance. Day length triggers hormones involved in flowering. Knowledge like this about plant responses to environmental changes is deliberately used in TREM to accomplish resource goals.

Tending activities vary in intensity and extent. For example, weeding is intensive, whereas burning is extensive. The effects on individual plants from tending—whether dividing roots, pruning branches, or removing bark—arise from the timing (season) and type (intensity) of actions, which are guided by the properties needed at harvest for some intended use. For example, indigenous people in Puget Sound extracted medicine and dye from the nettle plant and peeled it to make two-ply string for nets; they tended nettle patches by weeding and burning each autumn following harvest (White 1975). These practices enriched soil and simplified plant communities in patches dispersed amid areas with greater structural and compositional diversity.

Plants respond to tending practices based on (1) individual life histories, including reproductive traits, growth habits, distribution, and habitat requirements; (2) the part(s) removed; and (3) the timing and intensity of harvest(s). In long-lived perennials like trees and shrubs, for example, high adult mortality is more detrimental to population persistence than is periodic harvest of reproductive parts (flowers, fruits, and seeds). In addition to sexual reproduction by flowers and fruits, some plants are also capable of vegetative reproduction, a form of asexual reproduction. Roots of some plants will produce buds and stems that grow into new plants, and similar vegetative reproduction may also occur in stems. Runners are stems that grow along the ground surface, whereas rhizomes are underground stems that spread and produce new plants. Corms are stems in which food is stored; in contrast, bulbs are short stems covered by fleshy leaf bases containing stored food. Plants that reproduce primarily by vegetative reproduction have adapted to relatively unchanging environments and, with less variation in their population, would likely be less able to survive rapid changes in climate than would plants that benefit from the genetic diversity achieved through sexual reproduction. All of these plant attributes and more underlie TREM and, in combination, can contribute to desired forest structures and compositions.

Tending activities occur in association with several specific traditional uses (fig. 3.1) or in combinations, such as using plant-based dyes and fibers for ceremonial arts (plate 4A). By adopting a unique symbol for each species and then placing the symbol within a cell according to species growth habit (e.g., tree, shrub, herb), each cell in figure 3.1 illustrates the forest layers affected by various combinations of tending and use. The columns and rows of figure 3.1 are ordered alphabetically, starting at the upper left, where plants used for barter and tended by burning include herbs (beargrass and camas) and shrubs (hazelnut and huckleberry, plate 4B). Plants tended with fire to yield food include a fungus (matsutake, box 3.1; plate 3C), herbs like cattail and nettle, shrubs like salmonberry (plate 3B), plus the tanoak tree. For brevity, we have grouped tending activities that involve cutting plant tissues into one category, despite the gradient of possible intensity (from shaping to severing and killing). The variety of tending actions included in cutting and burning directly affects plants growing in every forest layer. Notwithstanding the limited plant sample and sources restricted to published literature, one obvious result of TREM made evident in figure 3.1 is its contribution to forest structural and floristic diversity. Historical records are dispersed but consistent, including information gathered for an Indian Claims Commission case about Skagit land in Puget Sound circa 1859 that

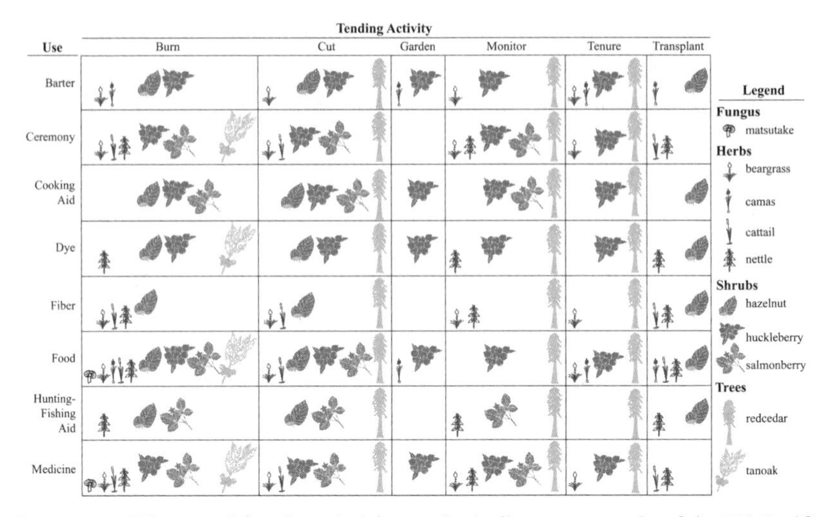

FIGURE 3.1. Plants and fungi tended for use by indigenous people of the US Pacific Northwest contribute to layered and diverse forests, as illustrated by a few species of different life-forms (fungi, herbaceous plants, shrubs, trees). Source materials on file.

explicitly named several species in figure 3.1 when describing forest composition (22 Ind. Cl. Comm. 28, 1976). Although the evidence for plants is direct and documented, it must be inferred for fungi and lichens.

The cultural preeminence of western redcedar is both well-documented and evident from figure 3.1, which shows that this tree is used for all the purposes listed here. Different parts of the cedar tree, harvested at each stage of its life, are used variously for barter, ceremony, dye, fiber, food, and medicine, and aids for cooking, hunting, and fishing (Blukis Onat 2002). Along the Northwest coast, working with the wood of cedar trees was traditionally a task for men, whereas women worked with the bark and roots. For millennia, humans have taken advantage of the regenerative capacities of trees to harvest their wood, branches, bark, resin, roots, leaves, and buds without killing the trees themselves. Living trees with evidence of past use can be found worldwide, including the Northwest coast of North America, Turkey, Australia, Africa, Polynesia, and Spain (Turner et al. 2009). Such *culturally modified trees* are classified by Turner et al. (2009) by the alteration purposes, including (1) incidental harvest results (e.g., collecting inner bark tissues for food, fiber for weaving); (2) intentional tending for future production (e.g., pruning and coppicing branches); and (3) intentional tending for marking (e.g., boundary, tenure/ownership, or as witness trees).

BOX 3.1. *TRICHOLOMA MAGNIVELARE*— AMERICAN MATSUTAKE OR TANOAK MUSHROOM

Mushrooms are the ephemeral reproductive structures (fruiting bodies) of longer-lived threadlike fungal networks that weave through forest floors. Fruiting typically begins soon after the advent of the rainy season. Mushrooms produce spores that germinate and fuse to establish new threadlike organisms. Many species of mushrooms, including matsutake, form a partnership with the tiny feeder roots of trees. The root colonized by the fungal threads (hyphae) is known as a mycorrhiza, and it's here that nutrients and water delivered by the fungal hyphae are exchanged for nourishing plant-produced sugars.

American matsutake is distributed in a variety of habitats in coniferous northern forests across Canada and in temperate forests of the Appalachians, Rocky Mountains, and Cascade and Pacific Coast Ranges (Hosford et al. 1997; plate 3C). It forms beneficial, mutualistic symbioses (mycorrhizae) with the roots of most conifers and some hardwood trees. In southwestern Oregon and coastal Northern California, it is particularly abundant in closed-canopy stands of mature tanoak and scattered Douglas-fir with a deep hardwood leaf-duff layer. It is also found in sandy coastal habitats with shore pine. In Mexico it is found in high-elevation pine and fir forests.

Accounts of wild mushroom use by indigenous peoples in the Pacific Northwest are few and varied depending on the tribe and location (Kuhnlein and Turner 1991; Richards 1997; Anderson 2013). In Northern California, mushrooms are thought to have been an important component of traditional cuisine (Richards 1997; Anderson 2013). Matsutake, commonly known as the tanoak mushroom among the Karuk, Yurok, and Hupa peoples, is regularly harvested in southern Oregon and Northern California, and elders in their 80s recall having picked tanoak mushrooms since childhood (Richards 1997). As the name suggests, tanoak mushrooms are commonly found beneath tanoaks, but they also occur among conifers and chinquapin. According to the Yurok and Karuk tribes, tanoak mushrooms were picked after tanoak acorns were gathered and are most sustainably harvested by gently pulling and twisting them from the ground to avoid disturbing the duff and soil (Richards 1997). Severe fire diminishes fruiting (Richards 1997).

In contrast to its shrubby appearance in dry forests, the tanoak grows into a tree in moist forests in the western hemlock–Sitka spruce zone or coast redwood zone. The large acorns, ground, leached, and then prepared as soup and bread, were traditionally an important food for indigenous people living within the range of tanoak in the Coast Ranges of south-

FIGURE 3.2. The leaves of beargrass create patterns in a basket being woven by Deanna Marshall. Photo by Frank K. Lake.

western Oregon and northwestern California, and these foods retain their ceremonial importance. Burning fosters tree growth and acorn production and helps keep this deciduous element in coniferous forests. Tending activities like cutting and burning stimulate sprouts and shape new branches in shrubs like hazelnut (Anderson 2013), which is used for making traps and blue dye and produces nuts used for food and barter (Moerman 1998). For shrubs like huckleberry, burning can enhance the production and quality of fruit harvested (plate 4B); moreover, the abundance of community associates may also be enhanced, such as the beargrass leaves used as fiber for traditional weaving (fig. 3.2). The timing of a fire or other disturbance, relative to plant life histories, affects the species and thus the community response. For example, after a disturbance, salmonberry can regrow from a buried seed bank and from a bud bank, but seedlings grow slower than shoots. The salmonberry bud bank occurs both on aboveground stems and on the rhizome, and all types of buds produce new shoots, which can quickly use sunlight and soil resources made available by disturbance (Zasada et al. 1994), thus outcompeting the seedlings of other species and boosting the dominance of salmonberry. This outcome would shift, however, if a fire were lethal to the bud bank and the plant instead reproduced from seed. A similar pattern exists in herbaceous plants like beargrass, which can sprout new leaves following fire unless the heat kills the rhizome.

Because TREM incorporates knowledge about species responses under varying conditions and at different levels of biological organization—a population of the same species, a community of interacting populations, an entire ecosystem—it is a useful source of information when ecological objectives include species diversification to support human uses. For example, planned and controlled burning can be used to target the season and intensity of fire to promote desired mortality and regeneration by species (plate 4B). It may take many years to accomplish the goals of burning. One outcome can be sites with diversified plant communities that provide for a variety of personal and commercial uses. There are promising examples of cooperative efforts using fire to accomplish ecological objectives for specific forest sites, but a mismatch exists in the geographic scale of current efforts versus the scale of historical indigenous burning in the region (Boyd 1999).

Multiple Types of Knowledge

People who have direct, tactile experience of plant tending and harvest can be expected to learn about their environment differently than people who do not, given what is now known about cortical plasticity of the human brain (Buonomano and Merzenich 1998). These differences can produce and reinforce different types of knowledge. The existence of multiple systems of learning and types of knowledge implies that not all information is accessible to everyone, because it is derived from, and reinforced by, different experiences. This means individuals think to ask certain questions or are disposed to observe certain relationships and responses based on a cultural lens that is reinforced cognitively. The dominant culture in a pluralistic society affects how knowledge systems are legitimized. A multitude of knowledge systems raises questions about whether translation among them is possible, and attention is being directed to relations between traditional knowledge and Western science (e.g., Alexander et al. 2011; Hummel and Lake 2015). One overlapping topic relevant here is plant phenology, which correlates the timing of biological phenomena (like annual flowering or breeding migrations) with climatic conditions. Traditional phenological knowledge often marks the co-occurrence of events—such as the maturing of salmonberry flowers and ripening of fruits occurring together with the return of adult sockeye salmon to freshwater (Lantz and Turner 2003)—that affect tending and harvest schedules. Phenological knowledge of forest plants is also derived through the methods of Western science, for example, to guide the seasonality of silviculture (e.g., Schmidt and Lotan 1980), but designed studies of phenological relationships among plants and animals are limited.

Blended Approaches in Forestry

The global climate and the political climate in the United States both make a case for conversations about incorporating traditional resource and environmental management practices to support innovations in forestry. Warmer and drier temperatures (chap. 16) change plant phenology. Recognition that seasonal changes are affecting plant-based industries and communities does exist, but human responses to this information vary. For example, changes in temperature and moisture affect agricultural systems. In some areas of the United States, this means farmers and ranchers can now sow and harvest one additional crop during the growing season, but there is no consensus on the source of, or responses to, the observed changes (box 3.2).

Conversations between American Indians and non-Indians about land and water, flora and fauna, occur in the context of tribal and federal law and personal and collective experiences of settlement history. Over time, laws and narratives are continually reinterpreted and power relations are renegotiated. Among several national events in the 1970s (Catton 2016) were two separate legal decisions in 1976 that combined to influence the modern role for TREM in forest management. The first was a unanimous

BOX 3.2. AS SEASONS CHANGE, RESPONSES VARY

Human responses to climate change information are mixed. People who harvest plants for commercial sale or personal use are particularly aware of changes in seasonal patterns of temperature and precipitation, factors directly affecting plant growth, harvest timing, and yields. Despite mutual interest in expected changes in climate, reactions to observed changes vary among farmers and ranchers, foresters, and indigenous people.

Adaptation management is defined by the IPCC (Intergovernmental Panel on Climate Change 2007) as "initiatives and measures to reduce the vulnerability of natural and human systems against actual or expected climate change effects." Adaptation practices have long been central to human agricultural systems worldwide. Indeed, successful farmers are adept at switching crop varieties, rotating cultivation intensities, and adjusting planting dates when local, annual effects on yield and quality are beneficial; doing otherwise can make the difference between storage and starvation. Similarly, adaptation is integral to indigenous tending practices of forests, but distinct differences in the life histories of key plants in forest ecosystems (long-lived) versus agricultural systems (short-lived) impart different views of time, hence influencing human calculations of risk. Moreover, levels of risk aversion and ecological understanding of risk-management strategies for conserving biological diversity differ (Hummel et al. 2009), so varied responses to climate change are unsurprising. In forestry, traditional tending practices typically focus on native species and schedules based on plant life histories, whereas agricultural models of forestry may use off-site planting stock and schedules tied to financial markets. Such differences in plant cultivation and harvesting by farmers, foresters, and indigenous people—and in the plant species themselves—are reflected in the varying responses of these three groups to climate change and in the general lack of consensus on the topic. The Society of American Foresters articulates how forests can contribute to reducing greenhouse gases (GHGs) (SAF 2008), which have been linked to changes in global climate. In contrast, the American Farm Bureau Federation opposes any regulation of GHGs from agricultural sources and questions the role of GHGs in climate change (AFBF 2015). No equivalent national organization represents tribal responses, which vary. Nonetheless, individual tribes and collaboratives have adopted climate adaptation plans (e.g., Swinomish Indian Tribal Community 2010). Absent agreement on human responses to climate change, the topic is subject to ongoing discussion, research, education, and debate (chap. 16).

decision by the US Supreme Court (426 U.S. 373) that state governments lack authority to assess taxes on the property of an Indian living on tribal land without specific congressional authority. This landmark case eventually contributed to a 1988 federal law establishing the Indian gaming industry, which earns billions of dollars annually. Steady growth in revenues from gaming is changing tribal influence over decisions about land and its management because there is more money to invest in communities, education, and infrastructure, and for protracted legal processes. The second decision was a US federal law governing the administration of national forests (P.L. 94-588) known as the National Forest Management Act (NFMA) of 1976. NFMA explicitly included renewable resources identified in the 1960 Multiple-Use Sustained-Yield Act (P.L. 86-517), like outdoor recreation, fish and wildlife, and wilderness, and provided for public involvement in agency preparation and revision of national forest plans. The two 1976 legal decisions set in motion separate chains of events that eventually led to legislation on tribal and federal relations over forests in 2004 (P.L. 108-278) and altered positions of responsibility and power.

Tribal and federal interests now appear to converge over the idea of forests with diverse species serving many uses. Differences in opinion and approach clearly remain and will require concerted effort to address. Tribal and federal discussions about land management and co-governance are distinct from discussions among emerging land-stewardship councils and collaborative groups (chap. 9). Whether conversations and decisions about innovations in forestry to sustain people and biodiversity are marked more by conflict and are decided in court, or by cooperation and are negotiated in meetings, depends on vision and leadership.

Literature Cited

AFBF (American Farm Bureau Federation). 2015. *Global climate change*. http://www.fb.org/issues/docs/climate15.pdf.

Alexander, C., N. Bynum, E. Johnson, and 10 coauthors. 2011. Linking indigenous and scientific knowledge of climate change. *BioScience* 61:477–484.

Anderson, M. K. 2013. *Tending the wild: Native American knowledge and management of California's natural resources*. Berkeley: University of California Press.

Blukis Onat, A. R. 2002. Resource cultivation on the northwest coast of North America. *Journal of Northwest Anthropology* 36:125–144.

Boyd, R., ed. 1999. *Indians, fire and the land in the Pacific Northwest*. Corvallis: Oregon State University Press.

Buonomano, D. V., and M. M. Merzenich. 1998. Cortical maps: From synapse to maps. *Annual Review of Neuroscience* 21:149–186.

Catton, T. 2016. *American Indians and national forests*. Tucson: University of Arizona Press.

Garibaldi, A., and N. Turner. 2004. Cultural keystone species: Implications for ecological conservation and restoration. *Ecology and Society* 9(3):1.

Hosford, D., D. Pilz, R. Molina, and M. Amaranthus. 1997. *Ecology and management of the commercially harvested American matsutake*. General Technical Report PNW-GTR-412. Portland, OR: USDA Forest Service, Pacific Northwest Research Station.

Hummel, S. 2003. Managing structural and compositional diversity with silviculture. Chap. 4. Pp. 85–119 in *Compatible forest management*. Edited by R. A. Monserud, R. W. Haynes, and A. C. Johnson. Dordrecht, the Netherlands: Kluwer Academic Publishers.

Hummel, S., and F. K. Lake. 2015. Forest site classification for cultural plant harvest by tribal weavers can inform management. *Journal of Forestry* 113:30–39.

Hummel, S., G. H. Donovan, T. A. Spies, and M. A. Hemstrom. 2009. Conserving biodiversity using risk management: Hoax or hope? *Frontiers in Ecology and the Environment* 7:103–109. doi:10.1890/070111.

Intergovernmental Panel on Climate Change. 2007. *Climate Change 2007: Mitigation. Contribution of Working Group III to the Fourth Assessment Report of the Intergovernmental Panel on Climate Change*. Edited by B. Metz, O. R. Davidson, P. R. Bosch, R. Dave, and L. A. Meyer. Cambridge, UK: Cambridge University Press.

Kroeber, A. L. 1939. *Cultural and natural areas of Native North America*. Vol. 38. University of California Publications in American Anthropology and Ethnology. Berkeley: University of California Press.

Kuhnlein, H. V., and N. J. Turner. 1991. *Traditional plant foods of Canadian indigenous peoples: Nutrition, botany, and use*. Amsterdam, the Netherlands: Gordon and Breach Science Publishers.

Lantz, T., and N. Turner. 2003. Traditional phenological knowledge of aboriginal peoples in British Columbia. *Journal of Ethnobiology* 23:263–286.

Lepofsky, D. 2009. The past, present, and future of traditional resource and environmental management. *Journal of Ethnobiology* 26:161–166.

Moerman, D. E. 1998. *Native American ethnobotany*. Portland, OR: Timber Press.

OTR (Office of Tribal Relations). 2015. *USDA Forest Service "Tribal Connections" map viewer user guide*. Washington, DC: USDA Forest Service, Office of Tribal Relations.

Richards, R. T. 1997. What the natives know: Wild mushrooms and forest health. *Journal of Forestry* 95:5–10.

Schmidt, W., and J. E. Lotan. 1980. *Phenology of common forest flora of the Northern Rockies—1928 to 1937*. Research Paper INT-RP-1980. Ogden, UT: USDA Forest Service, Intermountain Forest and Range Experiment Station.

SAF (Society of American Foresters). 2008. *Forest management and climate change: a position statement of the Society of American Foresters*. http://www.safnet.org/fp/documents/climate_change_expires12-8-2013.pdf.

Swinomish Indian Tribal Community. 2010. *Swinomish climate change initiative climate adaptation action plan*. La Conner, WA: Swinomish Indian Tribal Community, Office of Planning and Community Development. http://www.swinomish-nsn.gov/climate_change/Docs/SITC_CC_AdaptationActionPlan_complete.pdf.

Turner, N. J., Y. Ari, F. Berkes, I. Davidson-Hunt, Z. Fusun Ertug, and A. Miller. 2009. Cultural management of living trees: An international perspective. *Journal of Ethnobiology* 29:237–270.

Van Pelt, R. 2001. *Forest giants of the Pacific Coast*. Seattle: University of Washington Press.

White, R. 1975. Indian land use and environmental change: Island County, WA. *Arizona and the West* 17:327–338.

Zasada, J. C., J. C. Tappeiner III, B. D. Maxwell, and M. A. Radwan. 1994. Seasonal changes in shoot and root production and in carbohydrate content of salmonberry (*Rubus spectabilis*) rhizome segments from the central Oregon Coast Ranges. *Canadian Journal of Forest Research* 24:272–277.

Chapter 4

Wood-Products Markets, Communities, and Regional Economies

Richard W. Haynes, Claire A. Montgomery, and Susan J. Alexander

Moist coniferous forests have played a significant role in the economic and social well-being of the Pacific Northwest. The region's forest-products industry has evolved in the context of federal forest policy that changed from discussions about selling off federal land to consideration of (1) conservation and sustainable yield; (2) active management and planning; (3) multiple use and preservation; and (4) the present emphasis on ecosystem management and ecological forestry. As the dialogues changed, the alliance between forest industry and forest policy, which was once at the heart of Northwest forest management, also changed.

In this chapter, we describe the interactions of federal policy, forest industry, and rural communities as they developed in the continental Pacific Northwest from the pre-World War II era through the 1980s. We then discuss the current period, beginning in the 1990s with legal injunctions on US federal timber sales, when federal harvest declined to approximately one-tenth of the level targeted in the forest plans written in the 1980s, and the consequences for rural communities. We describe several economic issues related to the ongoing implementation of ecosystem management that contribute to the current public debate. We conclude by returning to the main theme that moist coniferous forests have been and continue to be a source of goods and services providing for and defining the social and economic conditions of forest-based communities. The moist forests of Alaska followed a different trajectory (box 4.1, 4.2).

BOX 4.1. HISTORICAL PERSPECTIVE FROM SOUTHEAST ALASKA

Southeast Alaska's economic well-being is closely tied to resource-dependent industries, including fishing, forestry, and mining. In the late 1800s, mining-related needs drove increased local timber harvest. Conservation concerns led to the creation of the Alexander Archipelago Forest Reserve in 1902. The Tongass National Forest, the largest national forest in the United States at 6.9 million ha (17 million ac), was created in 1907 by presidential proclamation.

Harvests in the pre-World War II period (1907–1942) were about 30.7 thousand cubic meters (~13 million board feet, MMBF) per year, later rising to more than 212 thousand cubic meters (90 MMBF) to support the war effort: 25-year contracts were the norm. The Tongass Timber Act of 1947 required harvest levels of 10.6 million cubic meters (4.5 billion board feet) per decade, ushering in a period with high timber harvests under 50-year contracts with two firms: Ketchikan Pulp and Paper Company and Alaska Lumber and Pulp Company. Sawmills and pulp mills in Southeast Alaska created wood products that constituted a major share of West Coast exports by the early 1970s.

The 1971 Alaska Native Claims Settlement Act established Native landownerships throughout Alaska. In Southeast Alaska, 13 Native corporations were entitled to select about 222,600 ha (550,000 ac) of land from the Tongass National Forest, reducing Tongass timber production lands by about 10%. The 1980 Alaska National Interest Lands Conservation Act designated 2.7 million ha (6.6 million ac) in Southeast Alaska as congressionally designated wilderness, national monument, or land-use designation II (managed to retain a roadless, wildland character), further reducing land available for timber harvest on the national forest. These protected areas contain about 0.7 million ha (1.7 million ac) of timber lands. On the entire Tongass, there are approximately 1.7 million ha (4.2 million ac) of "other forest land" and 2.3 million ha (5.8 million ac) of productive forest land, capable of producing ≥1.4 cubic meters per hectare (20 cubic feet per acre) per year of wood fiber, or having greater than 46.6 cubic meters per hectare (8,000 board feet per acre). Despite the reductions in the lands available for harvest, long-term contracts remained in place. Native corporations sold timber to the highest bidder, usually into the whole-log export market. In 1990, the Tongass Timber Reform Act repealed the required federal harvest levels of the Tongass Timber Act and replaced them with a requirement to "seek to meet . . . demand." Owing primarily to market forces, the long-term contracts ended in the 1990s, and the largest manufacturing facilities closed.

Before WWII

The forest-products industry was among the earliest manufacturing industries to emerge in the Northwest. In the nineteenth century, the indus-

BOX 4.2. CONTEMPORARY PERSPECTIVES ON ALASKAN TIMBER ECONOMIES

Until recently, Southeast Alaska produced most of the timber harvested in Alaska. In 2005, the Southeast supplied 74% of Alaska's total timber harvest, most (60.7%) coming from Native and private lands, with state and other public lands supplying 21.7%, and 17.5% coming from national forests. In contrast, most mill supply came from national forests (53%), followed by state and other public lands (38%) and private and Native lands (8%). Timber harvest in southwest Alaska, including Kodiak and Afognak Islands, has recently been increasing. Since 2007, harvests from private and Native corporation lands in southwest Alaska have increased from about 71 000 to 144 000 cubic meters (30 to 61 million board feet) per year. These logs are exported to foreign markets, primarily Asia.

The wood-products industry in Alaska currently consists of individual- and family-owned sawmills and independent logging businesses. An annual regional survey of the 20 largest or most-active sawmills began in Southeast Alaska in 2000. Annual surveys assess mill capacity, employment, production, and markets. In 2007, the 20 original mills became 22 (partial subdivision and sale of one mill). Of those 22 mills, 10 were active in 2012, 2 were idle, and 10 (primarily the largest facilities) had been decommissioned or were no longer in production. In 2011, sawmills in Southeast Alaska were operating at 9.6% of capacity. By comparison, sawmills in Idaho, Oregon, California, and Montana generally operate at more than 80% of their capacity unless there is a severe economic downturn. The capacity utilization rate of the last operating medium-sized sawmill in Southeast Alaska in 2011 was estimated at about 13%. Low utilization rates make it extremely difficult for sawmill owners to cover fixed costs, much less make a profit. Most communities in Southeast Alaska have declined in population since the early 1990s, particularly those in which wood manufacturing facilities closed; lack of stable employment is a persistent issue.

Alaskan wood-products markets are closely tied to those in North America and the Pacific Rim and are deeply affected by tight credit, low-cost-margin issues, and the slowly recovering domestic housing market. The future projection is cautious optimism, relative to price and demand. Timber supply in Southeast Alaska will most likely be a constraint, owing to high prices in log-export markets and the relatively limited availability of timber from state and federal lands.

try was based on processing large, primarily old-growth trees from coastal forests into lumber for a variety of markets. The late 1880s saw an expansion of railroads to access timber in more remote locations. The labor force lived in transient logging camps or in mill towns located on the margin of the forest and the oak chaparral, in the foothills of the Cascade Range and the coastal forests. Residents did not own property and there was little sense of community. Federal policy was dominated by the desire to shift land into private "productive" hands (Robbins 1997).

The industry rapidly expanded in the twentieth century with the development of railroads that served domestic and export markets. The industry further matured during this period as a result of capital investment in processing facilities and the beginning of *vertical integration*. These vertically integrated companies owned both forestland and processing capacity, a structure that came to dominate the wood-products sector through the 1980s (plate 5A). By 1930, mills in the Douglas-fir region accounted for one-third of all US lumber production, based on large, high-quality logs available from private landowners. Douglas-fir lumber commanded a price premium relative to other species because it was relatively knot free (Haynes 2008).

At the same time, a concern about the sustainability of timber harvest and the stability of communities rose to the surface of the federal forest policy discussion. This concern was fed by observations of timber boom-and-bust cycles in the Lake States, the southern Appalachian region, and in the Mid-South of the United States. These cycles left behind communities distressed or abandoned. The turn of the century saw the establishment of the federal forest reserves (now known as national forests) and the USDA Forest Service. The latter played a largely custodial role in protecting forest resources and watersheds from unsustainable timber-harvest practices and wildfire (Haynes et al. 2007).

A deliberate policy focus on community economics and social well-being, combined with a desire by the forest industry to protect its investment in mills from volatile fluctuations in wood supply, led to an alliance in the Northwest among the forest-products industry, communities, and counties that was rooted in the principles of sustained yield. Public forestry policy discussions in the 1920s were heavily informed by the belief that forest regulations designed to produce an even flow of timber would best benefit neighboring communities where processing facilities were located (SAF 1989).

Two framing concepts are critical to this history. The first is based in nineteenth-century silviculture, which was rooted in the emerging science of growing trees to produce timber and to generate a sustained yield of wood. The second concept arose from the early twentieth-century "conservationist" conviction that scientific management could bring about an accommodation between the concentration and accumulation of wealth on the one hand and broader social equity on the other (Nechodom et al. 2008). In addition to the benefits of stable employment, proponents of sustained-yield management on public timberlands also promised a steady stream of forest revenues to support school funds and local government expenditures. During this period several cooperative sustained-yield units—in which an industry partner committed some portion of their land base to be managed as a catchment with federal land under sustained-yield principles in return for monopoly access to timber in the catchment—were proposed. Because of the controversial monopoly access clause, only one cooperative sustained-yield unit was ever established: the Shelton Unit was formed with the Simpson Logging Company on the Olympic Peninsula, Washington, in 1946.

Throughout this period, technological innovation continued to transform the industry, diversifying products and expanding markets. Nevertheless, demand did not keep pace with the expanding supply, and because wood prices stagnated through the 1920s, there was little pressure on the national forests to sell timber.

After World War II

After the Great Depression and World War II, Americans turned to the task of rebuilding. In an effort to keep the American Dream attainable, housing affordability and stable communities became policy objectives, leading to a burst of housing construction. As a consequence, softwood lumber demand reached new heights and private timber supplies started to decline. From the late 1940s until the late 1980s, timber harvest in the Douglas-fir region increased by roughly 25% and wood prices rebounded (Haynes 2008).

The stability of rural communities became a formal policy priority for the US Forest Service in the Sustained Yield Act of 1944. This was manifested by the agency's shift from a custodial role to an active role in timber supply under sustained-yield management. Between 1945 and 1965, timber harvest on lands managed by the Forest Service in the western forests of Oregon and Washington rose from about 4.22 million cubic meters (745 million board feet, or 149 million cubic feet) to 22.8 million cubic meters (4,035 million board feet, or 807 million cubic feet) as federal land managers implemented sustained-yield practices (Haynes et al. 2007). The emphasis continued with the implementation of the National Forest Management Act of 1976 (P.L. 94-588, 16 U.S.C. §§ 1600-1614), which directed the Forest Service to identify timber-dependent communities where at least 10% of employment was involved in timber processing, and to consider the effect of federal actions on those communities when developing plans.

Although industry continued to support the idea of sustained-yield management on federal land, stable rural communities remained an elusive goal owing to the volatility and globalization of wood-products markets; labor-saving technological changes that led to a shift to larger, higher-capacity mills; and expansion of the products mix. These factors shifted labor demand away from small rural communities to more urban communities that were near ports, markets for final wood products, and labor pools.

Volatile Wood-Products Markets

Residential construction comprises the vast majority of solid-wood-product consumption in the United States. Because housing is a durable good, and investment in new housing can be postponed when income falls, fluctuations in the larger economy are amplified in wood-products markets. Timber-harvest fluctuations between 1965 and 1990 (fig. 4.1) closely tracked housing starts. Ironically, private harvest is less volatile than federal harvest in spite of sustained-yield policies. Purchasers of federal timber, to the extent that their contracts allow, harvest when wood prices are high, whereas industrial-forest landowners hold timber as a capital asset and respond to interest rates as well as prices, which tends to moderate the effect of the business cycle on private timber harvest. In Oregon and Washington, state forestlands that are managed for schools and other public services suffer from the same volatility in harvest levels as federal lands, for the same reason.

Globalization of Wood-Products Markets

The log-export trade grew rapidly as economic growth in Pacific Rim countries provided expanding markets for US timber. During the 1970s and 1980s, log exports accounted for roughly 20% of timber harvested. The rapid growth in harvests and competition between markets led to rising stumpage prices and a ban on exporting federal logs west of the 100th meridian (~Colorado, US) without further processing (36 CFR Ch. 11, sec. 223.162). Hence the region's private timberlands and state lands in Washington (from which logs were exportable until the 1990s) were the primary

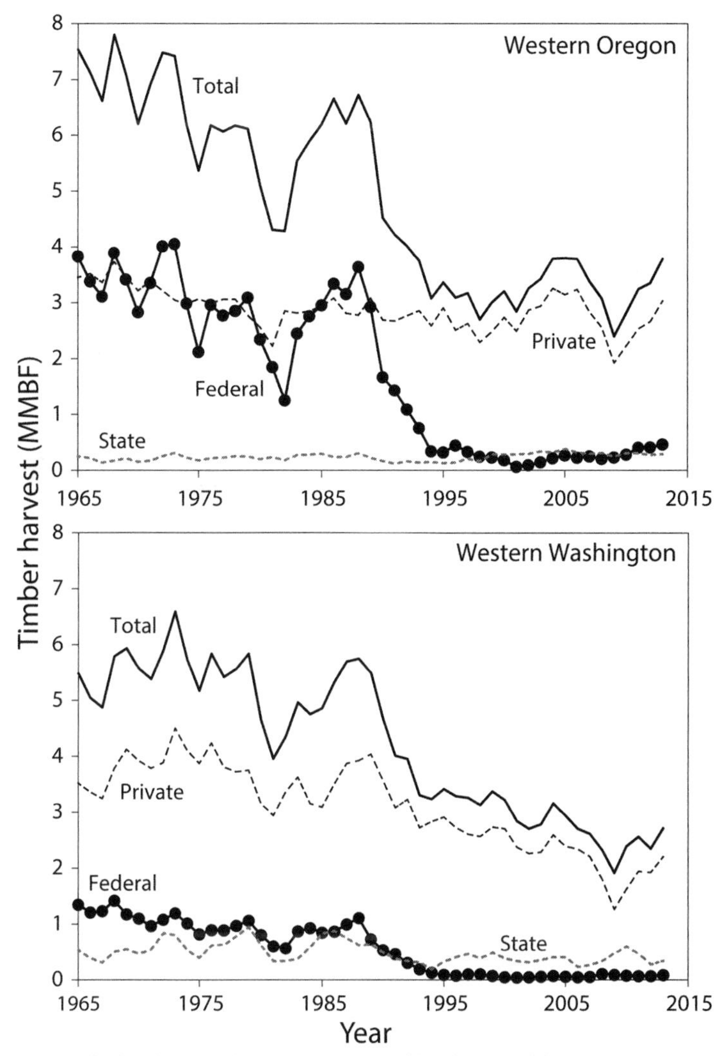

FIGURE 4.1. Timber harvest by major types of landownerships, western Washington and western Oregon (Zhou 2013).

source of logs for export. These forests tend to be located at lower elevations and closer to cities and ports relative to federal land. Because export markets favored larger, older, high-quality trees (e.g., Douglas-fir having straight stems, cylindrical boles, relatively small, infrequent branches, and high stiffness compared with other softwoods), higher log-export prices provided an incentive for some private landowners to manage for longer rotations. This had the ancillary ecological benefit of increasing the proportion of mature forests (older than 60 years) on some private lands, particularly nonindustrial private forestlands (Haynes 2008).

Expansion of the Product Mix

Industry diversification continued into the mid-1990s; the product mix expanded to include both lumber and products like paper that use residues from primary production. The proportion of stumpage used for lumber production fell from 76% in 1950 to 38% in 1979. By the late 1980s, the industry was bifurcated, with a growing segment of highly efficient mills that cut uniform sizes and grades of logs from mostly second-growth private timber for commodity markets, and some older, less-efficient mills that processed larger (and older) logs mostly from public timberlands for a range of markets, including high-quality domestic and export markets. The design of the older mills made them difficult to adapt to major changes that would soon shape the industry (Haynes 2008).

The Era of Ecosystem Management: 1991 and Beyond

Landmark changes started in 1991 with injunctions on the sale of federal timber in western Oregon and Washington that were addressed with the implementation of the Northwest Forest Plan. The Plan allowed federal sales to continue at roughly 20% of the 1980s level, yet actual sales rarely exceeded 10% of those levels. Plantation forests on industry land became the nearly exclusive source of wood in the region. The reductions in federal sales caused wood supplies to fall below existing log demand, and regional log prices spiked upward in 1993 as a result. This led to mill closures, particularly in rural communities with mills that were dependent on federal timber, and processing capacity shifted to other regions, particularly the US South and the interior Canadian provinces, where harvest levels increased (Haynes et al. 2007).

As the Japanese and other Asian economies collapsed at the start of the 1990s, the log-export market collapsed. Timber managers and landowners shifted formerly exported logs (annually, more than 10 million cubic meters [>2 billion board feet, log scale]) to the domestic market, helping the timber industry adapt to reduced federal harvest flows. This, and the shift of processing capacity and timber harvest to other regions, softened the impact on US consumers. However, the situation for rural communities in the region was aggravated by other developments in wood-products markets (Haynes et al. 2007), including reduced milling capacity for the changing resource base, continued labor-saving technological changes in both harvest and milling, and consolidation of processing capacity.

The region's timber industry restructured during the 1990s, becoming highly efficient but less product diverse, producing lumber from (~35- to 50-cm (14- to 20-inch) diameter logs (plate 5B) primarily for the domestic market with timber from private timberlands. By one measure, the product diversity that increased from 1950 to 1979 by 16% fell back to 1950 levels by 2002 (Haynes 2008). Little capacity remains to handle logs over 61 cm (24 inches) in diameter, but there is an evolving small-log industry using logs between 11 and 25 cm (4.5 and 10 inches) in diameter. Sawmills changed, with the development of both very large mills (producing ~600 to 800 cubic meters [300,000 to 400,000 board feet] per shift) and specialty mills, some of which are relatively small (less than ~100 cubic meters [50,000 board feet] per shift). Secondary processing has expanded into growing markets in engineered wood products to replace solid-wood products.

Changes in the economic structuring of forest industries began in the 1980s but dominated the 1990s and beyond. For many reasons, including changes in the federal tax code, the forest industry moved away from the vertically integrated structure that had dominated since the turn of the century. Formerly vertically integrated firms chose to sell forest landholdings to timber investment management organizations (TIMOs) or to restructure into separate manufacturing firms and landholding firms known as real estate investment trusts (REITs). By 2005, TIMOs and REITs held over half of the 23 million ha (58 million ac) that had been held by vertically integrated firms in the United States in 1980. Although predictions of how this restructuring would affect rural communities are largely speculative, there is some concern that TIMOs and REITs will be less likely to invest in relationship building in rural communities than the mill- and forestland-owning firms of the past (Clutter et al. 2005).

Since the mid-1990s, stumpage prices have been stable or declining. Relatively weak stumpage prices are leading to a shift toward forest management regimes that favor shorter rotations. Today, the economic incentive for all private landowners is to grow smaller, more uniform trees. One consequence of this is a divergence between ecological (early- versus late-seral) conditions on public and private timberlands.

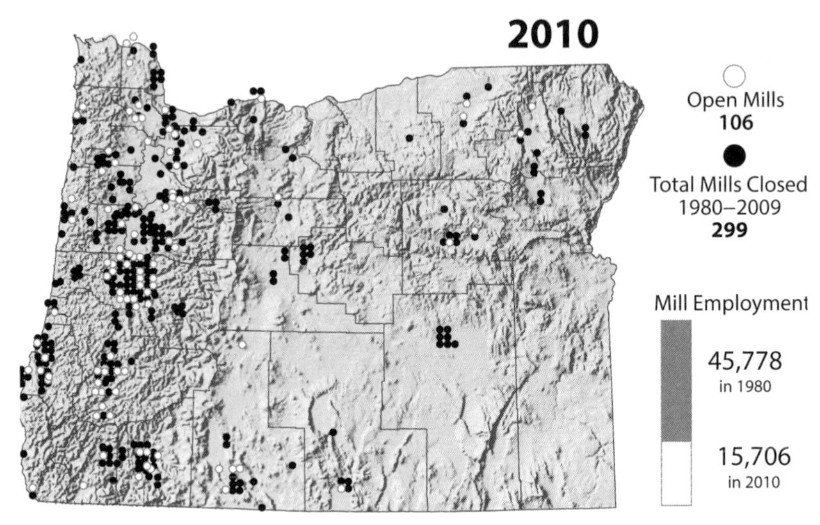

FIGURE 4.2. Operating sawmills in Oregon in 2010 and change from 1980 (adapted from OFRI et al. 2011, data source Ehinger 2010).

Implications for the Well-Being of Forest-Dependent Communities

The combined effect of these changes (restructuring of the timber industry, decline in federal timber sales, price volatility) on the rural economies of the region has been profound. Of the 405 mills operating in Oregon 1980, only 106 remained open in 2010. The surviving and new mills are in locations along main transportation corridors and close to private timberlands (fig. 4.2). Some rural areas formerly thought of as timber-dependent now have little local forest-products manufacturing, and logs harvested in the area are shipped to distant manufacturing centers, resulting in slightly lower stumpage prices than in the past and reduced employment in spite of relatively high harvests on private land.

The loss of employment in forest-dependent communities has contributed to increases in the rate of poverty in rural counties relative to metropolitan counties, particularly in Oregon. In Oregon, poverty rates were 1.3% higher in rural counties than in metropolitan counties in 1989; in 2010, they were 3% higher; in Washington, the discrepancy has been stable, about 2.5%. The impact of reduced federal timber harvest is greater in the five rural Oregon counties that received 75% of the payments in lieu of property taxes from the Oregon & California Railroad (O&C) lands, former railroad grant lands that are managed in trust for the counties by the Bureau of Land Management (BLM): poverty rates for those counties ranged from 15.7% to 19.5% in 2010 (metropolitan county average: 12.6%). Following implementation of the Northwest Forest Plan, harvest and revenue (in 1982 dollars) fell from an annual average of 2.779 million cubic meters (1,178 million board feet) and US$187 million between 1962 and 1990 to an annual average of 0.429 million cubic meters (182 million board feet) and US$78 million. These revenue losses have caused severe budget reductions for county governments where O&C revenues were a major source of public funding.

Beyond the poverty statistics, there are other indicators of growing socioeconomic distress in forest-based communities. According to the Clatsop County food bank, food distributed rose from 288 000 kg (635,000 pounds) in 2006–2007 to 607 000 kg (1,340,000 pounds) in 2012–2013. During 2012–2013, 25% of residents and 41% of children in Clatsop County qualified for food assistance (Betts et al. 2014).

Where federal timber harvest occurs, it attempts to be consistent with ecosystem management objectives, with socioeconomic trade-offs. Until the 1990s, stumpage prices for Oregon Department of Forestry, BLM, Washington Department of Natural Resources, and Forest Service timber were roughly equivalent. In the last two decades, stumpage prices for state and federal timber have diverged: federal timber brought an average of US$94 per cubic meter (US$195 per thousand board feet) and US$113 per cubic meter (US$234 per thousand board feet) less than state timber in Oregon and Washington, respectively. This divergence can be attributed to (1) ecosystem management strategies such as restrictions on harvesting larger trees; and (2) modified silvicultural systems that embrace broader management objectives, use different types of harvest methods (clearcut, uneven-age, thinning) and harvest systems (ground based, cable, and helicopter) that lead to greater variation in logging costs among timber sales. Studies report that patch-cut or group-selection methods can cost nearly 25% more than clearcut methods (Curtis et al. 2004).

In the midst of these changing timber policies and forest management objectives, rural community demographics also are changing. These changes include shifts in the proportion of employment in the forest sector relative to total employment, shifts in the economic diversity of communities, and strong in-migration (Donoghue et al. 2006). Through the 1970s, the forest-products industry was an economic mainstay of western Montana, Washington and Oregon, Northern California, and the Idaho Panhandle.

Often it provided the only manufacturing employment, the livelihood for dozens of small communities, and helped define the sense of place that frequently motivates residents "to struggle with each other for the future of the lands and homes they love" (Robbins 1997). It is no longer typical for all forest communities to be economically dependent on timber extraction. Some rural areas have been changed by urban flight and suburban sprawl that have created an interface where new development encroaches on open spaces and forests. Many communities that were once economically dependent on timber, mining, and ranching are now attracting amenity migrants as gateways to public land (Kruger et al. 2008). These incursions create challenges for the management of both private and public timberlands and affect community cohesiveness (Donoghue and Sturtevant 2008).

Research on economic transitions in resource-based communities in the West suggests that the ability of forest communities to attract and sustain new economies—such as service, value-added manufacturing, and knowledge-based industries—depends upon access to aesthetic amenities, transportation, and communication. New residents drawn to forest communities' clean air and water, scenery, and recreational opportunities provide economic growth opportunities, new tax bases, and demand for services. For many families, the mailbox and Internet have become sources of income as telework increases and as investment income, retirement benefits, or government payments increasingly account for much of the economy in forest communities. The harvest, sale, and processing of nontimber forest products, such as floral greens and wild edible fungi, also is a significant activity in forested landscapes (Alexander et al. 2011). Although these industries provide rural people with economic opportunities and additional food security (McLain et al. 2008), they serve only as income supplements and do not solve poverty.

Mitigating poverty and economic inequality in these communities requires building trust and addressing tensions among forest managers, long-time residents, and recent in-migrants (chap. 18; Harrison et al. 2016). Long-time residents harbor resentment about lost employment opportunities in the extractive industries of the past. Amenity migrants, who seek forest-based amenities, natural landscapes, and slower lifestyles, tend to be interested in protecting the natural integrity of landscapes and thus are unhappy with residual timber harvesting, ranching, and mining that detract from the visual landscape. New institutional arrangements for managing forested landscapes have developed in the last decade or so (Montgomery 2013) and may help to meet the needs of a broader array of stakeholders than in the past (collaboratives, chap. 9).

A Challenging Future

The goods and services coming from the moist coniferous forests of the Northwest largely define the social and economic conditions of forest-based communities in the region. At the broad scale, the public acts as if forested ecosystems are a set of "commons" capable of producing both market and nonmarket goods and services. Many of us expect that forest management will contribute to maintaining a broad set of environmental values, including timber production, wildlife habitat, aesthetics, biological diversity, water flows, ecological integrity, and recreation. Forest management, the purposeful management of timberlands to meet owner objectives, has evolved in the region and ranges from custodial (no active management) to intense management in which stands are managed for timber volumes or for particular species and sizes of trees.

Price incentives were a key component of sustainable forest management in this region, but in the last two decades reduced financial returns have threatened progress toward sustainable forest management. At the same time, population shifts and ongoing economic changes complicate the uneasy relation between public land-management agencies and privately funded enterprises that depend on public timber. This is also a region where the growing reliance on regulations and the changing nature of forest governance is challenging to many forest managers, who see regulations and shared decision making as impeding their management discretion. Nevertheless, the Northwest is a region where alliances among forest owners, the forest-products industry, various agencies, and interested public groups—all motivated by a shared desire to protect open spaces and the associated goods and services—led to early adoption of sustained-forest-based development. Our challenge now is to seek equilibrium among interests advocating environmental protection, employment that contributes to economic prosperity, public access, and social justice.

Literature Cited

Alexander, S. J., S. Oswalt, and M. Emery. 2011. *Nontimber forest products in the United States: Montreal Process criteria and indicators as measures of current conditions and sustainability*. General Technical Report PNW-GTR-851. Portland, OR: USDA Forest Service, Pacific Northwest Research Station.

Betts, M., B. Bourgeois, R. Haynes, S. Johnson, K. Puettmann, and V. Sturtevant. 2014. *Assessment of alternative forest management approaches: Final report of the Independent Science Review Panel*. Prepared with assistance from D. C. E. Robinson, A. W. Hall, and G. Stankey. Vancouver, BC, Canada: ESSA Technologies Ltd for Oregon Department of Forestry, Salem, OR.

Clutter, M., B. Mendell, D. Newman, D. Wear, and J. Greis. 2005. Strategic factors driving timberland ownership changes in the U.S. South. http://www.srs.fs.usda.gov/econ/pubs/southernmarkets/strategic-factors-and-ownership-v1.pdf.

Curtis, R. O., D. D. Marshall, D. S. DeBell, eds. 2004. *Silvicultural options for young-growth Douglas-fir forests: The Capitol Forest study—establishment and first results.* General Technical Report PNW-GTR-598. Portland, OR: USDA Forest Service, Pacific Northwest Research Station.

Donoghue, E. M., N. L. Sutton, and R. W. Haynes. 2006. *Considering communities in forest management planning in western Oregon*. General Technical Report PNW-GTR-693. Portland, OR: USDA Forest Service, Pacific Northwest Research Station.

Donoghue, E. M., and V. E. Sturtevant. 2008. *Forest community connections: Implications for research, management, and governance.* Washington, DC: Resources for the Future.

Ehinger, P. F. 2010. *Summary of mill closure data from 1990–2010*. Paul F. Ehinger and Associates. http://www.amforest.org/images/pdfs/Mill_Closures.pdf.

Harrison, J. L., C. A. Montgomery, and J. C. Bliss. 2016. Beyond the monolith: The role of bonding, bridging, and linking social capital in the cycle of adaptive capacity. *Society and Natural Resources* 29:525–539.

Haynes, R. W. 2008. *Emergent lessons from a century of experience with Pacific Northwest timber markets.* General Technical Report PNW-GTR-747. Portland, OR: USDA Forest Service, Pacific Northwest Research Station.

Haynes, R. W., D. M. Adams, R. J. Alig, P. J. Ince, J. R. Mills, and X. Zhou. 2007. *The 2005 RPA timber assessment update*. General Technical Report PNW-GTR-699. Portland, OR: USDA Forest Service, Pacific Northwest Research Station.

Kruger, L. E., R. Mazza, and M. Stiefel. 2008. Amenity migration, rural communities, and public lands. Pp. 127–142 in *Forest community connections: Implications for research, management, and governance*. Edited by E. M. Donoghue and V. E. Sturtevant. Washington, DC: Resources for the Future.

McLain, R., S. Alexander, and E. Jones. 2008. *Incorporating understanding of informal economic activity in natural resource and economic development policy*. General Technical Report PNW-GTR-755. Portland, OR: USDA Forest Service, Pacific Northwest Research Station.

Montgomery, C. A. 2013. Institutional environments and arrangements for managing complex aquatic ecosystems in forested landscapes. *Forest Policy and Economics* 35:50–56.

Nechodom, M., D. R. Becker, and R. W. Haynes. 2008. Evolving interdependencies of community and forest health. Pp. 91–108 in *Forest community connections: Implications for research, management, and governance*. Edited by E. M. Donoghue and V. E. Sturtevant. Washington, DC: Resources for the Future.

OFRI (Oregon Forest Resources Institute), Oregon Department of Forestry, and Oregon Forest Industries Council. 2011. *Oregon forest product manufacturing, 1980–2010*. Narrated slide show. https://www.youtube.com/watch?v=QWfJtTxiKwc.

Robbins, W. G. 1997. *Landscapes of promise. The Oregon story, 1940–2000*. Seattle, WA: University of Washington Press.

SAF (Society of American Foresters). 1989. *Report of the Society of American Foresters National Task Force on Community Stability*. SAF 89-06. Bethesda, MD: Society of American Foresters.

Zhou, X. 2013. *Production, prices, employment, and trade in Northwest forest industries, all quarters 2012*. Resource Bulletin PNW-RB-265. Portland, OR: USDA Forest Service, Pacific Northwest Research Station.

Chapter 5

An Ecosystem Services Framework

Dale J. Blahna, Stanley T. Asah, and Robert L. Deal

Ecosystem services are the full range of social, ecological, and economic benefits that people obtain from nature (Millennium Assessment 2003; Smith et al. 2011). These services include both biophysical (e.g., water, food, and fiber) and intangible (e.g., cultural or health) benefits. The concept originated in ecological economists' attempts to assign monetary valuations to the goods and services humans receive from naturally functioning ecosystems, so that the full array of direct and indirect benefits are captured in environmental policy, management, and decision making (Westman 1977). The importance and value of ecosystem services are being recognized internationally (Farley and Costanza 2010; Muradian et al. 2010), as illustrated by the Intergovernmental Platform on Biodiversity and Ecosystem Services (IPBES 2012), which is currently supported by 124 nations. Many US state and federal natural resource agencies have adopted policies that include analyses of ecosystem services in planning and decision making. The US Environmental Protection Agency, US Geological Survey, and US National Park Service all have new initiatives regarding the identification and mapping of ecosystem services. Specific to national forestlands, the US Forest Service's new planning rule (USDA 2012) requires all 175 national forests to report key ecosystem services for forest plan assessments and revisions.

The general purposes of ecosystem services analyses have expanded dramatically since the late 1990s (Kline et al. 2013). What began as a "simple" process of identifying monetary values of natural capital has evolved into large and diverse sets of metrics, analyses, and frameworks. Nevertheless, the fundamental purposes of accounting for ecosystem services are conceptually straightforward: (1) to measure and describe how humans benefit from forest ecosystems (assessment purpose); and (2) to use service and benefit measures to evaluate trade-offs of alternative forest-management decisions (decision-support purpose). Although the methods for conducting ecosystem services analyses may be complex, some relatively specific principles can be applied to help focus and frame analyses. In this chapter, we provide background on ecosystem services assessments and trade-off analysis. Ecosystem services are discussed more specifically relative to Pacific Northwest moist-forest landownerships in chapter 6.

Ecosystem Services Assessments

The concept of ecosystem services is inherently anthropocentric. The overarching goal is to describe environmental services that a landscape provides in terms of human benefits or outcomes so that the value of natural systems for people can be understood in decision making (Costanza et al. 1997). Understanding both sides of the environmental-service/human-benefit equation is required for an ecosystem services assessment. When thinking of clean water as a forest ecosystem service, the description of human benefits will be very different if they are measured as the market value of drinking water, the importance of clean water for salmon habitat, or the value of water as an indispensable requirement for human life. Although it is impossible to describe all service-benefit combinations for a forested landscape (Vatn and Bromley 1994), understanding the target human benefits that will frame an assessment is as important as understanding environmental conditions.

Ecosystem services assessments are inherently complex because services are numerous and interlinked, and there can be many different ways to describe overlapping and intersecting services and benefits (Fisher et al. 2009). Even an apparently simple human benefit from a forest, such as mushroom harvesting, can incorporate indirect social and cultural benefits like an appreciation for a unique sense of place in the forest, social group interaction, and a sense of community among mushroom harvesters (Asah et al. 2014). And while some benefits, especially those traded on the open market, can be described in relatively specific (and agreed upon) measures and valued in monetary terms, many cultural benefits like sense of place and social cohesion are subjective and difficult to monetize or quantify (Vatn and Bromley 1994; Bockstael et al. 2000; Kline and Mazotta 2012; Kline et al. 2013). Related to this, there have long been philosophical concerns that any attempts to quantify intrinsic values of nature are misleading (Chan et al. 2012a; Silvertown 2015). Furthermore, human benefits derived from ecosystems are variable and subject to shifting norms and values over time. Although salmon habitat has been considered a valued ecosystem service of Northwest forests, spanning Native to contemporary cultures, benefits from the provision of northern spotted owl habitat have increased dramatically in recent years as biologists recognize the importance of old-growth forests for owl survival. Yet northern spotted owls may be a valued ecosystem service only for some people, whereas other people concerned about the human benefits of the timber industry may consider these owls a negative value. So the human benefits provided by salmon and owls are not just related to food and beauty, but also to scarcity, political capital, and traditional and intrinsic values of nature. Thus it is important to understand the underlying human values that help explain how ecosystem services and benefits are linked (Chan et al. 2012a).

It is also important to recognize that benefits derived from ecosystem services are context dependent; they may shift with geographic, temporal, and sociopolitical context. For example, it is not enough to identify the market cost of an individual salmon and then tally up the total "value" based on the estimated number of salmon. The value of the last known salmon in a vanishing salmon run will be *much* higher than that of one salmon in a larger run, owing to scarcity, irreversibility, political image, and other intangible values. Likewise, there is no absolute measure of ecosystem services benefits; they are based to a large extent on public perceptions (Vatn and Bromley 1994; Fisher et al. 2009). As a result, consensus is emerging that it is essential to incorporate decision makers, managers, scientists, and the public in ecosystem services assessments (Asah et al. 2012; Nahlik et al. 2012; Kline et al. 2013).

One approach for dealing with the complexity of ecosystem services assessments is to focus analysis on *final* ecosystem goods and services. Rather than measuring and summing all interrelated services and intermediate services, emphasis is placed on the identification of a smaller number of benefits resulting from direct human interactions with ecosystems (Boyd and Banzhaf 2007; Nahlik et al. 2012; Wong et al. 2015). Ringold et al. (2013) defined final services as "biophysical features that people perceive as being directly related to their well-being" and identified a six-step process for working with beneficiaries to identify ecosystem attributes, metrics, and indicators of final services. For catch-and-release anglers on rivers in 12 western states, they identified size and abundance of different fish species, aesthetic importance of the stream setting, and usability and safety of sites as indicators of fishing quality (classified as high, medium, or low). The intermediate ecosystem services that were not quantified in this simpler analysis included ecosystem attributes, functions, or processes that were essential to delivering those final services, such as the food webs that supported fish to achieve desired numbers and sizes, and requisite habitat conditions, including water quality and riparian interactions. So while the quantification of final ecosystem goods and services is more tractable, their delivery depends on intermediate services that are important to recognize.

Another consideration for ecosystem services assessments is the quality and availability of data (Tallis et al. 2008). In the preceding example, Ringold et al. (2013) used existing biophysical data from the US Environmental Protection Agency's Environmental Monitoring and Assessment Program to quantify angler preferences. For some ecosystems, there may be a wealth of biophysical resource data, but there may not be a match between the data and relevant ecosystem services. Also, compared with biophysical data, social and cultural data for estimating human values of ecosystem services are often scarce (Daniel et al. 2012). As a result, only a few types of human-benefit measures are possible using existing data; primary social-science data need to be collected.

Socioeconomic methods for monetary valuation of ecosystem services focus on *contingent valuation* (a survey-based technique for the valuation of nonmarket resources in which people are asked to state their value for a service contingent on a specific scenario given), willingness to pay, and willingness to accept (Bockstael et al. 2000; Champ et al. 2003). There is also an expanding literature on methods for estimating more intangible ecosystem services values, with methods drawn from such diverse fields as metaphor analysis, landscape aesthetics, outdoor recreation, and cultural heritage, spiritual, and religious studies (Daniel et al. 2012; Raymond et al. 2013). However, the selection, use, and effectiveness of specific market and nonmarket valuation methods depend upon the management context and may need to be interpreted within the scope of the purpose of the analysis, which rarely is simply conducting an assessment of ecosystem services.

A complete accounting of human benefits of any ecosystem is difficult, if not impossible. Hence, managers need to focus on the benefits that are important to their decisions, based on the ecological, social, and economic context (Fisher et al. 2009). Rather than starting with a detailed assessment of existing ecosystem attributes and services, applied ecosystem services studies may be conceptually reverse-engineered to first identify and prioritize key ecosystem services and human-benefit measures that are most important in addressing management and policy issues (Wong et al. 2015; Penaluna et al. 2016).

Ecosystem Services for Decision Support

Rarely does an ecosystem services assessment by itself meet forest planning or management needs. The primary purpose for conducting an ecosystem services analysis is to provide an understanding of the trade-offs in human benefits that will result from different policy or management actions (Fisher et al. 2009; Chan et al. 2012b; Kline and Mazzotta 2012; Wong et al. 2015). Simply conducting a descriptive assessment of existing ecosystem services and related human benefits conveys very little about the changes in human welfare that may result from different management scenarios. There is no single place to fit an ecosystem services analysis into a broader planning or decision-making framework, but many examples of specific protocols are available (Raphael et al. 2007; de Groot et al. 2010; Thompson et al. 2013; US Environmental Protection Agency 2015; Wong et al. 2015; NESP 2016). In this section we provide a framework to help conceptualize the general stages for applying an ecosystem services approach to planning and trade-off analysis.

Six general steps comprise an applied ecosystem services analysis (fig. 5.1). Steps 1, 2, 5, and 6 represent the standard assessment process discussed above. Steps 1 and 2 define current ecosystem conditions and relevant ecosystem services by applying ecological and socioeconomic production functions (Kline et al. 2013; Wong et al. 2015). Steps 2 and 5 (connected by dashed line) represent the identification of final ecosystem services (or other approach to focus the assessment) and the indicators and metrics used to measure the human benefits. However, there is no direct or logical link between steps 5 and 6 without a *process* for linking management objectives to potential actions that will form the basis for analyzing trade-offs (step 3) and for prioritizing key ecosystem services (step 4) that best address those trade-offs (Chan et al. 2012b; Kline et al. 2013; Wong et al. 2015). Prioritizing key services helps narrow the scope of analysis to a manageable and measureable set of outcomes.

An iterative problem-framing process (Clark and Stankey 2006; Thompson et al. 2013) bridges ecosystem services to management and policy and guides the development of steps 3 to 6. This process starts by identifying the specific focusing issue or problem statement(s) that will guide

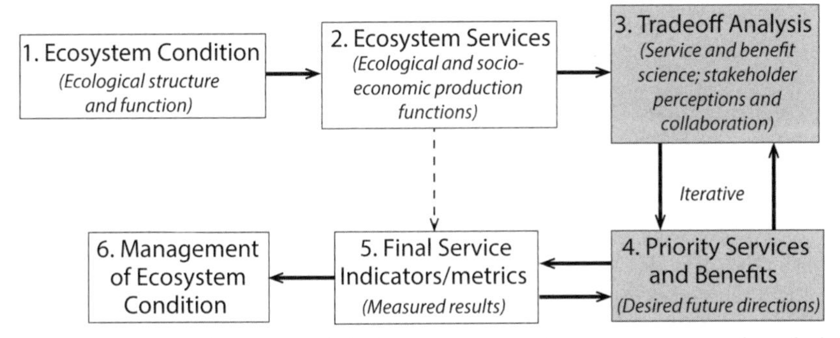

FIGURE 5.1. Step-by-step model of ecosystem services assessment and analysis process.

the analysis (Blahna and Yonts-Shepard 1990), after which scale of analysis, stakeholders, and other analysis sideboards can be developed (Raphael et al. 2007; Fisher et al. 2009; Thompson et al. 2013). Studies of ecosystem service trade-off analyses include examples of issue statements, but most do not explain how or why the focusing issues were developed. The guiding issues can be based on one or more of the following three factors.

Project Purpose or Outcome—The primary legal or administrative reasons, goals, or objectives for implementing the project or management actions can guide analyses (Wong et al. 2015). For example, Ricketts et al. (2004) estimated the benefits of wild pollinators as a forest ecosystem service for economically important coffee production in Costa Rica. In another case, Wong et al. (2015) conducted an assessment for construction of six new lakes and restoration of a large wetland area in the Yongding River Ecological Corridor in Beijing, China. They evaluated ecosystem services that specifically addressed five policy goals that made the project a funding priority: water purification, local climate regulation, water storage, dust control, and aesthetics.

Environmental or Social Problems—Many ecosystem services studies are designed to provide insights into different approaches for addressing or combating specific environmental problems. For example, US Geological Survey scientists conducted a study of six ecosystem services of the Puget Sound basin, Washington, USA, that were important for understanding the effects of regional population growth and urbanization (Bagstad et al. 2014). They evaluated ecosystem services related to carbon storage and sequestration, flood regulation in 100-year floodplains, sediment regulation for reservoirs, scenic viewsheds for homeowners, and open-space proximity for homeowners. Forest condition and structure played a pivotal role in each of these ecosystem services.

Stakeholder Conflicts—Conflict issues often are ignored or avoided in planning and decision making (Blahna and Yonts-Shepard 1990; Endter-Wada et al. 1998). Yet avoiding conflicts has had the unanticipated effect of exacerbating controversies in the long run, resulting in the more recent focus on collaborative approaches to management (Keough and Blahna 2006). By addressing specific environmental conflicts, ecosystem services analyses can help identify solutions that address the goals of multiple entities. For example, Foley et al. (2014) conducted an ecosystem services analysis for a wetland restoration project on the Deschutes National Forest, Oregon, USA. The restoration project was being developed to reduce ecosystem threats from "stand-replacing fire, beetle infestations, and unmanaged recreation impacts." Although this is an example of addressing environmental problems, the study also included a detailed analysis of the effects on matsutake mushroom harvesting. Compared with other ecosystem services that were evaluated, like wildlife habitat and cultural values of the wetland, mushroom picking was a very specific benefit that was included because of past mushroom harvester conflicts with the agency and recreationists.

Guiding issues should be based on a realistic assessment of the decision context and the time and funding available for conducting a detailed multidisciplinary ecosystem services analysis. Each issue requires a separate analysis. The guiding issues should also be specific and concise so that relevant management actions are evident (Blahna and Yonts-Shepard 1990; Clark and Stankey 2006). The next task for framing the analysis is to use the guiding issue or problem statements to develop objectives that will result in desired future outcomes (both environmental and social) and potential management actions to meet those objectives (Thompson et al. 2013). From these objectives and potential actions, a type of triage process can be conducted to identify priority ecosystem services to be used in detailed trade-off comparisons.

Identifying priority ecosystem services and collecting the human-benefit data that will inform the detailed analysis of trade-offs is not just an analytic or scientific process but includes multiple opportunities for public engagement and interaction among scientists, managers, and stakeholders (Ringold et al. 1996; Chan et al. 2012b; Nahlik et al. 2012). Guiding issues also can be used to help clarify the public engagement opportunities necessary to perform service-benefit trade-off analyses for each issue. Thus, steps 3 through 6 (fig. 5.1) are repeated for each issue or problem statement. However, it is not a clear, linear process but may vary depending on the individual issue and decision context. One issue may be clearly defined, easily quantified, and move rapidly through steps 3 through 6. Other issues may be very difficult, complex, or controversial, requiring iterative loops back and forth between steps 3 and 4 or even steps 3, 4, and 5.

The final goal of the process is to show the total changes in human benefits for each alternative management action, along with the distribution of positive and negative outcomes of each action for each stakeholder group (Vatn and Bromley 1994; Fisher et al. 2009; Kline and Mazzotta 2012). Given the diversity of potential benefits to be summarized per issue and set of trade-offs, including monetary values for provisioning resources and qualitative estimates of intangible benefits, it is unlikely that outcomes can be displayed using a single metric (e.g., dollar values), and benefit measures often will be supplemented with professional judgment. Kline and Mazotta (2012) recommended using tables, graphics, or narratives to visually summarize ecosystem services outcomes of different management scenarios.

Hence a well-planned and strategic approach is needed for conducting an effective ecosystem services analysis. A coarse-filter, rapid, descriptive assessment of existing data about ecosystem services will not be a useful decision aid but can be valuable in providing initial input to the issue-framing process needed to design steps 3 to 6 (fig. 5.1). In a broader summary of the process (fig. 5.2), the six steps have been collapsed into three very general steps: ecosystem conditions (combined steps 1 and 6, fig. 5.1); ecosystem services analysis (steps 2 and 5); and decision context and framing (combined steps 3 and 4). The arrow widths in fig. 5.2 symbolize the relative amount of effort and data used in the analysis. The top arrows illustrate a rapid, coarse-filter assessment of existing data and stakeholders that feeds into the framing process. After the initial problem framing—in which decisions are made about guiding issues, key ecosystem services, and human-benefit metrics—more detailed fine-filter data (illustrated by the thicker arrows, fig 5.2) are used in a *targeted analysis of ecosystem service trade-offs*, leading to the final ecosystem management policy or decisions (fig. 5.2, bottom arrows).

Future Applications in Forest-Management Planning

The global application of the ecosystem services concept is expanding dramatically. In April 2015, the International Union of Forestry Research Organizations (IUFRO 2015) established a task force on the Contribution

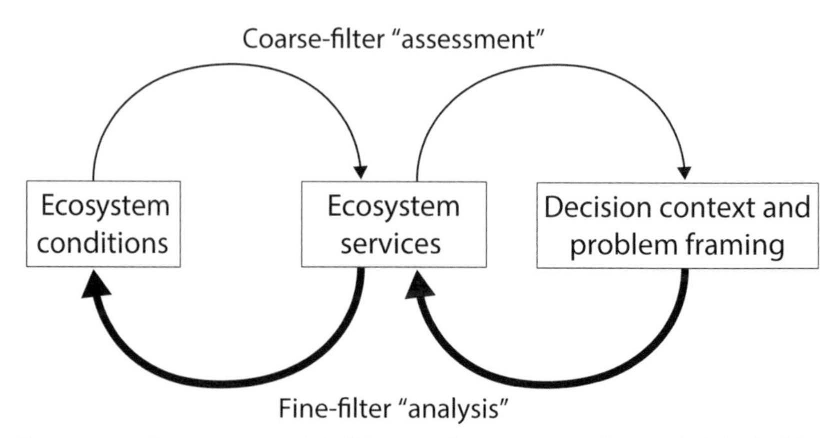

FIGURE 5.2. A summary model of the use of ecosystem services analysis in decision support.

of Biodiversity to Ecosystem Services (CoBES) to illustrate use of the concept in forest management and policy. The main goals are to provide policy makers and forest managers with information to improve the sustainable management of forests, identify research needed on the relationship between biodiversity and ecosystem services, and inform policy makers of the cross-sectoral importance of forest ecosystem services. During its first term (2015–2017), CoBES is developing several products, including a database of relationships between forest biodiversity and ecosystem services, a suite of publications, and international conferences.

The US Forest Service is developing strategies to address new requirements for including ecosystem services in the Forest Service planning rule (USDA 2012) and to apply the concept in operations and management decisions. Also, in response to the growing interest and potential applications of ecosystem services in decision making, the US Forest Service established the National Ecosystem Services Strategy Team in 2013 (NESST). The purpose of this team is "to collaboratively develop national strategy and policy around ecosystem services and integrate it into Forest Service programs and operations" (Deal et al. 2014).

The guidelines for implementing the Forest Service planning rule provide a specific example of how the ecosystem services framework presented above can be used (USDA 2013). The guidelines call for a three-phase forest plan revision process: assess, revise, and monitor. Ecosystem services data are included in the assessment phase for each national forest, using existing data. This roughly corresponds to a coarse-filter assessment (fig. 5.2). How ecosystem services data can be used to help evaluate management trade-offs is not specified. Since the number and type of ecosystem services provided by a national forest is very large, it will be imperative to focus on a few key planning issues to conduct the detailed ecosystem services analyses (fine-filter analysis, fig. 5.2). The targeted ecosystem services then can be used to form the basis for a practical approach for monitoring forest plans and for using the results for adaptive management, based on changing ecosystem-services benefits over time.

Next Steps

Ecosystem services analyses can provide a useful approach for estimating changes in human benefits under different forest policy and management scenarios. As a decision-support tool, the purpose and structure of an ecosystem services analysis must first be clarified by decision makers and analysts. Although the first (descriptive assessment) type of ecosystem services analysis is primarily an empirical, analytic process with some supplemental input from stakeholders (e.g., to identify human benefits and values), the second type of analysis (decision-support) is essentially a *social and political process that is informed by science*. Even the most ardent critics of ecosystem services, those opposed to the fundamental anthropocentric basis of the concept, recognize the importance of identifying and communicating how healthy ecosystems contribute to human well-being, and the need for using democratic processes to evaluate and make final decisions about forest management and conservation (Silvertown 2015).

It is important to note that these democratic processes are topic, context, and scale dependent. In short, there is no magic bullet (or single framework) for structuring and using an ecosystem services analysis. However, a key element is developing clear and concise issue or problem statements to help focus analysis. Also, thinking backward—that is, starting from management objectives and possible actions—may help structure both the detailed ecosystem services assessment and the process for using the benefit analysis in planning and decision making (Penaluna et al. 2016). To provide decision support, ecosystem services analyses need to be context dependent and integrated fully into the entire policy or management planning process. This includes a systematic review of management needs and issues, collaborative processes for identifying objectives and management options for each issue, and the prioritization of key services and benefits. This requires transparent, iterative, and collaborative planning and decision processes.

Literature Cited

Asah, S. T., D. J. Blahna, and C. M. Ryan. 2012. Involving forest communities in identifying and constructing ecosystem services: Millennium Assessment and place specificity. *Journal of Forestry* 110:149–156.

Asah, S. T., A. D. Guerry, D. J. Blahna, and J. J. Lawler. 2014. Perception, acquisition, and use of ecosystem services: Human behavior, and ecosystem management and policy implications. *Ecosystem Services* 10:180–186.

Bagstad, K. J., F. Villa, D. Batker, J. Harrison-Cox, B. Boigt, and G. W. Johnson. 2014. From theoretical to actual ecosystem services: Mapping beneficiaries and spatial flows in ecosystem service assessments. *Ecology and Society* 19(2):64. http://dx.doi.org/10.5751/ES-06523-190264.

Blahna, D. J., and S. Yonts-Shepard. 1990. Preservation or use? Confronting public issues in forest planning and decision making. Chap. 13. Pages 161–176 in *Outdoor recreation policy: Pleasure and preservation*. Edited by J. Hutcheson, F. Noe, and R. Snow. New York: Greenwood.

Bockstael, N. E., A. M. Freeman, R. J. Kopp, P. R. Portney, and V. K. Smith. 2000. On measuring economic values for nature. *Environmental Science and Technology* 34:1384–1389.

Boyd, J., and S. Banzhaf. 2007. What are ecosystem services? The need for standardized environmental accounting units. *Ecological Economics* 69:616–626.

Champ, P. A., K. J. Boyle, and T. C. Brown. 2003. *A primer on nonmarket valuation*. New York: Springer Science and Business Media.

Chan, K. M. A., A. D. Guerry, P. Balvanera, and 14 coauthors. 2012b. Where are *cultural* and *social* in ecosystem services? A framework for constructive engagement. *BioScience* 62:744–756.

Chan, K. M. A., T. Satterfield, and J. Goldstein. 2012a. Rethinking ecosystem services to better address and navigate cultural values. *Ecological Economics* 74: 8–18.

Clark, R. N., and G. H. Stankey. 2006. *Integrated research in natural resources: The key role of problem framing*. General Technical Report PNW-GTR-678. Portland, OR: USDA Forest Service, Pacific Northwest Research Station.

Costanza, R., R. d'Arge, R. de Groot, and 10 coauthors. 1997. The value of the world's ecosystem services and natural capital. *Nature* 387:253–260.

Daniel, T. C., A. Juhar, A. Arnberger, and 19 coauthors. 2012. Contributions of cultural services to the ecosystem services agenda. *Proceedings of the National Academy of Sciences of the United States of America* 109:8812–8819.

Deal, R., E. Weidner, and N. Smith. 2014. Integrating ecosystem services into U.S. Forest Service programs and operations. In *Federal resource management and ecosystem services guidebook*. Durham, NC: National Ecosystem Services Partnership, Duke University. Available at: www.nespguidebook.com/cms/wp-content/uploads /2014/11/FRMES-CE-10-FULL-PDF.pdf.

de Groot, R. S., R. Alkemade, L. Braat, L. Hein, and L. Willemen. 2010. Challenges in integrating the concept of ecosystem services and values in landscape planning, management, and decision making. *Ecological Complexity* 7:260–272.

Endter-Wada, J., D. Blahna, R. Krannich, and M. Brunson. 1998. A framework for understanding social science contributions to ecosystem management. *Ecological Applications* 8:891–904.

Farley, J., and R. Costanza. 2010. Payments for ecosystem services: From local to global. *Ecological Economics* 69:2060–2068.

Fisher, B., R. K. Turner, and P. Morling. 2009. Defining and classifying ecosystem services for decision making. *Ecological Economics* 68:643–653.

Foley, T., J. Bowles, N. Smith, and P. Calligiuri. 2014. An ecosystem services approach to management of a complex landscape: The Marsh project. Case 7 in *Federal resource management and ecosystem services guidebook*. Durham, NC: National Ecosystem Services Partnership, Duke University. http://www.nespguidebook.com/cms/wp-content/uploads/2014/11/FRMES-CE-7-FULL-PDF.pdf.

IPBES (Intergovernmental Platform on Biodiversity and Ecosystem Services). 2012. Resolution on the Intergovernmental Science-Policy Platform on Biodiversity and Ecosystem Services. http://www.ipbes.net/images/Resolution%20establishing%20IPBES_2012.pdf.

IUFRO (International Union of Forest Research Organizations). 2015. *Contribution of biodiversity to ecosystem services in managed forests*. http://www.iufro.org/science/task-forces/biodiversity/.

Keough, H. L., and D. J. Blahna. 2006. Achieving integrative, collaborative ecosystem management. *Conservation Biology* 20:1373–1382.

Kline, J. D., and M. J. Mazzotta. 2012. *Evaluating tradeoffs among ecosystem services in the management of public lands*. General Technical Report PNW-GTR-865. Portland, OR: USDA Forest Service, Pacific Northwest Research Station.

Kline, J. D., M. J. Mazzotta, T. A. Spies, and M. E. Harmon. 2013. Applying the ecosystem services concept to public land management. *Agricultural and Resource Economics Review* 42:139–158.

Millennium Ecosystem Assessment. 2003. Introduction and conceptual framework. Chap. 1 in *Ecosystems and human well-being: A framework for assessment*. Millennium Ecosystem Assessment. Washington, DC: Island Press.

Muradian, R., E. Corbera, U. Pascual, N. Kosoy, and P. H. May. 2010. Reconciling theory and practice: An alternative conceptual framework for understanding payments for environmental services. *Ecological Economics* 69:1202–1208.

Nahlik, A. M., M. E. Kentula, M. S. Fennessey, and D. H. Landers. 2012. Where is the consensus? A proposed foundation for moving ecosystem service concepts into practice. *Ecological Economics* 77:27–35.

National Ecosystem Services Partnership (NESP). 2016. *Federal resource management and ecosystem services guidebook*. https://nespguidebook.com/.

Penaluna, B. E., D. H. Olson, R. L. Flitcroft, M. Weber, J. R. Bellmore, S. M. Wondzell, J. B. Dunham, S. L. Johnson, and G. H. Reeves. 2016. Aquatic biodiversity in forests: A weak link in ecological and ecosystem service resilience. *Biodiversity and Conservation*. doi:10.1007/s10531-016-1148-0 .

Raphael, M. G., R. Molina, C. H. Flather, R. S. Holthausen, R. L. Johnson, B. G. Marcot, D. H. Olson, J. D. Peine, C. H. Sieg, and C. S. Swanson. 2007. A process for selection and implementation of conservation approaches. Chap. 12. Pp. 334–362 in *Conservation of rare or little-known species*. Edited by M. G. Raphael and R. Molina. Washington, DC: Island Press.

Raymond, C. M., G. G. Singh, K. Benessaiah, J. R. Bernhardt, J. Levine, H. Nelson, N. J. Turner, B. Norton, J. Tam, and K. M. A. Chan. 2013. Ecosystem services and beyond: Using multiple metaphors to understand human-environment relationships. *BioScience* 63:536–546.

Ricketts, T. H., G. C. Dailey, P. R. Ehrlich, and C. D. Michener. 2004. Economic value of tropical forest to coffee production. *Proceedings of the National Academy of Sciences of the United States of America* 101:12579–12582.

Ringold, P. L., J. Alegria, R. L. Czaplewski, B. S. Mulder, T. Tolle, and K. Burnett. 1996. Adaptive monitoring design for ecosystem management. *Ecological Applications* 6:745–747.

Ringold, P. L., J. Boyd, D. Landers, and M. Weber. 2013. What data should we collect? A framework for identifying indicators of ecosystem contributions to human well-being. *Frontiers in Ecology and the Environment* 11:98–105.

Silvertown, J. 2015. Have ecosystem services been oversold? *Trends in Ecology & Evolution* 30:641–648.

Smith, N., R. Deal, J. Kline, D. Blahna, T. Patterson, T. Spies, and K. Bennett. 2011. *Ecosystem services as a framework for forest stewardship: Deschutes National Forest overview*. General Technical Report PNW-GTR-852. Portland, OR: USDA Forest Service, Pacific Northwest Research Station.

Tallis, H., P. Kareiva, M. Marvier, and A. Chang. 2008. An ecosystem services framework to support both practical conservation and economic development. *Proceedings of the National Academy of Sciences of the United States of America* 105:9457–9464.

Thompson, M. P., B. G. Marcot, F. R. Thompson, S. McNulty, L. A. Fisher, M. C. Runge, D. Cleaves, M. Tomosy. 2013. *The science of decision-making: Applications for sustainable forest and grassland management in the National Forest System*. General Technical Report WO-GTR-88. Washington, DC: USDA Forest Service.

US Environmental Protection Agency. 2015. *National Ecosystem Services Classification System (NECSCS): Framework design and policy application*. EPA-800-R-15-002. Washington, DC: US Environmental Protection Agency.

USDA (US Department of Agriculture, Forest Service). 2012. National Forest System land management planning. Final rule and record of decision. 36 CFR Part 219. *Federal Register* 77:21162–21267. http://www.fs.usda.gov/Internet/FSE_DOCUMENTS/stelprdb5362536.pdf.

——. 2013. National Forest System land management planning directives. *Federal Register* 78(39):13316–13319. http://www.fs.usda.gov/Internet/FSE_DOCUMENTS/stelprdb5411235.pdf.

Vatn, A. and D. W. Bromley. 1994. Choices without prices without apologies. *Journal of Environmental Economics and Management* 26:129–148.

Westman, W. E. 1977. How much are nature's services worth? *Science* 197:960–964.

Wong, C. P., B. Jiang, A. P. Kinzig, K. N. Lee, and Z. Ouyang. 2015. Linking ecosystem characteristics to final ecosystem services for public policy. *Ecology Letters* 18:108–118.

SECTION 2

Dynamic Systems as a New Paradigm

A long history of anthropogenic and natural processes has shaped the dynamics of moist forest systems in the US Pacific Northwest. Some forest-management changes on both private and public lands occurred rapidly following the implementation of the federal Northwest Forest Plan in 1994, whereas others evolved more slowly because of difficulties in reconciling differences in management objectives among landownerships. Unplanned changes have stemmed from a variety of other factors, including fire, pests and disease, invasive species, and climate change, as well as the changing human demographics of the region. Learning processes, adaptive management, and the increasing role of multiple stakeholders in cross-ownership land-management decisions have improved planning and management outcomes.

Using the conceptual framework of ecosystem services introduced in chapter 5, chapter 6 explores how the rapidly developing value system of ecosystem services can help to mold a common purpose in forest planning. Both urban and rural communities value societal goods and services from these forests, including commodity production, recreational experiences, water, carbon, and biodiversity. These values are matched to forest management priorities that differ among landownerships, setting up a spatial inequality of service provisioning across the landscape. This leads to uncertainties regarding long-term sustainability of both societal and ecological services. In chapter 7, forest landscape changes in the US Northwest are documented, showing an evident footprint of prior land-management decisions and large-scale disturbances, including the loss of regional old-growth forests in the last century. Chapter 8 focuses on the importance of learning processes and adaptive management in improving management for multiple objectives across this complex landscape. Emerging changes in collaborative stakeholder roles in forest planning decisions are presented in chapter 9, with examples of collaborative groups from the broader Pacific Northwest region.

Chapter 6

Ecosystem Services with Diverse Forest Landowners

Robert L. Deal, Paul E. Hennon, David V. D'Amore, Raymond J. Davis, Jane E. Smith, and Eini C. Lowell

Pacific Northwest moist coniferous forests provide a wide array of globally important goods and services, including water, carbon sequestration, wood products, fish and wildlife habitat, cultural values, and world-class recreation. These forests are owned and managed by a mix of public, private, and tribal landowners (plates 6, 7), however, who often have different forest-management objectives. Overall, this diverse landownership provides a highly variable forest landscape with forest-management objectives ranging from intensive management on industrial forestlands, to longer rotations on state and tribal lands, to an emphasis on preservation and restoration of late-successional forests to support endangered species and water quality on federal lands. In this chapter, we synthesize some of the objectives of different landowners in the region and the potential opportunities and challenges of integrating goods and services (ecosystem services) into forest management. We show how broad assessment of ecosystem services can be used to plan management activities and to evaluate trade-offs of managing public and private lands to provide a suite of goods and services.

Forestlands and Management Practices

The temperate rain forests of the northwest coast of North America are one of the major timber regions of the United States, widely recognized for their productivity, well-organized timber markets, and large-scale timber-processing industries (Haynes et al. 2003) (chap. 4). In the conterminous US portion of this region about 44% of moist-forest land is owned and managed by the federal government; 29% is corporately owned and managed industrial forest; 13% is family owned; 11% is owned by state and local government; and the remaining 3% is tribal forestland (Hewes et al. 2014) (plate 6). Private lands in the Northwest have relatively few but large industrial forestlands and thousands of small family forestlands and woodlots. In Southeast Alaska, forests are primarily owned and managed by the federal government, interspersed with large blocks of Alaska Native corporation forestland (plate 7).

Forest management objectives often differ among ownership classes. Private industrial forestlands have become more intensively managed, with increasingly shorter rotations and silvicultural practices that focus on Douglas-fir wood production (Haynes 2007). State, tribal, and family forestlands often have longer rotations and a focus on more diverse forest management objectives. On federal lands, we see a fairly recent emphasis on restoration of older forests, multiple uses, ecosystem resilience, and endangered species (Spies et al. 2007).

A major principle of contemporary sustainable forest management in the Pacific Northwest and worldwide is assuring that forests are not only sustainably managed for timber and other economic products, but also for balancing responsible stewardship of other forest and human values (fig. 6.1). The shift of forest-management priorities to a more multiple-use approach is especially apparent on public lands, and has been mandated by federal and state laws since 1994. However, a strong signature of the past emphasis on timber production is visible at landscape scales (chap. 7), and has been used as evidence supporting corporate profit and greed (Vaillant 2005). The legacy of extensive clearcut harvests has left many public lands, now prioritized for late-successional forest reserves and other ecological benefits, as uniform young stands with closed canopies. The realization of multiple-use objectives over these landscapes will take many decades to attain, with active management of younger stands, light thinning of developing stands, and old-growth preserves with little or no management (chap. 10).

After more than two decades of management under new sustainability standards and guidelines of the US federal Northwest Forest Plan (USDA and USDI 1994), the area of older forests is still decreasing, but at a slower rate than in the decades preceding the Plan (Davis et al. 2015) (chap. 7). This had been anticipated, owing to the necessary time lag before ecological sustainability could be achieved on federal lands, but consequently, plan-

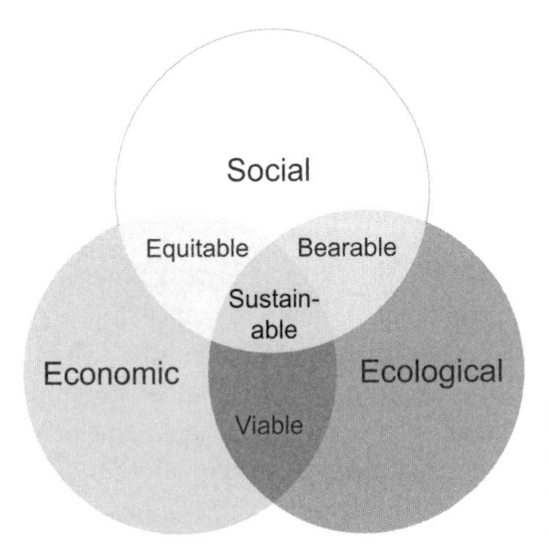

FIGURE 6.1. Sustainability Venn diagram showing the intersections among social, economic, and ecological sustainability.

ning for socioeconomic sustainability from federal forest management has lagged as well. In response to timber markets and global economic forces, forest-management intensity has increased on nonfederal lands (chap. 4).

Managing forests to enhance ecosystem services and assure long-term sustainability of forest products, biodiversity (chap. 13), and other resources is a goal shared around the world. Sustainability includes balancing ecological, economic, and social costs and benefits. Achieving this goal would ensure that we consume water, forest products, and energy at a sustainable rate (ecological sustainability); that we use these resources efficiently and responsibly (economic sustainability); and that we maintain the social well-being of communities or regions in the long term by serving the common good (social sustainability). Integration of ecological, economic, and social sustainability will ensure that resource use is viable, achievable, and equitable (fig. 6.1). Yet there is considerable uncertainty around how to attain these three key objectives simultaneously and how to recognize success. The mixed ownership and variation in forest-management approaches in the Pacific Northwest region confound the already difficult problem of integration. On this landscape, such broader goals are achieved regionally, but not at finer spatial scales. Unfortunately, regional-scale accounting of ecosystem services results in losses of services at finer scales. For example, local community socioeconomic sustainability from timber resources can be lost when harvests are patchy in time or space, and locally valued ecosystem services such as endemic species of concern can be lost with increased harvest intensities. Three constructs are currently developing that may aid in resolving this conflict: (1) broader recognition of the importance of ecosystem services (chap. 5); (2) application of adaptive-management strategies (adjusting actions as experience is gained and lessons are learned—chap. 8); and (3) collaborative group governance of forested watersheds and landscapes (chap. 9).

Ecosystem Services as a Framework for Describing Forest Benefits

The concept of ecosystem services has developed considerably since 1994 (chap. 5) and helps frame and describe the comprehensive set of benefits that people receive from forests. Commonly recognized benefits include timber, climate regulation, and water purification. Yet, there also are aesthetic, spiritual, and cultural benefits such as recreation opportunities and retention of special places for immersion in nature. Ecosystem services are formally recognized in the US Forest Service planning rule (USDA 2012) and may be a useful means to help manage multiple resources (Smith et al. 2011). Managing for ecosystem services involves understanding the broad suite of benefits a forest provides and clarifying relationships between the quantity and quality of services provided and the condition of ecosystems that provide them.

Timber continues to be an important foundational *provisioning* ecosystem service from actively managed forestlands. Understanding the relationship between timber harvest and other ecosystem services is essential to inform decision making about where, when, and to what extent logging is sustainable, or even beneficial to forest integrity (e.g., by improving habitat and watershed values or reducing fuel hazards). Gains and losses to other *cultural* services, like recreation and aesthetic values, merit consideration, as do the short- and long-term timing of multiple-service provisioning and the geographic context at small and large scales. Active forest restoration can contribute to *regulating* ecosystem services that are relatively new to forest management. For example, a viable forest industry can provide capacity to mitigate greenhouse gases by undertaking forest restoration activities that ultimately produce wood products that store carbon (chap. 12).

Biodiversity, as characterized by the Millennium Ecosystem Assessment (MEA 2005), provides an essential underpinning of ecosystem health and function with subsequent effects on *final ecosystem services* such as food, clean water, clean air, and aesthetic pleasure. Biodiversity can therefore be seen as a complex of *intermediate ecosystem services*: the diversity of the plant, animal, and microbial species (e.g., fungi, box 6.1) within an ecosystem that influence critical ecological functions and processes. Such functions

BOX 6.1. FUNGI: LITTLE-KNOWN TAXA WITH MULTIPLE ECOSYSTEM SERVICES

Fungi are a diverse taxon, with a wealth of species serving functional roles in food webs and nutrient cycling that promote forest health and biodiversity, as well as pathogenic species perceived as a type of disturbance. The wet, temperate, locally variable forest environment and high richness of tree and understory plant species in moist forests are conducive to the growth of a wide variety of fungi nested in complex webs of species interactions. Some species are a major food source for small mammals, which in turn provide primary prey species for northern spotted owls and other predators. Belowground mycorrhizal fungi are particularly important because they form mutually beneficial symbioses with roots of most conifers and many hardwood trees and shrubs. These fungi sustain the life of their plant hosts by providing essential nutrients in exchange for photosynthetically produced sugars. Pathogenic fungi contribute to a fine-scale disturbance in moist coniferous forests. Stem decay fungi of live trees are key drivers of tree death through stem breakage. Laminated root rot affecting Douglas-fir, mountain hemlock, and western hemlock is a fungal root disease caused by *Phellinus sulphurascens* (formerly known as and closely related to *P. weirii* [Larsen et al. 1994]) that results in a considerable reduction in timber production (e.g., loss of ~4.4 million cubic meters [1.86 billion board feet] annually, Thies and Sturrock 1995). Pathogenic fungi are ecosystem engineers, creating habitats and affecting ecological processes.

Fungi also are a provisioning ecosystem service as a special forest product for humans. Some edible fungi were collected by Native tribes of the Pacific Northwest (chap. 3; box 3.1; plate 3C), and a wider variety of species are harvested today; the magnitude of the commercial harvest generates concern for long-term species viability.

Despite their profound diversity and ecological roles, fungi have often been overlooked relative to management activities. Although root rot fungi have been addressed in management activities to reduce their effects (development of resistant strains, planting away from disease pockets, removing infected roots and stumps), there is a developing concern that other common fungi could become rare because of habitat loss or large-scale logging (Molina et al. 2001). Some fungi may be associated with late-successional forest habitat conditions (Smith et al. 2002), and fragmentation of these habitats may threaten locally endemic species.

and processes include plant productivity, soil fertility, water quality, nutrient cycling, pollution and waste reduction, biomass accumulation, resistance to disease and disturbance, and other environmental conditions that in turn affect final services and human welfare.

In addition to supporting a more integrated approach to decision making, the language of ecosystem services can help managers highlight connections between forests and people and strengthen relationships with the public. Ecosystem services, when considered as natural capital, can lead landowners, and managers, and the public to regard landscapes as natural assets by ensuring that people who rely on these services know their value and the cost of losing them (Kline 2006; Collins and Larry 2008). An ecosystem services framework can clarify relationships between the quantity or quality of services provided by forests and the condition of forest ecosystems (Daily 1997). This approach can bring attention to the functions performed by healthy ecosystems, distinguish between those that are low and high functioning, lead to investigations about causes of ecosystem degradation or impeded function, and identify where restoration or other actions are needed. Hence, managing for forest ecosystem services should ultimately help us to attain forest ecosystem resilience.

Managing to sustain functions and processes also encourages a broad-scale perspective and serves dual objectives of enhancing land stewardship while providing public benefits. This perspective can help highlight functions and processes in a decision-making framework, clarify priorities and management needs, and support the design and implementation of projects with clearly articulated goals and results. This approach also can assist managers with analysis of the impacts of projects across resource areas by considering potential trade-offs among ecosystem services provided, rather than focusing on just one objective or working with independent and isolated programs. By providing a clear framework for describing these relationships, the ecosystem services paradigm can promote collaborations among interest groups that share stewardship goals. Set in an adaptive-management context, land managers for both public and private land can better evaluate the trade-offs inherent in management activities when presenting their objectives in terms of ecosystem services outcomes.

Disturbances and Ecosystem Services in Moist-Forest Ecosystems

Effective forest management must account for disturbances and their effect on ecosystem services. The moist coniferous forest region of the Northwest possesses distinct climatic and geographic characteristics that drive natural disturbance patterns (chaps. 2, 7, 16), forest productivity (chap. 11), and species diversity (chap. 13). For example, fire occurs at varying intervals and severities based on inherent properties of the biophysical ecoregion, forming temporal patterns known as fire regimes (chap. 7). The wettest areas of the coastal region generally remain wet throughout the year, especially in Southeast Alaska, and these ecosystems generally lack fire. Here, windstorms, floods, and landslides are dominant disturbances (Deal et al. 1991). Finer-scale or gap-phase disturbances maintain multi-aged forests in an old-growth condition in wind-protected landscapes (Alaback and Juday 1989; Lertzman et al. 1996). Canopy gaps can be created when trees are uprooted or suffer stem breakage or when dead trees remain as standing snags as a result of different abiotic (e.g., ice, wind) and biotic factors (Spies et al. 1990; Hennon and McClellan 2003) (pathogenic fungi, box 6.1). These fine-scale forest disturbances contribute to structural heterogeneity and ecological processes, including the creation of different structures for final ecosystem services such as biodiversity and timber products, as well as their supporting ecosystem services such as species habitat, soil development, and nutrient cycling.

In the face of planned and unplanned disturbances, management to retain biodiversity as an important ecosystem service can be challenging and needs to consider the natural range of variability, also termed historical range of variability (HRV; Keane et al. 2009) of species assemblages and habitat conditions on the landscape. Large and extreme disturbances may reduce species abundances or occurrences, but some types and patterns of disturbance may be required to maintain particular species. For example, moist forests provide habitat for both late- and early-seral species; the former may require vigilance for habitat retention and connectivity, and the latter may require disturbance.

Retaining both early- and later-seral species on the landscape may be a complex endeavor requiring oversight. For instance, after an era of intense timber management, concern for species associated with old-growth forests led to the ecosystem-management framework of the federal Northwest Forest Plan (USDA and USDI 1994), including late-successional reserve land-use allocations. Yet, if natural disturbances occur over broader scales than anticipated (e.g., fire) and those affected areas are patch reserves devoted to habitat protection for late-seral species, these species' refugia may be lost, and reevaluation of habitat preserves, restoration, and connectivity may be needed (chap. 7). Owing to current landownership priorities, the burden for much of the delivery of the ecosystem services related to biodiversity concerns relies heavily on public landholdings. Hence, although the disturbance ecology of the moist coniferous forests may be an essential framework both to understand its heterogeneity and for the development of management approaches to maintain valued ecosystem services that depend on site-to-landscape variability (chaps. 14, 15), its utility as a paradigm for forest management varies with landownership, in alignment with the services it delivers. That is, it could be useful for public lands where biodiversity sustainability is a priority, but less useful across all lands unless a more collaborative landownership paradigm were to develop (chaps. 9, 20).

Reliance on public lands to maintain biodiversity as an ecosystem service can fall short, especially for those taxa that are listed as federally threatened and endangered. For example, Pacific salmonids have emerged as a key group of forest-dependent species relying on both public and private landownerships for sustainability. Fish populations are an ecosystem service to which all lands contribute (chaps. 13, 14, 15), with decades of fish habitat degradation stemming from disturbances across all lands. Timber harvests across the region initially focused on wood production, and effects on water quality (e.g., stream temperature, sedimentation); aquatic-riparian habitats (e.g., instream wood, stream channel configuration); and aquatic species were discounted. Today, a signature of clearcut harvesting and splash dams (box 6.2) can be detected on stream habitats and in some species distributions. Photographs of historical Pacific salmon fisheries are sobering, and such productivity has not been recovered in the region owing to the combined effects of timber harvests coupled with a host of other factors, including roads, dams, mining, grazing, and ocean conditions. Moreover, aquatic-riparian systems are especially susceptible to climate conditions affecting water temperatures, timing and level of streamflows, and likelihood of landslides. Landslides also affect stream sedimentation and are related to site geomorphology (e.g., steep topographies) and other disturbances, including past timber-harvest practices and roads crossing streams.

Today, stream maintenance and restoration are managed by creating streamside buffer zones (chapter 14), an approach used on both private and public landownerships. However, although buffer widths vary by jurisdiction, with wider buffers often prescribed for public lands. In summary, current regional forest-management objectives range from intensive management on private industrial lands to an emphasis on preservation and restoration of late-successional forests on federal lands. Because they are managed separately, there is an apportionment of ecosystem service de-

BOX 6.2. SPLASH DAMS

Splash damming was a common method of log transport in the Pacific Northwest from the 1880s through the 1950s. Logs were stored in pools behind splash dams constructed across streams, then released in large freshets to downstream mills. In addition, log drives down stream systems also occurred. Downstream obstacles such as large boulders and natural logjams were sometimes removed with dynamite to facilitate log transport. The practice stripped stream channels to bedrock and scoured aquatic habitat. Legacy effects on stream geomorphology, substrates, pools, and channel complexity can persist on splash-dammed streams for 50 to 130 years (Miller 2010). In western Oregon, Miller (2010) located and mapped 232 splash-dam sites and 213 log-drive sites, and found that splash-dammed streams experienced flood-disturbance regime shifts for up to 40 years, with floods that were more frequent and of greater magnitude. Splash dams have had a long-lasting effect on salmonid habitats, and their legacy is an important consideration for the restoration of stream habitats.

livery by ownership. Although many regional services can be provided this way, those with place-based restrictions may be vulnerable to losses. For example, if high-quality habitat for sensitive salmonids occurs in stream reaches with reduced habitat protections, they may be vulnerable to losses that may not be compensated for by protections in low-quality habitat (Reeves et al. 2016). Similarly, local human communities reliant on timber revenue may be unsustainable if they are adjacent to newly created forest reserves. Objectives of diverse landowners in the region continue to present opportunities and challenges to combining social, economic, and ecological objectives into forest-management. The ecosystem services concept can help to integrate the broad suite of services provided by all lands, including values that are often overlooked in traditional forest-management decisions. An expanding bioeconomy (biomass uses) may provide another avenue to simultaneously address national energy concerns, climate change, and economic development. Forest management that promotes environmental sustainability and values ecosystem services could, if coupled with qualified industrial specialties, provide opportunities for stable employment and economic growth. This topic is expanded later (chaps. 19, 20), after additional concepts of forest and societal dynamics, science lessons learned, and key future considerations have been presented.

Literature Cited

Alaback, P. B., and G. P. Juday. 1989. Structure and composition of low elevation old-growth forests in research natural areas of Southeast Alaska. *Natural Areas Journal* 9:27–39.

Collins, S., and B. Larry. 2008. Caring for our natural assets: An ecosystems services perspective. Pp. 1–11 in *Integrated restoration of forested ecosystems to achieve multi-resource benefits: Proceedings of the 2007 National Silviculture Workshop.* Edited by R. L. Deal. General Technical Report PNW-GTR-733. Portland, OR: USDA Forest Service, Pacific Northwest Research Station.

Daily, G. C. 1997. *Nature's services*. Washington, DC: Island Press.

Davis, R. J., J. L. Ohmann, R. E. Kennedy, W. B. Cohen, M. J. Gregory, Z. Yang, H. M. Roberts, A. N. Gray, and T. A. Spies. 2015. *Northwest Forest Plan—the first 20 years (1994–2013): Status and trends of late-successional and old-growth forests*. General Technical Report PNW-GTR-911. Portland, OR: USDA Forest Service, Pacific Northwest Research Station.

Deal, R. L., C. D. Oliver, and B. T. Bormann. 1991. Reconstruction of mixed hemlock-spruce stands in coastal Southeast Alaska. *Canadian Journal of Forest Research* 21:643–654.

Haynes, R. W. 2007. Integrating concerns about wood production and sustainable forest management in the United States. Pp. 1–18 in *Sustainable forestry management and wood production in a global economy*. Edited by R. L. Deal, R. White, and G. Benson. Binghamton, NY: Haworth Press.

Haynes, R. W., R. A. Monserud, and A. C. Johnson. 2003. Chap. 1. Compatible forest management: Background and context. Pp. 3–34 in *Compatible forest management*. Edited by R. W. Haynes, R. A. Monserud, and A. C. Johnson. Dordrecht, the Netherlands: Kluwer Academic Press.

Hennon, P. E., and M. H. McClellan. 2003. Tree mortality and forest structure in temperate rain forests of Southeast Alaska. *Canadian Journal of Forest Research* 33:1621–1634.

Hewes, J. H., B. J. Butler, G. C. Liknes, M. D. Nelson, and S. A. Snyder. 2014. *Map of distribution of six forest ownership types in the conterminous United States*. [Scale 1: 10,000,000, 1: 34,000,000.] Research Map NRS-6. Newtown Square, PA: USDA Forest Service, Northern Research Station.

Keane, R. E., P. F. Hessburg, P. B. Landres, and F. J. Swanson. 2009. The use of historical range of variability (HRV) in landscape management. *Forest Ecology and Management* 258:1025–1037.

Kline, J. D. 2006. *Defining an economics research program to describe and evaluate ecosystem services*. General Technical Report PNW-GTR-700. Portland, OR: USDA Forest Service, Pacific Northwest Research Station.

Larsen, M. J., F. F. Lombard, and J. W. Clark. 1994. *Phellinus sulphurascens* and the closely related *P. weirii* in North America. *Mycologia* 86:121–130.

Lertzman, K. P., G. D. Sutherland, A. Inselberg, and S. C. Saunders. 1996. Canopy gaps and the landscape mosaic in a coastal temperate rain forest. *Ecology* 77:1254–1270.

MEA (Millennium Ecosystem Assessment). 2005. *Ecosystems and human well-being synthesis*. Washington, DC: Island Press.

Miller, R. R. 2010. *Is the past present? Historical splash-dam mapping and stream disturbance detection in the Oregon Coastal Province*. Master's thesis. Corvallis, OR: Oregon State University. http://hdl.handle.net/1957/18998.

Molina, R., D. Pilz, J. Smith, S. Dunham, T. Dreisbach, T. O'Dell, and M. Castellano. 2001. Conservation and management of forest fungi in the Pacific Northwestern United States: An integrated ecosystem approach. Pp. 19–63 in *Fungal conservation: Issues and solutions*. Edited by D. Moore, M. M. Nauta, S. Evans, and M. Rotheroe. Cambridge, UK: Cambridge University Press.

Reeves, G. H., B. R. Pickard, and K. N. Johnson. 2016. *An initial evaluation of potential options for managing riparian reserves of the Aquatic Conservation Strategy of the Northwest Forest Plan*. General Technical Report PNW-GTR-937. Portland, OR: USDA Forest Service, Pacific Northwest Research Station.

Smith, J. E., R. Molina, M. Huso, D. Luoma, D. McKay, M. A. Castellano, T. Lebel, and Y. Valachovic. 2002. Species richness, abundance, and composition of hypogeous and epigeous ectomycorrhizal fungal sporocarps in young, rotation-age, and old-growth stands of Douglas-fir (*Pseudotsuga menziesii*) in the Cascade Range of Oregon, U.S.A. *Canadian Journal of Botany* 80:186–204.

Smith, N., R. Deal, J. Kline, D. Blahna, T. Patterson, T. A. Spies, and K. Bennett. 2011. *Ecosystem services as a framework for forest stewardship: Deschutes National Forest overview*. General Technical Report PNW-GTR-852. USDA Forest Service, Pacific Northwest Research Station.

Spies, T. A., J. F. Franklin, and M. Klopsch. 1990. Canopy gaps in Douglas-fir forests of the Cascade Mountains. *Canadian Journal of Forest Research* 20: 649–658.

Spies, T. A., K. N. Johnson, K. M. Burnett, J. L. Ohmann, B. C. McComb, G. H. Reeves, P. Bettinger, J. D. Kline, and B. Garber-Yonts. 2007. Cumulative ecological and socioeconomic effects of forest policies in coastal Oregon. *Ecological Applications* 17:5–17.

Thies, W. G., and R. N. Sturrock. 1995. *Laminated root rot in western North America*. General Technical Report PNW-GTR-349. Portland, OR: USDA Forest Service, Pacific Northwest Research Station. In cooperation with: Natural Resources Canada, Canadian Forest Service, Pacific Forestry Centre, Victoria, BC, Canada.

USDA (US Department of Agriculture, Forest Service). 2012. National Forest System land management planning. Final rule and record of decision. 36 CFR Part 219. *Federal Register* 77:21162–21267. http://www.fs.usda.gov/Internet/FSE_DOCUMENTS/stelprdb5362536.pdf.

USDA and USDI (US Department of Agriculture and US Department of the Interior). 1994. *Record of decision for amendments to Forest Service and Bureau of Land Management planning documents within the range of the northern spotted owl* [plus Attachment A: Standards and Guides]. [Place of publication unknown]: US Department of Agriculture and US Department of the Interior. http://www.reo.gov/library/reports/newroda.pdf.

Vaillant, J. 2005. *The golden spruce: A story of myth, madness and greed*. New York: W. W. Norton.

Chapter 7

Patterns of Change across the Forested Landscape

Raymond J. Davis, Andrew N. Gray, John B. Kim, and Warren B. Cohen

The scope and extent of past natural disturbances and human-derived changes to the forest landscape often provide the historical context for management but are often insufficiently accounted for in forest planning. In particular, static components of many management plans are not easily adapted to unforeseen system dynamics. For example, when the Northwest Forest Plan was designed in 1993, the inherently dynamic nature of the forest ecosystem and landscape was acknowledged, but there was a general lack of scientific information about the ecological processes that would shape forests of the future. The expectation was that both management and natural disturbances would influence change in the forested landscape, but how management would then adapt to these altered conditions was not clear. At the time, climate change was not well understood and was just beginning to be discussed in relation to forests.

Considering the influence of climate on many processes that shape forests, historical context (while informative) can take on a very different light. Furthermore, changes in industrial forest management, socioeconomic conditions, and political events are additional sources of complex dynamics that are not well accounted for in forest planning. Management within the historical range of variation may no longer be possible at some point in the future, given the scope, extent, and complex dynamics of cumulative changes (Wiens et al. 2012).

In this chapter we explore how past natural disturbances and human management have left their spatial signatures on the moist coniferous forest landscape in the conterminous US Northwest. We also examine recent regional changes in forests, climate, and socioeconomic conditions that have come to light with the wealth of information from monitoring, research, and new technologies.

Agents of Change

Forests are dynamic, and their composition, structure, and extent change through time in response to a range of natural and human causal agents. The landscape patterns of Northwest moist forests have been and continue to be shaped by swift disturbances and unhurried forest succession. The pace and extent of these forces change with the climate, both environmentally and sociologically. In general, forest disturbances are much easier to detect quickly, as they often operate on shorter time scales. For instance, the effects of a large, hot-burning wildfire on forest patterns can be seen days after the smoke has cleared. However, observing patterns of forest succession takes patience. The changes that follow such a fire, including forest regrowth, will not be readily apparent for years, decades, or even centuries.

Climate

As recorded by written human history and by geological and biological records, the climate is always changing, usually slowly and gradually, but sometimes abruptly. We are now living in a warm period between glaciations, about 10,000 years after glaciers retreated from much of North America. As the Northwest warmed and the glaciers retreated, many lifeforms that were not suited for the Ice Age climate colonized the region. Today's climate is strongly influenced by temperature variations in the Pacific Ocean, such as the Pacific Decadal Oscillation and the El Niño–Southern Oscillation. The shifting patterns of climate shape the distribution, survival, and productivity of forest tree species.

Climate drives the natural disturbance regimes reflected in the frequency and severity of wildfires and the extent and intensity of windstorms, ice storms, floods, and landslides. Drier conditions and thunderstorms bring wildfires to inland mountain ranges every summer. Nearer to the coast, large wildfires are much less frequent than inland, and constant fog and dew promote fungal diseases. In the highest mountain elevations, extreme winter temperatures, snowpack extent, and shallower soils create a difficult environment for tree growth. Winter climate patterns dictate storm tracks, which affect forest disturbance through windstorms, floods, and landslides. Winter temperature and precipitation patterns govern the accumulation of snowpack and the timing of snowmelt, affecting streamflow and groundwater flow. As climate changes, the forests and their disturbance patterns change along with it.

Human-induced climate change has begun and has already been observed in the Northwest (Abatzoglou et al. 2014; Mote et al. 2014). An analysis of weather observation and climate simulation data for the last hundred years shows a long-term warming of the region at a rate similar to global observations. As for precipitation, only a small increase in the spring has been detected. If human activities around the globe continue as before without significant mitigation action, the climate of the Northwest is projected to warm by 5°C (9°F) by the end of this century, thereby increasing the frequency and intensity of wildfires, altering the hydrologic cycles of forest ecosystems, and shifting the distribution of forest tree species. Chapter 16 discusses climate change in greater depth.

Disturbances

Natural disturbances and those that are human caused are not necessarily separate, and their interactions can be complex. For example, drought can increase the frequency of fire starts by both lightning and human activities. Sometimes a natural disturbance will be followed by human actions such as timber salvage or replanting, which in turn may influence future "natural" disturbances (e.g., wildfire). In addition to natural disturbances directly related to climate conditions, other natural disturbances in the region include the rare earthquake, tsunami, or volcanic eruption.

Under the contemporary climate, wildfires have been the primary natural shaper of forest patterns in the conterminous US Northwest, whereas windstorms predominated in the wetter forests of Alaska and British Columbia. In the past, wildfires occurred infrequently yet were moderate to severe in their consequences (e.g., killing most or all of the live trees). At the beginning of the twentieth century, Henry Gannett of the United States Geological Survey wrote in his report *The Forests of Oregon* that the most startling feature noted on land-classification maps of the state was the extent of the burned areas in the Coast Range and the northern half of the western Cascade Range (Gannett 1902), although human ignition cannot be ruled out as causing many of these fires. He also observed that wildfires in the moister forests resulted in the "nearly or quite complete destruction" of timber, whereas in drier forests where rainfall was lower and timber less dense, fires resulted in only "partial destruction." The spatial signatures of these fires are still apparent in today's forest patterns. In fact, these patterns sometimes played significant roles in shaping the national and state forests of today.

By far the leading human-caused disturbance agent shaping forest patterns of the recent past and contemporary times has been timber harvesting. In addition, agriculture, urbanization, and associated infrastructures such as highways, roads, railroads, power lines, gas lines, and dams for hydropower and flood control have had widespread impacts and have significantly changed forest patterns in localized areas. Forest patterns have been further affected by mining, grazing, and introduction of invasive species. Timber harvesting has had a strong spatial signature over time. It began in areas surrounding major settlements and shipping ports, then expanded from urban areas and along spreading transportation systems to the margins of private and public forestlands upon which there were "almost unbroken stands of old-growth Douglas-fir" covering the foothills and lower slopes of the Cascade Range for most of the length of the state of Oregon (Andrews and Cowlin 1940).

Socioeconomic Factors

Human land use in the twentieth century was based in large part on the template established by Native Americans in prior centuries. Highly productive marine and riverine ecosystems made native tribes in the region among the wealthiest in North America before Old World diseases decimated their numbers. Tribes maintained extensive prairies and montane meadows through burning for game and food plants (e.g., camas and tanoak, chap. 3). Many of those extensive valley prairie systems (e.g., Willamette, Puget Sound, and Fraser) later became population and agricultural centers. The abundant and highly productive timber resources of the region fueled the socioeconomic development of these settlements, and they began to expand and spread into the surrounding forested landscapes. By 1909, it was estimated that the forested areas in Oregon had declined by 12% (Kellogg 1909), and by 1936 Andrews and Cowlin (1940) estimated that 19% of the original moist forests in Oregon and Washington had been harvested.

On the remaining forestlands, private timberlands were managed to produce revenue from timber harvest, and forest industries were focused on maximizing profits through intensive forest management of their lands. The rate of harvest of large, merchantable, late-successional forests on private timberlands became a concern in the 1920s (Peavy 1929), with speculations that privately owned timber in western Oregon would be exhausted by the mid-1900s. Indeed, by the middle of the twentieth century, productive older moist forests on private lands became scarce. This led to an increased rate of timber harvesting of older forests on federal lands to sustain the economic industry that had developed from the harvesting of private lands and to meet the increased public demand for timber following World War II (chap. 4).

The inroads into federal timber first focused on easily accessible areas and fueled a program of road building to access more. As the rate of timber harvesting on federal lands accelerated, social demands for the establishment of federal laws for managing public forests for long-term multiple uses, including conservation of native fish and wildlife species, eventually put the brakes on the rate of federal forest harvest. These new laws required the development of forest-management plans by the federal government and eventually culminated in the development and implementation of the Northwest Forest Plan in 1994.

These forest landscape changes are illustrated in three maps showing different time periods (fig. 7.1). At the beginning of the twentieth century, we estimate there were ~7.7 million ha (19.0 million ac) of older moist forests in Oregon and Washington. By 1940, these forests had decreased by 24%, owing to timber harvesting, wildfire, and land development. By 2012, our latest estimate shows a 49% decline since the early 1900s, accounting for some older forest recruitment. Looked at another way, older forests of today currently cover about half the area they did a century ago.

Ownership Factors

The ownership patterns in this region were shaped by competing social, political, and economic forces, often with much controversy. Most notable among these was the establishment of the forest reserves beginning in 1891 and ending in 1907. These reserves later become national parks and national forests managed by the National Park Service and the US Forest Service. Their spatial signatures remain easily visible on maps based on remotely sensed images of today's forests (plate 8C), along with the checkerboard pattern resulting from restoration of the Oregon and California Railroad (O&C) lands to public ownership under federal management in 1916.

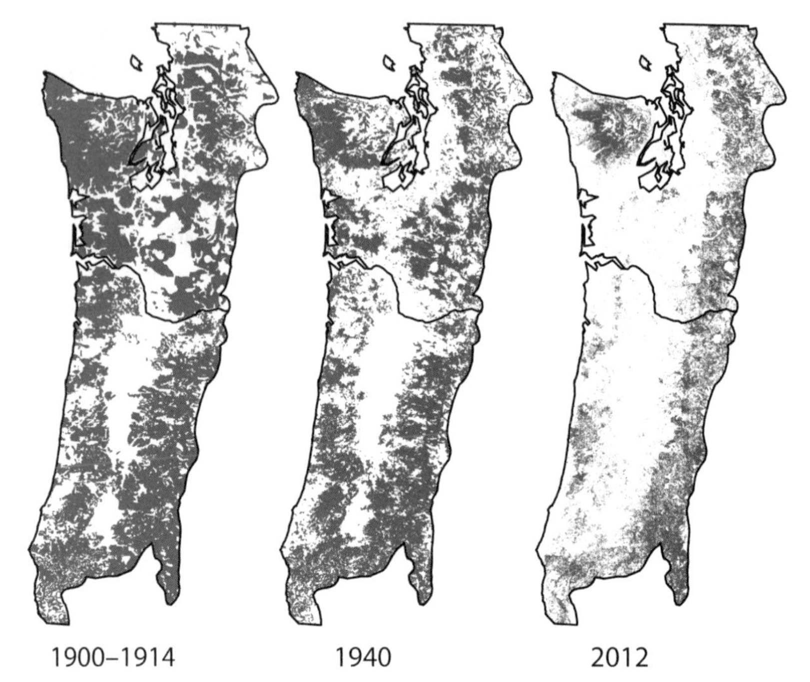

FIGURE 7.1. Landscape pattern changes of older forest over the last century.

Some of these reserves (and also state forests) were delineated around the perimeters of the large wildfires of the 1800s, because these areas were less desirable than unburned forests and had not been claimed by private individuals and companies.

In the conterminous United States, the federal government's current management approach for Northwest moist forests focuses primarily on restoring a deficit of older forest in the region that resulted from timber harvesting on both private and public lands over the last century. Of federal forestland in the conterminous US Northwest, 32% is congressionally withdrawn from management for timber, and another 30% is administratively designated for maintenance and restoration of old-forest structure, which precludes timber harvesting in older forests. The contrast between federal and private industrial management approaches has resulted in a forested landscape that is visibly bimodal, with highly divergent forest patterns and dynamics.

Monitoring Change

Broad-scale changes in forest patterns reflect the past but also inform the future. Keeping track of these changes and understanding their cause and effect are important for informed forest management. One of the first examples of broad-scale forest monitoring was the hand-drawn forest maps in *The Relation of Geography to Timber Supply* (Greeley 1925) that showed changes in forest patterns from early settlement times to [then] current conditions. The technology of broad-scale forest monitoring has come a long way since then. Today we rely heavily on a network of forest inventory and analysis plots (Bechtold and Patterson 2005) as well as space-borne instruments to detect and record forest changes (Cohen et al. 2002; Healey et al. 2008). Modern mapping techniques are designed to take advantage of dense time-series of satellite data, which allow us to capture more subtle disturbances such as insects and disease, and partial burning or timber harvesting (Kennedy et al. 2012). This is important because forest management and climate are changing and the moist forests of this region are experiencing more nuanced disturbances.

Monitoring population change and its influence on forests is important. The population in western Oregon and Washington quadrupled between 1940 and 2010 from 2.1 to 8.6 million people, and increased by 1.3% per year from 2000 to 2010. Population growth has resulted in conversion of privately owned forest and agricultural lands to urban land uses (table 7.1). Since 1936, there has been an overall decline of 6% in forestlands and 36% in agricultural land across the area, with a concomitant increase in development. In western Oregon, between 1961 and 1986, forestlands under private and state ownerships declined at a rate of 0.2% per year owing to road building or conversion into urban or agricultural uses (MacLean 1990). Such conversion affects not only the land it covers, but the characteristics and management of surrounding forests.

Over time, patterns of disturbance and recovery become apparent (plate 9), and we see that the majority of recent natural- and human-caused disturbances (from 1990 to 2012) have occurred on nonfederal lands (plate 9C). Although recovery occurs across all forest ownerships following disturbance, forests showing stable or recovering signatures are mostly on federally managed land (plate 9B). As a result of new technologies, we can see more nuanced changes in forest patterns over the course of the last few decades. We now have a good handle on coarse patterns of forest extent and change and some indicators of fragmentation and degradation, and

TABLE 7.1. Estimates of area (million ha) by land-use type in western Oregon and Washington, USA (data from Andrews and Cowlin 1940; Lettman et al. 2011; Gray et al. 2013).

	Area in land-cover type, by year (million ha)			
Land cover type	*1936*	*1975*	*2010*	*Change*
Forest	11.74	11.43	11.07	-0.67
Agriculture	1.89	1.38	1.21	-0.68
Other nonforest	0.49	1.31	1.84	1.35
Total land	14.12	14.12	14.12	

recent enhancements in forest monitoring may help to provide more comprehensive information on changes in forest composition, structure, and function.

The Big Picture Today

In the US Pacific Northwest, landscape patterns in moist coniferous forests are now closely tied to human actions, perhaps even more so today than they were a few hundred years ago, given the sheer increase in human densities and technological advances. However, natural disturbances still play an important part in shaping these forests, while forest succession carries on at a slower pace. Disturbances and conversion of Pacific Northwest coastal moist forests in the conterminous United States largely reflect established landownership patterns, a result of over 150 years of social, economic, and political decisions. Those decisions were not made randomly, however; they were often influenced by environmental conditions and forest disturbance history. Large wildfires that left immense tracts of land covered by young, unmerchantable forests were overlooked by the timber industry, which sought productive, accessible forestlands with large saw logs that lay in close proximity to settled areas. The more remote forests, where access was difficult and transportation costs high, were less desirable and often wound up as public domain. Similarly, lower-elevation lands where timber harvesting operations were unimpeded by winter conditions and could continue year-round incurred more harvesting than higher elevations where logging operations were tied to the seasons.

When considering older forests, the forest landscape pattern most prominently reflects the late nineteenth- and early twentieth-century decisions by the US government to set aside millions of acres of forest land as forest reserves. In addition, the decision to place the O&C lands in the public trust added millions more acres under federal management. These reserves and public trust lands were designed to benefit the local economy and society by providing a long-term, sustainable supply of timber and other resources. In contrast, private timber lands were managed to maximize profits, and as history shows, the large timber resource on those lands was largely depleted by the mid-twentieth century. The results of these decisions are reflected in the landscape pattern of older forests seen today (plate 8).

Current management of federal forests under the Northwest Forest Plan includes a network of older forest reserves across the entire breadth of the federal forest landscape (plate 9A). In essence, these are contemporary reserves within the larger historic reserves. If the current strategy is maintained, perhaps the forest landscape patterns of the future will reflect this decision, and thus these reserve patterns, just as today's forest reflects the pattern of forest reserves established over a century ago. Current monitoring results demonstrate that the pattern of human-caused forest disturbance is concentrated on nonfederal lands and particularly on industrial timberlands. The pattern of stable or recovering forests is concentrated on federal forestlands, consistent with the current management guidelines (plate 9). The recovery trajectory is not instantaneous, however, and new patterns are projected to emerge over the next several decades.

Over the last century, not only has the pattern of the forest changed, so too have the dynamic forces that have caused these changes. Although timber harvesting currently constitutes most of the disturbance on private forest land, wildfires account for most of the recent disturbances that are shaping forest patterns on federal forests. Interestingly, most large wildfires have been occurring on reserved forestlands. Some argue that wildfire is a natural ecosystem process that should be allowed to function as it did over the last few centuries. However, the current forest landscape is a novel one that is split into two distinctly different segments based on ownership and interlaced with human infrastructure. It will be challenging to allow "natural" disturbances to play their important roles in the ecosystem in the context of this novel forest landscape.

If reserves continue to diminish precipitously, affecting the ecosystem services they were designed to provision, a new look at their spatial distribution and trajectory may be worthwhile. How climate change might affect this dynamic is a further unknown: will the expected increase in frequency of large wildfires challenge the current federal forest reserve design? Given the dynamics of forest landscapes, how this question will be addressed remains to be seen.

Literature Cited

Abatzoglou, J. T., D. E. Rupp, and P. W. Mote. 2014. Seasonal climate variability and change in the Pacific Northwest of the United States. *Journal of Climate* 27:2125–2142.

Andrews, H. J., and R. W. Cowlin. 1940. *Forest resources of the Douglas-fir region*. Miscellaneous Publication No. 389. Washington, DC: USDA Forest Service, Pacific Northwest Forest and Range Experiment Station.

Bechtold, W. A., and P. L. Patterson. 2005. *The enhanced Forest Inventory and Analysis Program: National sampling design and estimation procedures*. General Technical Report SRS-80. Asheville, NC: USDA Forest Service, Southern Research Station.

Cohen, W. B., T. A. Spies, R. J. Alig, D. R. Oetter, T. K. Maiersperger, and M. Fiorella. 2002. Characterizing 23 years (1972–1995) of stand replacement disturbance in western Oregon forests with Landsat imagery. *Ecosystems* 5:122–137.

Davis, R. J., J. L. Ohmann, R. E. Kennedy, W. B. Cohen, M. J. Gregory, Z. Yang, H. M. Roberts, A. N. Gray, and T. A. Spies. 2015. *Northwest Forest Plan—the first 20 years (1994–2013): Status and trends of late-successional and old-growth forests*. General Technical Report PNW-GTR-911. Portland, OR: USDA Forest Service, Pacific Northwest Research Station.

Gannett, H. 1902. *The forests of Oregon*. Professional Paper No. 4, Series H, Forestry, 1. Washington, DC: US Geological Survey.

Gray, A. N., D. L. Azuma, G. J. Lettman, J. L. Thompson, and N. McKay. 2013. *Changes in land use and housing on resource lands in Washington state, 1976–2006*. General Technical Report PNW-GTR-881. Portland, OR: USDA Forest Service, Pacific Northwest Research Station.

Greeley, W. B. 1925. The relation of geography to timber supply. *Economic Geography* 1:1–14.

Healey, S. P., W. B. Cohen, T. A. Spies, M. Moeur, D. Pflugmacher, M. G. Whitley, and M. Lefsky. 2008. The relative impact of harvest and fire upon landscape-level dynamics of older forests: Lessons from the Northwest Forest Plan. *Ecosystems* 11:1106–1119. doi:10.1007/s10021-008-9182-8.

Kellogg, R. S. 1909. *The timber supply of the United States*. Circular No. 166. Washington, DC: USDA Forest Service.

Kennedy, R. E., Z. Yang, W. B. Cohen, E. Pfaff, J. Braaten, and P. Nelson. 2012. Spatial and temporal patterns of forest disturbance and regrowth within the area of the Northwest Forest Plan. *Remote Sensing of Environment* 122:117–133.

Lettman, G. J., A. A. Herstrom, D. Hiebenthal, N. McKay, and T. J. Robinson. 2011. *Land use change on non-federal land in Oregon 1974–2009*. Salem, OR: Oregon Department of Forestry.

MacLean, C. D. 1990. *Changes in area and ownership of timberland in western Oregon: 1961–86*. Resource Bulletin PNW-RB-170. Portland, OR: USDA Forest Service, Pacific Northwest Research Station.

Mote, P., A. K. Snover, S. Capalbo, S. D. Eigenbrode, P. Glick, J. Littell, R. Raymondi, and S. Reeder. 2014. Chap. 21: Northwest. Climate change impacts in the United States. Pp. 495–521 in *The Third National Climate Assessment*. Edited by J. M. Melillo, T. C. Richmond, and G. W. Yohe. US Global Change Research Program. http://nca2014.globalchange.gov/report/regions/northwest.

OSBF (Oregon State Board of Forestry). 1914. *Map of the state of Oregon*. Compiled by Theodore Rowland under the direction of F. A. Elliott, State Forester. Salem, OR: Oregon State Board of Forestry.

Peavy, G. W. 1929. *Oregon's commercial forests: Their importance to the state*. Bulletin No. 2. Salem, OR: Oregon State Board of Forestry.

Plummer, G. H., F. G. Plummer, and J. H. Rankine. 1902. *Map of Washington showing classification of lands*. Map. Washington, DC: US Geological Survey.

Wiens, J. A., G. D. Hayward, H. D. Safford, and C. Giffen. 2012. *Historical environmental variation in conservation and natural resource management*. Oxford, UK: Wiley and Sons.

Chapter 8

Learning to Learn: The Best Available Science of Adaptive Management

Bernard T. Bormann, Byron K. Williams, and Teodora Minkova

Of all the dynamic processes in the human-forest ecosystem that determine long-term sustainability, learning and learning-based planning and decision making (adapting) are among the most important. In the framework of human-forest ecosystem sustainability (chap. 1), adaptive management is a social ecosystem process to be considered equally with ecological processes such as energy capture and flow, nutrient and water cycling, and diversity- and structure-based resilience to disturbance.

People play a critical role in the ecosystem by applying what they know to forest management, but they must also have the humility to recognize that they do not always know enough to manage complex ecosystems successfully over time without a strong learning element. Learning is essential simply to find out if management objectives were met and has a vital longer-term role as well. For example, maximum sustained yield, especially in fisheries, proved to be largely unsustainable in many cases (Ludwig et al. 1993). Learning about, and responding to, needed changes soon enough to avoid irreparable damage are the keys to sustaining high levels of ecological and community benefit in the long term. Here we briefly review basic concepts of adaptive management and evaluate major advances and challenges over the last 20 years, with an eye to guiding future efforts, especially for public forest lands.

Adaptive Management—Background, Context, and Definition

Adaptive management, an approach that involves the dual pursuit of management and learning, has been a part of natural resources management for many decades. For example, Beverton and Holt (1957) described an adaptive approach in fisheries without calling it adaptive management. Later, Holling (1978) and Walters and Hilborn (1978) provided a conceptual framework for adaptive resource management, and Walters (1986) followed up with a more complete technical treatment of adaptive decision making. Lee (1993) expanded the context by providing a comprehensive exposition of social and political dimensions. These pioneering efforts sparked a growing interest in adaptive management as a practical means of addressing landscape-scale natural resource issues.

Adaptive management is based on the recognition that we have only a partial understanding of natural resource systems and their responses to management and that there is value in tracking resource conditions and using what is learned as those resources are being managed (Williams 2011). Learning in adaptive management occurs through the practice of management itself, with adjustments to decision making occurring as understanding improves. For many resource-management problems, using management in an experimental, learning-oriented context is optimal for gaining the understanding needed to manage more effectively (Williams et al. 2007).

Since adaptive management was first introduced, an emphasis on uncertainty and how to manage for it has developed, and science-based learning tools have become integrated into decision-making frameworks. In the ongoing adaptive process, learning informs management, and management supports learning about the resource system (Williams and Brown 2014). In some cases, management interventions can be viewed as experimental "treatments" that are implemented according to a management design. The ultimate focus of adaptive decision making is on management, and learning is valued in terms of its contribution to improving management (Walters 1986).

Unfortunately, many people think adaptive management is monitoring or changing management direction in the face of failed policies—it is not. Adaptive management includes structured learning based in part on the scientific method (box 8.1), including (1) exploring alternative ways to meet management objectives; (2) predicting outcomes of alternatives based on the current state of knowledge; (3) implementing one or more of these alternatives; (4) monitoring to track effects of management actions; and (5) using results to update knowledge and adjust management actions (Murray and Marmorek 2004). Adaptive management also focuses on the dual processes of learning and adapting, often through partnerships of managers, scientists, and other stakeholders learning together how to create and maintain sustainable resource systems (Bormann et al. 2006).

Applications of adaptive management typically involve systems characterized by change and uncertainty. Common features of adaptive management of ecosystems are (1) the natural resource being managed changes through time in response to fluctuating environmental conditions and management actions; (2) environmental variation is only partially predictable, and induces an element of apparent randomness in biological and ecological processes, which in turn leads to unpredictable system behaviors; (3) periodic and potentially time-specific management interventions directly or indirectly influence system behaviors; and (4) effective management is limited by uncertainty about resource conditions and processes and the influence management has on them. Reducing this uncertainty can lead to improved management actions (Williams and Brown 2012). It is worth recognizing that there are few ecosystems left that are not characterized by change and uncertainty as climate change unfolds.

Notwithstanding its theoretical and practical advantages, adaptive management is infrequently applied in the real world of natural resource management, even though many management plans, policies, and procedures assert otherwise. The context for resource management continues to evolve, and the value of managing adaptively continues to be questioned by many managers. If adaptive management makes so much sense in concept, a reasonable question to ask is, why has it not been implemented more frequently and successfully?

Despite the conceptual advantages of adaptive management, a number of impediments have been identified (table 8.1). Given these and other challenges for natural resource management (overlapping jurisdictions; conflicting priorities among scientists, decision makers, and stakeholders), it is not surprising that adaptive management has been viewed with skepticism (McLain and Lee 1996; Walters 1997; Rogers 1998). Many of these issues constitute barriers to good management, whether adaptive or not. Effective and efficient resource management ultimately requires approaches that can overcome such barriers while institutionalizing bridges between science discovery and its application to resource management. Adaptive management can contribute to these bridges.

BOX 8.1. ADAPTIVE MANAGEMENT AND THE SCIENTIFIC METHOD

Science is a structured learning process to continually refine our knowledge. Science-based adaptive management is an application of the scientific method to facilitate learning while managing in order to improve decisions (Williams et al. 2007). Decisions about applying adaptive management can then be said to rely on the best available science of how to do adaptive management. Inexperience with science may lead to misinterpretation of adaptive management.

Generally, the scientific method identifies and examines hypothesized causes for observed phenomena. This usually involves comparing predictions against data, where a match supports the hypothesis (Hempel 1965). A more detailed treatment includes theories, hypotheses, predictions, observations, and comparisons.

Theories. A relevant theoretical framework provides a conceptual basis on which observations can be targeted, facts can be discerned, and data interpreted.

Hypotheses. In natural resource systems, hypotheses represent different views about how resource components are related, resource processes are structured and parameterized, or resources are influenced by management interventions. From hypotheses come testable predictions.

Observations. Scientific investigation in natural resources usually involves collecting some form of data that are pertinent to hypothesis-based predictions. Using statistically sound surveys, experiments, and other efforts with inferential power is key to successful data collection. Observations serve as "ground-truth," against which predictions can be measured.

Comparisons. The confrontation of hypotheses (and their predictions) against observations defines the scientific method in natural resources. Alternative hypotheses about natural resources compete with each other in an arena of observed evidence, in which the most appropriate hypothesis gains support through repeated testing by its compatibility with statistically sound observations.

We emphasize that scientific method as described here is fully complementary with the traditional goals and objectives of natural resource management. Many of the presumptive causes of resource patterns and dynamics are recognized from observations made during the course of management decision making. In some instances, management concepts are directly applied in research designs to help improve understanding of the causes for resource patterns and dynamics. The link between resource management and scientific assessment, in which management both supports and is supported by science, exemplifies adaptive management. It is for this reason that adaptive management is so often described as science- or learning-based management and is held by most practitioners to be a model for the marriage of science and management.

TABLE 8.1. Commonly cited impediments to effectively implementing adaptive-management processes (McLain and Lee 1996; Walters 1997; Gregory et al. 2006; Williams and Brown 2012).

Impediments to Adaptive Management
1. A lack of technical expertise or support for an existing framework for decision making and up-front costs and institutional changes required for development of a new framework.
2. Resistance to acknowledging uncertainty.
3. The belief that needed management actions are known, and that follow-up monitoring and assessment are unnecessary and consume resources that could be put to better use.
4. The belief that adaptive management is being used, when it is not. For example, monitoring by itself is not enough to make a project "adaptive," nor is implementation of changes without a deliberate process of reasoning that characterizes adaptive management.
5. Risk aversion leading to strategies with little or no opportunity for learning. A culture that conceives of failures as unproductive has little chance of successfully applying an adaptive approach. Making uncertainties and lack of understanding explicit provides a rationale for a prescribed course of action and identifies likely outcomes.
6. Approaches focused on near-term gains and losses that devalue long-term management benefits and costs (e.g., given fiscal constraints, organizational capacity, and accountability measures). If the future is heavily discounted, there is little incentive to use adaptive management to learn how to manage better in the future.
7. Insufficient stakeholder engagement. Stakeholders continue to be disillusioned with management practices, withhold support for a project, mount legal challenges, turn to insular thinking, and stifle innovation.
8. Lack of institutional commitment to follow through with monitoring and assessment (sampling design, data collection and summarization, database management, and data assessment) after an initial start-up of adaptive decision making.

Adaptive Management in US Pacific Northwest Forests

Given the west coast Canadian roots of adaptive management (Holling 1978; Walters 1986; box 8.2), it should not be surprising that its first regional-scale forestry application appeared in the Northwest's moist

BOX 8.2. APPLICATIONS OF ADAPTIVE MANAGEMENT

Here we highlight three examples of adaptive management from private and state lands in the Pacific Northwest.

1. The MacMillan Bloedel Ltd. forest-products company started an adaptive-management program in 1998 in British Columbia, Canada, on leased lands currently managed by Western Forest Products, Inc. The goal was to learn how to sustain biodiversity in managed forests while assessing variable-retention harvest as an alternative to clearcutting (Bunnell and Dunsworth 2009). The program also examined cost, operational feasibility, and social acceptance. After 5 years, variable-retention harvest became the order of business on 1.1 million ha (2.7 million ac) of coastal forest. The program's strength was evident as the lease changed through four companies. This success is attributed to continued support by senior management, corporate commitment, dedicated program champions, a rigorous research and monitoring program with external scientific advisers, and thoughtful engagement of stakeholders from the early stages of the process (Bunnell and Dunsworth 2009; Smith 2009).

2. Since 2000, an adaptive-management program has been implemented by the Washington Board of Forest Practices to assess the effectiveness of the rules for private and state lands under the Forest and Fish Agreement (WADNR 2013a). Program stakeholders include state and federal agencies, forest landowners, county and tribal governments, and environmental organizations. Very specific administrative procedures were adopted for the process, roles, and responsibilities of the program participants, but the program did not specify particular ecological thresholds and triggers for rule changes. This open-ended approach was viewed as necessary to bring stakeholders together. New research and monitoring information since 2000 has often reduced uncertainty around existing rules rather than changing them, in part because the science used to develop the original agreement was fairly robust.

3. Working with the Olympic Natural Resource Center (University of Washington), the Washington Department of Natural Resources committed to adaptive management in 1992 by setting aside 110 000 ha (~270,000 ac) as the Olympic Experimental State Forest. The habitat conservation plan (WADNR 1997) allowed flexibility to test bold ecosystem-based management strategies to integrate revenue production (timber harvest) and ecological values (habitat conservation) using a disturbance-ecology-based approach. Harvested areas are interspersed with lightly managed or unmanaged areas to create landscape mosaics of forest structure and seral stages, and specialized harvest methods are being tested to create forest structural complexity. Although the management "experiment" was implemented, the learning aspect and its application to management have been hard to achieve, largely owing to the lack of an effective decision-making framework with technical expertise and sustained agency support (WADNR 2013b).

coniferous forests in the form of the federal Northwest Forest Plan (USDA and USDI 1994). The Plan wrestled with how to manage federal forests after a legal injunction in 1991 halted all timber harvest on federal lands in the range of the northern spotted owl. Federal and university scientists developed the overarching conceptual framework of this Plan (FEMAT 1993) after a call for action from President Clinton at the Forest Conference on April 2, 1993. After debate, the scientific team decided to apply ecosystem management based on principles of conservation biology (using a fixed-area, land-use allocation model) rather than disturbance ecology (e.g., creating a shifting mosaic of successional stages). Getting the right spatial distribution of late-successional reserves was deemed more important than the argument that management could emulate complex natural disturbance patterns through time without fixed reserves. Within the principles adopted, numerous additional fine- to coarse-scale mitigations addressed myriad natural resource issues (USDA and USDI 1994). High initial uncertainties about the approach and about specific mitigations were clearly recognized, and adaptive management was chosen as the way the Plan would evolve over time as more information was gathered.

In 1994, adaptive management of forests had never been attempted at large spatial scales. The theories were in place, but collective experience was limited—the science of adaptive management was in its infancy, with few analyses of its effectiveness. Many adaptive management impediments (table 8.1) and successes were observed over the first years of Plan implementation (Stankey et al. 2003) and were documented in a 10-year (1994–2003) post-Plan monitoring report (Bormann et al. 2006). The two greatest lessons learned from impediments over the first 10 years were that (1) formalizing adaptive steps and committing to monitoring worked better than allocating land to adaptive management areas; and (2) problems implementing this new strategy would have been reduced if the Plan had called for even more focused feedback.

One major success was the first closing of the adaptive-management loop when agency executives officially endorsed three of the report recommendations: (1) redesign approaches to adaptive management to be more systematic and rigorous; (2) develop more active ways of reducing fuels in fire-prone, late-successional reserves; and (3) consider changes to the monitoring program (Bormann et al. 2007). In 2007, a new, more detailed adaptive-management framework was developed and adopted by regional agency executives (fig. 8.1).

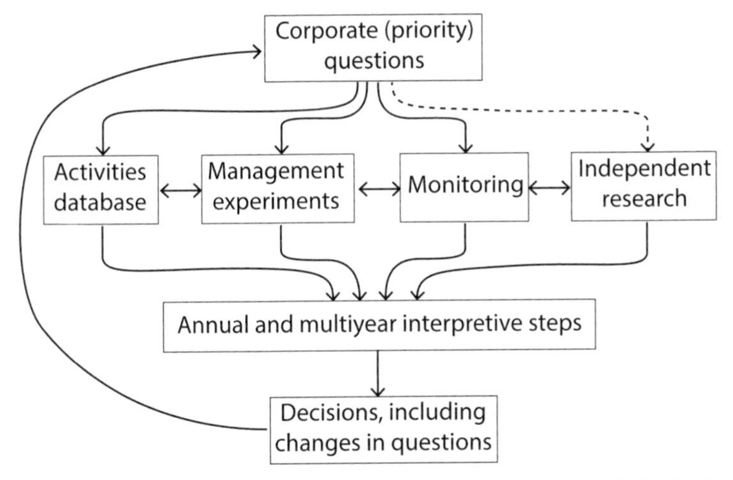

FIGURE 8.1. The adaptive-management framework chosen by US federal Northwest Forest Plan agency executives in 2007. Learning methods (second row) are fitted to questions and then results are interpreted collectively to inform decisions. Management experiments were seen as likely the best way to determine effectiveness yet have been the least-employed method.

Management Experiments

To evaluate adaptive management success during years 10 to 20 of Plan implementation (2004–2013), we first examined the use and outcomes of the 2007 framework (fig. 8.1). Agency executives held a workshop in 2008 to identify priority (*corporate*) questions about Plan efficacy, including information needs that are best handled using management experiments—a major learning method in the adaptive-management framework in which managers and researchers apply and compare different viable strategies using simple design principles at management scales. Three questions emerged: (1) how should we manage after catastrophic wildfire; (2) how can we best reduce fuel loads; and (3) how can we deal with owl habitat losses in dry forests? These questions were likely influenced by a series of large wildfires, beginning with the ~200 000-ha (500,000-ac) Biscuit Complex fire in southwestern Oregon in 2002.

The Pacific Northwest Research Station of the US Forest Service, the US Fish and Wildlife Service, the Bureau of Land Management, and the regional National Forest System of the US Forest Service developed a management-experiment study plan to address the question of managing after wildfire (Bormann et al. 2007). This study plan was a template to develop several management experiments (fig. 8.1). For example, one management experiment addressed how active management in and near northern spotted owl habitat on dry forests might slow habitat loss resulting from the unexpectedly frequent wildfires in the area. However, the postwildfire study plan was never fully developed, in part over concerns that planning for management after wildfire—with or without an experiment—could take away from appropriations designated for planning wildfire-hazard abatement in unburned forests. There also were concerns about potential lawsuits. More generally, a lack of funding and support affected completion of this and other management experiments.

Other investments in science-based projects with potential adaptive-management outcomes also were reduced. For example, existing management experiments in western Oregon on the Siuslaw and Rogue River–Siskiyou National Forests continued, but with few resources for the monitoring outlined in their study plans (USDA 2001, 2004). Large-scale research studies—different from management experiments because they were developed and monitored by researchers—received some continued support from management and research agencies, although they too experienced a drop in priority (e.g., the Demonstration of Ecosystem Management Options Study and the Long-Term Ecosystem Productivity Study; http://www.fs.fed.us/pnw/research/lsse/index.shtml).

Declining field monitoring of experiments in general is counterproductive, since assessments of outcomes would produce knowledge valuable to revisions of federal forest plans by jurisdiction (e.g., per national forest), as well as the larger Northwest Forest Plan. An apparent disenfranchisement between science and management can be tracked economically, but perhaps socially as well, as personnel turnover resulted in loss of legacy knowledge of the Plan and its implementation intricacies, partnerships, and ownership established by the previous generation of the workforce. A lesson learned here is that both scientific and management-based studies reliant on multiple-year or decadal monitoring were not socioeconomically grandfathered into the structure of the organizations, and with fluctuating budgets and personnel they lost priority.

The 2012 Planning Rule

Use of adaptive-management frameworks for forest management was bolstered in 2012 by a new US-wide national forest planning rule (USDA

2012) that established the means by which forest-management decisions will be made. Numerous attempts at revision of the 1982 planning rule, nearly all headed by a committee of outside scientists, had failed by lawsuit or political caveat, and the rule had always reverted back to the original 1982 version. The Department of Agriculture was aware of these problems and had asked the Forest Service to develop a new rule collaboratively with stakeholders. As they undertook this effort through a series of roundtables across the country, stakeholders repeatedly insisted that science-based adaptive management be incorporated into the rule. This pressure was effective. The rule even adopted a "planning and learning" focus and mandated question-driven effectiveness monitoring and monitoring programs with designs that would be coordinated with the regional forester and the regional research station director. These were unprecedented changes with a potential to leapfrog the adaptive-management doldrums in Northwest forests that had existed from 2008 to 2012.

Lessons since 2003

The lesson that social processes can make structured learning more difficult reemerged in the second decade following the 1994 Plan (after the 10-year monitoring reports). In a number of cases, individual members of a collaborative group effectively vetoed an experimental design because it included a treatment they could not support. It has been problematic for some environmental activists to agree to treatments supported by groups seeking to benefit rural economies that rely on timber products: the ecology-first versus people-first dichotomy expressed in a study design. It may appear to some participants that management activities designed to provide new information will lead to a slippery slope back to the old days of intensive timber harvest. Although this opposition may appear to outsiders as antilearning or intolerant, it also may be viewed as a failure in the design of the collaborative process and management projects. Collaborators may need mechanisms that allow them to agree to an experiment without appearing to endorse all the treatments. Such mechanisms may include greater clarity in the collaborative operating principles or a commitment by decision makers to address key questions without full agreement by all the collaborators.

Some of the Plan-mandated regional effectiveness monitoring has continued, especially tracking northern spotted owl and marbled murrelet populations, late-successional and old-growth forest conditions, and aquatic-riparian indicators. Several 20-year (1994–2013) Plan monitoring reports were completed (e.g., Davis et al. 2016a, 2016b; Dugger et al. 2016; Falxa et al. 2016), and individual national forest plan revisions may emerge based on those and other science-based reviews and interpretations. If so, these would represent partial success for adaptive management. It is worth noting that the topic of adaptive management was included in the 10-year report but has yet to be addressed in the 20-year reports. The appropriate question to ask during plan revisions is, Will closing the adaptive-management loop through periodic syntheses and subsequent management changes be enough to achieve long-term sustainability without a concerted effort to ask and answer critical questions using science-based adaptive-management tools such as management experiments?

The major lessons from the second decade of Northwest Forest Plan implementation of adaptive management appear to be (1) it has yet to be implemented to address the original uncertainty about whether to use fixed reserves or shifting mosaics as a basis for management; (2) continued funding problems and frequent executive turnover hampered development of learning-based programs; (3) improved frameworks still have a high probability of failure without solid institutionalization in both management and research communities; and (4) the recent stakeholder-driven inclusion of science-based adaptive management in developing the new planning rule highlights that it is easier to find support among external collaborators than inside federal agencies. These lessons suggest that the expanded institutionalization created in the 2012 planning rule, and more important, how it is implemented, will determine the success of adaptive management in US Northwest forests.

A learning-based, long-term sustainability approach provides a way to reassess the modern purposes of managing moist conifer-dominated forests and putting all the pieces together to make it work as a system. Adaptive management has developed into a young science discipline that can credibly provide "best science" for those trying to apply it. Work remains on how much adaptive management can be implemented directly with collaborative groups (chap. 9), but collaborative application offers advantages, because citizen-driven, science-based adaptive management would likely be the most effective way to assure that long-term sustainability is successfully achieved. Collaboration at this level still represents a big change for management agencies; however, genuine collaboration could accelerate and broaden this next phase of the evolution of adaptive management.

Literature Cited

Beverton, R. J. H., and S. J. Holt. 1957. *On the dynamics of exploited fish populations*. London: Her Majesty's Stationery Office.

Bormann B. T., R. W. Haynes, and J. R. Martin. 2007. Adaptive management of forest ecosystems: Did some rubber hit the road? *BioScience* 57:186–191.

Bormann, B. T., D. C. Lee, A. R. Kiester, D. E. Busch, J. R. Martin, and R. W. Haynes. 2006. Adaptive management and regional monitoring. Chap. 10 in *Northwest Forest Plan—the first ten years (1994–2003): Synthesis of monitoring and research results*. Edited by R. W. Haynes, B. T. Bormann, and J. R. Martin. PNW-GTR-651. Portland, OR: USDA Forest Service, Pacific Northwest Research Station.

Bunnell, F. L., and G. B. Dunsworth, eds. 2009. *Forestry and biodiversity: Learning how to sustain biodiversity in managed forests*. Vancouver, BC, Canada: University of British Columbia Press.

Davis, R. J., B. Hollen, J. Hobson, J. E. Gower, and D. Keenum. 2016a. *Northwest Forest Plan—the first 20 years (1994–2013): Status and trends of northern spotted owl habitats.* General Technical Report PNW-GTR-929. Portland, OR: USDA Forest Service, Pacific Northwest Research Station.

Davis, R. J., J. L. Ohmann, R. E. Kennedy, W. B. Cohen, M. J. Gregory, Z. Yang, H. M. Roberts, A. N. Gray, and T. A. Spies. 2016b. *Northwest Forest Plan—the first 20 years (1994–2013): Status and trends of late-successional and old-growth forests.* General Technical Report PNW-GTR-911. Portland, OR: USDA Forest Service, Pacific Northwest Research Station.

Dugger, K. M., E. D. Forsman, A. B. Franklin, and 35 coauthors. 2016. The effects of habitat, climate, and barred owls on long-term demography of northern spotted owls. *The Condor: Ornithological Applications* 118:57–116.

Falxa, G. A., and M. G. Raphael, tech. coords. 2016. *Northwest Forest Plan—the first 20 years (1994–2013): Status and trends of marbled murrelet populations and nesting habitat*. General Technical Report PNW-GTR-933. Portland, OR: USDA Forest Service, Pacific Northwest Research Station.

FEMAT (Forest Ecosystem Management Assessment Team). 1993. *Forest ecosystem management: An ecological, economic, and social assessment. Report of the Forest Ecosystem Management Assessment Team*. Portland, OR: US Department of Agriculture; US Department of the Interior [and others].

Gregory, R., D. Ohlson, and J. Arvai. 2006. Deconstructing adaptive management: Criteria for applications to environmental management. *Ecological Applications* 16:2411–2425.

Hempel, C. G. 1965. *Aspects of scientific explanation*. New York: Free Press.

Holling, C. S., ed. 1978. *Adaptive environmental assessment and management*. Chichester, UK: Wiley.

Lee, K. N. 1993. *Compass and gyroscope: Integrating science and politics for the environment*. Washington, DC: Island Press.

Ludwig, D., R. Hilborn, and C. Walters. 1993. Uncertainty, resource exploitation, and conservation: Lessons from history. *Science* 260:17.

McLain, D., and R. G. Lee. 1996. Adaptive management: Promises and pitfalls. *Environmental Management* 20:437–448.

Murray, C., and D. R. Marmorek. 2004. Adaptive management: A science-based approach to managing ecosystems in the face of uncertainty. In *Making ecosystem-based management work: Proceedings of the Fifth International Conference on Science and Management of Protected Areas, Victoria, BC, May, 2003*. Edited by N. W. P. Munro, T. B. Herman, K. Beazley, and P. Dearden. Wolfville, NS, Canada: Science and Management of Protected Areas Association. http://www.essa.com/downloads/AM_paper_Fifth_International_SAMPAA_Conference.pdf.

Rogers, K. 1998. Managing science/management partnerships: A challenge of adaptive management. *Conservation Ecology* 2:2. http://www.consecol.org/Journal/vol3/iss1/art10.

Smith, A. 2009. Lessons learned from adaptive management practitioners in British Columbia. In *Adaptive management: A practitioner's guide*. Edited by C. Allan and G. H. Stankey. New York: Springer Science–Business Media.

Stankey, G. H., B. T. Bormann, C. Ryan, B. Shindler, V. Sturtevant, R. N. Clark, and C. Philpot. 2003. Adaptive management and the Northwest Forest Plan: Rhetoric and reality. *Journal of Forestry* 101:40–46.

USDA (US Department of Agriculture, Forest Service). 2001. *Five Rivers landscape management project, final environmental impact statement*. Corvallis, OR: USDA Forest Service, Siuslaw National Forest Service.

———. 2004. *Biscuit Fire recovery project, final environmental impact statement*. Medford, OR: USDA Forest Service, Rogue-Siskiyou National Forest.

———. 2012. National Forest System land management planning. Final rule and record of decision. 36 CFR Part 219. *Federal Register* 77:21162–21267. http://www.fs.usda.gov/Internet/FSE_DOCUMENTS/stelprdb5362536.pdf

———. 2016. Eastside restoration. http://www.fs.usda.gov/detail/r6/landmanagement/resourcemanagement/?cid=stelprdb5423597.

USDA and USDI (US Department of Agriculture and US Department of the Interior). 1994. *Record of decision for amendments to Forest Service and Bureau of Land Management planning documents within the range of the northern spotted owl* [plus Attachment A: Standards and Guides]. [Place of publication unknown]: US Department of Agriculture and US Department of the Interior. http://www.reo.gov/library/reports/newroda.pdf

Walters, C. J. 1986. *Adaptive management of renewable resources*. Caldwell, NJ: Blackburn Press.

———. 1997. Challenges in adaptive management of riparian and coastal ecosystems. *Conservation Ecology* 1(2):1. http://www.consecol.org/vol1/iss2/art1.

Walters, C. J., and R. Hilborn. 1978. Ecological optimization and adaptive management. *Annual Review of Ecology and Systematics* 9:157–188.

WADNR (Washington Department of Natural Resources). 1997. *Final habitat conservation plan*. Olympia, WA: Washington Department of Natural Resources.

———. 2013a. Section 22. Guidelines for adaptive management program. In *Forest Practices Board manual*. Olympia, WA: Washington State Department of Natural Resources. http://file.dnr.wa.gov/publications/fp_board_manual_section 22.pdf.

———. 2013b. *Revised draft environmental impact statement for the Olympic Experimental State Forest (OESF) Forest Land Plan*. Olympia, WA: Washington State Department of Natural Resources.

Williams, B. K. 2011. Adaptive management of natural resources—framework and issues. *Journal of Environmental Management* 92:1346–1353.

Williams, B. K., and E. D. Brown. 2012. *Adaptive management: US Department of the Interior applications guide*. Washington, DC: US Department of the Interior.

———. 2014. Adaptive management: From more talk to real action. *Environmental Management* 53:465–479.

Williams, B. K., R. C. Szaro, and C. D. Shapiro. 2007. *Adaptive management: U.S. Department of the Interior technical guide*. Washington, DC: US Department of the Interior.

Chapter 9

The Emergence of Watershed and Forest Collaboratives

Rebecca L. Flitcroft, Lee K. Cerveny, Bernard T. Bormann, Jane E. Smith, Stanley T. Asah, and A. Paige Fischer

Recent decades have seen the emergence of collaborative organizations for forest governance in landscape-scale management. A *collaborative* is defined as an organized collection of landowners, stakeholders, resource agencies, tribes, or other organizations that come together to address common issues and resolve problems through deliberation, consensus building, and cooperative learning (Goldstein and Butler 2010). Collaboratives are designed to be transparent, diverse, and inclusive (Wondolleck and Yaffee 2003). Although some collaboratives are community driven or place based, with the goal of protecting local interests or access, others are sparked by land managers who seek integrated solutions to multiple interests and objectives.

Collaboratives vary in form and include public advisory committees, watershed councils, stewardship groups, community-based organizations, and other stakeholder associations. Recently government agencies have become central players. Public land-management agencies like the US Forest Service have adopted collaborative models as standard practice, especially around landscape-scale issues such as fire management, climate change, and forest restoration (box 9.1; Wondolleck and Yaffee 2003). Collaboration is an integral part of agency dialogue with the public, and ideally, collaborative groups play an important role in fleshing out projects, details, and specifications; establishing zones of agreement; weighing alternatives; and providing feedback on actions. Collaborative approaches may provide a more culturally acceptable means of working with tribes involved in government-to-government negotiations.

BOX 9.1. LEGAL BASIS FOR COLLABORATIVES AND US FOREST POLICY

The US federal policy context for collaboration begins with the Administrative Procedures Act of 1946 (APA), which requires federal agencies to

- Keep the public informed of their organization, procedures, and rules.
- Provide for public participation in the rule-making process.
- Establish uniform standards for the conduct of formal rule making and adjudication.

Judicial review under the APA allows federal courts to set aside agency action by concluding that the regulation is "arbitrary and capricious, an abuse of discretion, or otherwise not in accordance with the law" (5 USC § 706[2][A]). The National Environmental Policy Act of 1970 (NEPA) applies this act for the main purpose of assuring the quality of the human environment when making decisions. The NEPA law established the Council of Environmental Quality (CEQ), which writes regulations so that NEPA can bring together federal, state, and local governments and other stakeholders to "balance environmental, economic, and social values" to create a "protective harmony between humans and the human environment" (42 U.S.C. § 4321).

Through the National Forest Management Act of 1976 (NFMA), the Forest Service implements APA, NEPA, and CEQ regulations at national forest scales with its national planning rule. Before 2011, the Forest Service used its 1982 planning rule mandating public involvement geared to notices of intent to write environmental documents (environmental assessments and environmental impact statements), disseminate scoping documents, and allow specific comment periods for people to raise questions about a proposed action. The 2012 planning rule (USDA 2012) extends public involvement to broaden forest planning activities.

Ultimately, most collaboratives seek win-win management outcomes in which decisions are made that are good for both ecological and human systems. In the past, arguments in favor of economic growth have been positioned in opposition to sustainable ecological systems. Collaboratives seek the ground between these two goals through enhanced group learning, stakeholder integration, and innovative management solutions. When collaborative members develop trust (chap. 18) and gain experience working together, they have a balanced, long-term approach that results in solutions that may be ecologically sustainable, economically viable, and socially acceptable (Conley and Moote 2003; Lauber et al. 2008).

How well collaboratives can achieve their goals is still an open question. Although collaboratives can be successful at bringing diverse voices to the table, difficulties are sometimes encountered. Like any formal organization, collaboratives face challenges in leadership, attrition, fatigue, internal conflict, and achieving consensus. They typically are not very representative of the broad array of pertinent stakeholders and tend to include those who may benefit or lose from collaborative processes and outcomes. A lack of social capacity, resources, skills, time, and funds to get work done can dampen collaborative effectiveness. Despite these concerns, collaboratives are touted as a highly appropriate governance approach to address socioecological problems at the landscape scale. Understanding both the strengths and weaknesses of collaboratives can advance the discussion of how forests can be managed for the well-being of ecological and human systems now and in the future.

Need for Collaboratives

The proliferation of collaborative forest landscape management has been spurred by increased interest in citizen engagement with public resource management. Also, public agencies need to expand capacity by working in partnerships, and it is increasingly recognized that socioecological changes occur at scales that transcend traditional landownerships and require cooperation among multiple landowners and interested stakeholders.

Many social and ecological problems become tractable only at large, multiownership scales. For instance, broad-scale disturbances such as wildfire, flooding, windthrow, climate change, disease, and invasive species cross ownership boundaries. Land-use changes on neighboring private lands, such as intensive residential development, agriculture, commercial development, or resource-based industrial development, can affect broader-scale ecosystem services such as water and air quality as well as aquatic and terrestrial habitat, including wildlife corridors on public lands. Simply put, the scale of the threat does not match that of the management action from a single landowner's jurisdiction.

The conterminous US Pacific Northwest (Oregon and Washington, USA) has a storied history of conflict around the management of natural resources that has been characterized both by polarization (pro-environment versus pro-development) and by a lack of trust in federal land management agencies (chap. 18). As a result, litigation and appeals are frequent responses to environmental decisions reached by land managers. Past methods for involving concerned publics in federal forest management, usually driven by regulations, have proved unsatisfactory to people who did not think their views were given fair consideration, and projects were then appealed and litigated, devolving the spirit of collaboration. Moreover, the National Environmental Policy Act (NEPA) requires agencies to demonstrate that proposed projects or actions will not have an adverse effect on the biophysical, economic, or sociocultural environment. Requirements of NEPA, along with other regulatory measures, have slowed the process of environmental decision making as agencies seek to create bullet-proof decisions that will hold up in court. In the case of large-scale forest restoration, collaboration is perceived as a means to reduce the burden of the NEPA process and decrease the potential for litigation or appeal by fostering diverse engagement in the deliberation phase.

Optimism for a new century of forest planning is emerging from the shadow of earned distrust that US federal agencies can manage land and resources with a sustainable and conservation-oriented approach. However, this optimism coincides with obvious contradictions, including a growing number of people who see national forests as places for recreation, sources of clean water and fish and wildlife habitat, and sinks to store carbon rather than for wood and mineral extraction, yet still demand greater use of wood products to build homes for the ever-increasing human population.

Although studies have not definitively demonstrated the success of collaboratives in building trust and diminishing lawsuits, early anecdotal accounts suggest that the model has promise. In the Blue Mountains region of Oregon, for example, the number of new legal cases and appeals filed against US national forests declined steadily after the formation of the Blue Mountain Forest Partners, a collaborative focused on the Malheur National Forest (McLain et al. 2014). In this case, a long-time environmental advocate began working alongside representatives of the timber industry to try new ways of addressing resource challenges that did not embrace a win-lose paradigm. This collaborative has expanded its scope to include other resource challenges beyond forest restoration.

The emphasis on collaboration has been supported by the highest levels of the US Forest Service and was woven into the goals of the 2012 National Forest planning rule (USDA 2012), which, when fully implemented in the subsequent round of forest plan revisions, will make collaboration even more ubiquitous (box 9.1). The expectations and roles of collaboratives will broaden from current engagement in NEPA, stewardship contracting, multiparty monitoring, and other planning and assessment efforts to include forest planning.

Here we examine a variety of forest management collaboratives since 1994 in the US Northwest. We address the composition of their socioecological networks and the ecological frameworks behind them, and whether

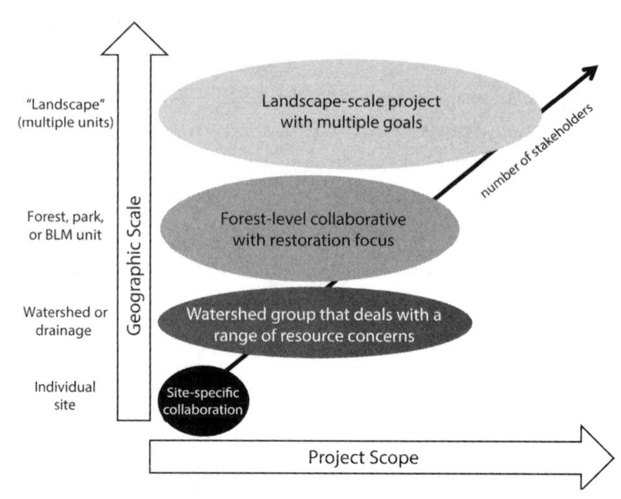

FIGURE 9.1. As the group focus expands to encompass new resource issues and increases in geographic extent, there are likely to be more stakeholders to engage.

they have resulted in policy shifts or emerging governance structures. Finally, we explore preliminary evidence for their effectiveness to inform their further evolution. Some examples draw from dry forest landscapes but illustrate key concepts and lessons learned in collaborative forest management.

Types of Collaboratives

The form of a collaborative varies based on the spatial extent of the unifying topic, the scope addressed by the group, the complexity of ecological and social issues, and the sociopolitical landscape (fig. 9.1; box 9.2). Collaborative approaches occur at many geographic extents and can be reflected in programs ranging from an individual site or resource area to a watershed (e.g., watershed planning groups); to a national forest, park, or other management unit (e.g., natural resource advisory committees, forest collaboratives); up to the landscape scale, which can include multiple national forests; other federal, state, and tribal lands; and private landowners (e.g., the Collaborative Forest Landscape Restoration Program).

In 2009, in recognition of the benefits from conservation partnerships at the landscape scale, US Department of the Interior secretary Ken Salazar issued Secretarial Order No. 3289 to establish the Landscape Conservation Cooperatives (LCCs), a network of 22 individual self-directed conservation

BOX 9.2. COLLABORATIVES ACROSS GEOGRAPHIC SCALES AND PROJECT SCOPES

Collaboratives have evolved in response to a variety of management goals, necessitating different geographical extents of cooperation and project scopes (fig. 9.1). At a site or individual project scale, multiple groups may collaborate to design or implement work. These types of collaborative actions have limited geographic scope and are designed to fix an obvious problem, such as the replacement of barrier culverts that block movement of aquatic organisms in a watershed. At the scale of an individual watershed, groups such as watershed councils may organize and coordinate restoration actions that occur within their specific areas.

In Oregon, watershed councils provide a nexus for multiple agencies, nonprofits, and local citizens to jointly plan and implement diverse restoration actions. At the forest scale, collaboratives work across jurisdictions to improve ecosystem health. For example, the Tapash Sustainable Forest Collaborative (Yakima, Washington) is working to improve ecosystem health and natural functions of the Washington landscape from the high eastern Cascade Range to the Columbia River basin. This coalition of federal, state, and tribal partners is focused on restoring fire-adapted ecosystems by exploring stewardship contracting, ecosystem-services markets, and cellulosic-ethanol production from forest biomass.

An example of a landscape-scale design for collaboratives is the legislatively established Collaborative Forest Landscape Restoration Program (CFLRP). CFLRP's projects take an all-lands approach to forest restoration, working across jurisdictions to manage at the landscape scale. The Forest Service collaborates with tribes, other landowners, communities, local organizations, conservation interests, and other agency partners on 10-year projects that promote forest restoration and enhanced community resilience.

areas covering the United States, including Pacific and Caribbean islands, as well as parts of Canada and Mexico. The intent was to provide a collaborative framework that could deliver the scientific information needed for effective management and catalyze conservation planning and actions across multiple jurisdictions through partnerships that include the US Forest Service.

The Forest Ecosystem Management Assessment Team (FEMAT 1993) that produced the US federal Northwest Forest Plan is an example of a regional collaborative group consisting of federal agencies along with public input processes. The team developed a governance structure for ~10 million ha (24 million ac) of largely moist coniferous forest (USDA and USDI 1994).

Collaborative approaches also may be community based (e.g., Fire Safe Councils), bringing local actors together to address shared resource concerns or socioeconomic needs. Collaborative groups can focus on a finite set of issues, such as water quality, wildlife habitat, fire protection, or travel management, or they can adopt a specific role or set of functions, such as stewardship contracting, environmental assessment, or strategic planning.

How Do Collaboratives Work?

Collaboratives assume many organizational forms and use a variety of governance mechanisms in their operations. *Governance* refers to the rules, forms, and norms that guide decision making and may include formal and informal institutions, social groups, processes, interactions, and practices (Reed and Bruyneel 2010). The forms and structures used by collaborative groups may vary depending on the geographic scale being considered; the number of years the group has been in operation; the operating environment of forest governance, such as the issues they address and activities they conduct (Conley and Moote 2003); or the individual preferences and experiences of internal leaders. The capacity of local collaborative groups to organize themselves, manage their projects, acquire necessary skills and resources, and achieve planned outcomes is critical to their success. There is an emerging need for information that evaluates the capacity of collaboratives to achieve their goals.

Oregon appears distinct from other US states because it has both federal support from the US Forest Service and legislated state investment in collaboratives. Across the state, 24 known collaboratives currently operate (Oregon Solutions 2013). The state sponsors Federal Forest Health Collaborative Capacity Assistance Grants, which can be used for facilitation, organizing meetings and field tours, travel costs, monitoring and evaluation, and communication. Statewide, numerous nonprofit organizations, foundations, universities, and state extension entities have emerged to aid in the genesis of new collaboratives and to provide ongoing technical and administrative support as they take root, grow, evolve, change leadership, and tackle a wider array of interlocking issues. Many collaboratives in Oregon have benefited from the presence of long-term facilitators with expertise in group process. In addition, opportunities are available for training and assistance in leadership, administrative management, grant writing, and other skills. Other states in the Northwest, including Idaho, Washington, and Alaska, have seen a steady growth of collaboratives. However, the statewide collaboratives network in Oregon is more advanced owing to the coalescence of state funding and broader institutional support.

Studies of collaboratives reveal a range of organizational forms. Formal mechanisms include official incorporation as a nongovernmental organization, hired staff, trained facilitation, use of formal subcommittees or standing committees, an operating charter, ground rules, and specific norms for deliberation and decision making. Many enduring collaboratives are far less organized, with fewer explicit rules, but have informal mechanisms that lead to successful outcomes. Informal unifying mechanisms may include social relations, long-term engagement by active participants, local knowledge, trust, institutional memory, strong communication networks, and clear but implicit goals (Conley and Moote 2003; Lauber et al. 2008). A rapid assessment of Oregon community collaboratives found that more than half of the groups had dedicated facilitation, a steady pattern of meetings, two or more administrative policies or documents, and a standing leadership committee (Davis et al. 2015). The study revealed that many long-standing collaboratives had few of the formal mechanisms.

In recent years, there has been some effort for collaborative groups to organize into collectives around common geographic areas. In Oregon, a new Coalition of Collaboratives is made up of five collaborative groups in eastern Oregon that focus on large-scale restoration efforts in cooperation with the US Forest Service Eastside Restoration Strategy (USDA 2016), in order to improve coordination and communication around eastside resource issues. In addition, several statewide meetings of Oregon collaboratives have occurred, in which leaders assemble to learn about how they can maximize their impact by working together. These meetings offer information, training, and social networks that can aid collaboratives. Such efforts to improve coordination and communication among collaboratives in a particular region may be an important aspect of the longevity and viability of the collaborative governance model.

Effectiveness of Collaboratives

The intent of collaboration-based management at landscape scales encompasses both social and ecological goals. Evaluation of the efficacy of collaboratives at achieving their goals is challenging and requires monitoring. At minimum, ecological, social, and economic monitoring requires the

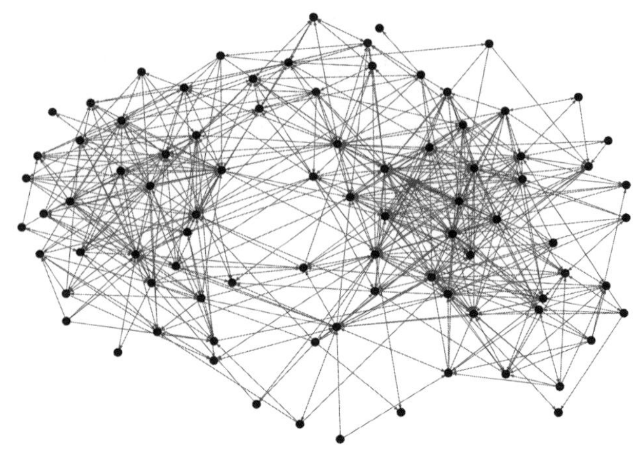

FIGURE 9.2. Graphic representation of the connections among 87 organizations concerned with increasing wildfire risk in the study area and the ties the organization representatives reported with the organizations they work with. The configuration of the network shows two separate groups with some level of connection between the groups (box 9.3; Fischer et al. 2016).

clear identification of management-related environmental indicators and metrics that are tracked over time.

Clearly defined ecological goals that can become part of ongoing research and monitoring, or that can be developed into programs that assess the specific effect of management action on the desired ecological condition, will benefit the effectiveness of collaboratives. Quantification of restoration-driven activities (such as the number of hectares [acres] thinned, or the number of road-stream culverts replaced) is simpler than assessment of the ecological effects of those activities. The direct effect of changes in management on the ecological integrity of rivers and upslope habitats has been explored by a variety of researchers. Large-scale and long-term monitoring programs have been put into place to evaluate the effects of forest management under the US federal Northwest Forest Plan (for northern spotted owls, marbled murrelets, late-successional and old-growth forest conditions, aquatic-riparian systems, socioeconomics; USDA and USDI 2015). However, some lessons learned from these efforts show that even statistically rigorous and continuously funded research programs designed to link ecological conditions to management actions have difficulty making these connections (Kershner and Roper 2010). The complexity of ecological interactions at landscape scales can confound even well-designed

BOX 9.3. SOCIAL-ECOLOGICAL NETWORKS IN FORESTED SYSTEMS

Social capital theory suggests that actors in cohesive social networks are positioned to build trust (chap. 18) and mutual understanding of problems and to act collectively to address these problems. Further, actors engaged with diverse partners are positioned to access new information and resources that are important for innovation and complex problem solving. Patterns of interaction within a network of organizations involved in forest and wildfire management in Oregon were investigated by Fischer et al. (2016) for evidence of structural conditions that create opportunities for collective action and learning. They found that tendencies to associate with others with similar management goals, geographic emphases, and attitudes toward wildfire strongly shaped network structure (fig. 9.2), which potentially constrained interactions among organizations with diverse information and resources and limited opportunities for the learning and complex problem solving needed for adaptation. In particular, Fischer et al. found that organizations with fire protection and forest restoration goals comprised distinct networks despite shared concerns about the problem of increasing wildfire risk.

sampling programs. This ecological complexity similarly makes it difficult to evaluate the effects of collaboratives on forest management.

Goals of collaboration-based management that address social issues include inclusiveness (ensuring that membership in the collaborative reflects the diversity of stakeholders and community members); reduction in legal challenges; building trust among stakeholders; identifying actions with both social and ecological benefits; and enabling the emergence of networks that may facilitate other collaborative initiatives (fig. 9.2; box 9.3). McDermott et al. (2011) defined nine factors that influence the success of collaboratives: (1) external sources of support, including local political actors and agency officials; (2) legal authority and supportive laws and policies; (3) community involvement and legitimacy; (4) adequate and reliable funding; (5) personnel or human capacity to do the work; (6) access to and exchange of information and research; (7) effective leadership; (8) trust between participants; and (9) social capital (networks of social relations among people and groups that enable them to coordinate and cooperate for mutual benefit). Evaluation of each of these factors can be assessed with appropriate statistical rigor. For example, Bergman et al. (2004) determined that stakeholders want to be confident that working relationships will be productive before investing in collaboration, and Flitcroft et al.

(2009) showed that networks among trusted neighbors lead to the expansion of stakeholder participation.

Challenges of Working Collaboratively

To be effective, the challenges and shortcomings of the collaborative movement must be recognized. Many collaboratives have a consensus operating principle in which everyone effectively retains veto power over any decision, citing lack of trust. This can hinder management because the only action that can be accomplished is the least controversial and may not solve the most important problems. This situation may be particularly problematic in instances in which participants in a collaborative bring with them latent agendas that may be related to past negative experiences, including the pursuit of old battles. Stakeholders with unspoken agendas may undermine proposed solutions by arguing that the solutions are not ecologically valid or that management cannot be trusted to implement them, or by orienting deliberations toward dead ends such as the perceived undesirable characteristics of other stakeholders rather than the main issues (Asah et al. 2012a).

For a collaborative to be successful, all members must be open to new paths toward the future, with the goal of an improved and integrated community and ecological system. Success of forest collaboratives may hinge on the acceptance of nontraditional ideas and approaches to management that may be achieved by focusing on shared organizational learning and mutual discovery. Such learning and discovery may include efforts to uncover the latent agendas that might be shaping the uncompromising positions that some stakeholders may hold (Asah et al. 2012b). Developing protocols for group learning, system monitoring, and incorporating feedback is a critical component of an adaptive-management approach to collaboration (Berkes 2009). If this cannot be done successfully, alternatives to a consensus approach may facilitate action and learning.

Although they often attempt to be diverse and encompassing in terms of membership, organized collaboratives have not been the right solution for everyone. Funding may be inconsistently available across regions or between years, causing some collaboratives to stall, whereas others appear more successful. Lack of access to collaborative meetings (due to geographic distance or time of meeting) can be a constraint, as well as challenges associated with integrating new participants who may be unfamiliar with the regulatory processes or with the intricacies of NEPA assessments and stewardship contracting. In some cases, potential collaborators have information or needs that are proprietary, sensitive, or culturally significant, and they do not wish that information to become widely dispersed in a collaborative model (e.g., traditional knowledge held by indigenous people). Collaboratives also often do not adequately accommodate stakeholders who are not directly involved, such as ordinary citizens who benefit from ecosystem services. Thus, collaboratives may often include only representatives of key direct stakeholder groups that are typically well organized and have well-articulated and direct interests along with the resources to engage in face-to-face meetings. Alternative mechanisms for engaging with the voices of those who are less directly involved would enhance the sociopolitical license of collaboratives.

Collaboratives pose other difficulties. The burden of work in a collaborative may fall on a small group of people, and the energy to maintain the collaborative over the long term may be compromised as individuals burn out. Collaboratives may be off-putting to some because of the actors or institutions involved. In other cases, the focus of the collaboratives may be too broad or too narrow to meet the needs of every organization with an interest in natural resource governance. Studies have found that collaboratives lacked participation from tribes, recreation organizations, and ranching organizations, among others (McLain et al. 2014) or have not met the needs of private landowners (Fischer and Charnley 2012).

New collaboratives have emerged with specialized focus or with different sets of members that may focus on a geographic area also encompassed by another forest collaborative. This overlap in the areas of interest by collaboratives may cause tension between groups or further complicate on-the-ground decision making. This is further compounded by the conflicting scales of policies (i.e., national policy, local applications) that provide legal frameworks for land management.

Future Considerations

Some cultural change may be necessary among collaborative participants so that they are truly able to listen and to share decision making. If participants perceive that some landowners, managers, and other particularly interested stakeholders come in with an agenda already formed, with the purpose of "bringing the public along," the collaborative effort will fail. Effective partnerships will also require that individual forest managers routinely seek and incorporate the input from collaboratives in forest management.

Regardless of any potential shortcomings, collaboration is becoming a necessary element of a socioecological framework of broad-scale forest management. People who are knowledgeable and care about forests and local communities may help others learn about the complex human processes behind community well-being and ecological sustainability. The more collaboratives contribute to formal processes, including planning, science-based adaptive management, and decision making, the more participants will appreciate the complexities of forest-based socioecological systems. Flexibility in work and consultation timelines allows collaboratives and associated management agencies to work more closely together.

Collaborative approaches are required for management strategies to align with the number and broad spatial extent of environmental stressors. Collaboratives provide an avenue to participation in the active management of forests for individuals and groups that have been previously excluded. The perceived and real exclusion of outside interests from management of federal forests has bred discontent and distrust. Collaboratives offer forest managers an opportunity to develop trust with local, regional, and national partners in the management of the forested commons. At their best, collaboratives offer a path toward management that is both ecologically sound and socially accepted.

Literature Cited

Asah, S. T., D. N. Bengston, K. Wendt, and K. C. Nelson. 2012a. Diagnostic reframing of intractable environmental problems: Case of a contested public land-use conflict. *Journal of Environmental Management* 108:108–119.

Asah, S. T., D. N. Bengston, K. Wendt, and L. DeVaney. 2012b. Prognostic framing of stakeholders' subjectivities: A case of all-terrain vehicle management on state public lands. *Environmental Management* 49:192–206.

Bergman, W. J., R. R. Bliss, C. A. Johnson, and G. G. Kaufman. 2004. *Netting, financial contracts, and banks: The economic implications*. Working Paper No. 2004-02. Chicago: Federal Reserve Bank of Chicago.

Berkes, F. 2009. Evolution of co-management: Role of knowledge generation, bridging organizations and social learning. *Journal of Environmental Management* 90:1692–1702.

Conley, A., and M. A. Moote. 2003. Evaluating collaborative natural resource management. *Society & Natural Resources* 16:371–386.

Davis, E. J., L. K. Cerveny, M. Nuss, and D. Seesholtz. 2015. Oregon's forest collaboratives: A rapid assessment. *Oregon State University Scholar's Archive*. http://ir.library.oregonstate.edu/xmlui/handle/1957/55791.

FEMAT (Forest Ecosystem Management Assessment Team). 1993. *Forest ecosystem management: An ecological, economic, and social assessment.* Report of the Forest Ecosystem Management Assessment Team. Portland, OR: US Department of Agriculture; US Department of the Interior [and others].

Fischer, A. P., and S. Charnley. 2012. Risk and cooperation: Managing hazardous fuel in mixed ownership landscapes. *Environmental Management* 49: 1192–1207.

Fischer, A. P., K. Vance-Borland, L. Jasny, K. E. Grimm and S. Charnley. 2016. A network approach to assessing social capacity for landscape planning: The case of fire-prone forests in Oregon, USA. *Landscape and Urban Planning* 147:18–27.

Flitcroft, R. L., D. C. Dedrick, C. L. Smith, C. A. Thieman, and J. P. Bolte. 2009. Social infrastructure to integrate science and practice: The experience of the Long Tom Watershed Council. *Ecology and Society* 14(2):36.

Goldstein, B. E., and W. H. Butler. 2010. Expanding the scope and impact of collaborative planning: Combining multi-stakeholder collaboration and communities of practice in a learning network. *Journal of the American Planning Association* 76:238–249.

Kershner, J. L., and B. B. Roper. 2010. An evaluation of management objectives used to assess stream habitat conditions on federal lands within the Interior Columbia Basin. *Fisheries* 35:269–278.

Lauber, T. B., D. J. Decker, and B. A. Knuth. 2008. Social networks and community-based natural resource management. *Environmental Management* 42:677–687.

McDermott, M. H., M. A. Moote, and C. Danks. 2011. Effective collaboration. Pp. 81–111 in *Community-based collaboration: Bridging socio-ecological research and practice.* Edited by E. F. Dukes, K. E. Firehock, and J. E. Birkhoff. Charlottesville and London: University of Virginia Press.

McLain, R., K. Wright, and L. K. Cerveny. 2014. *Who is at the forest restoration table? Final report on the Blue Mountains Forest Stewardship Network, Phase I*. Portland, OR: Portland State University. https://www.pdx.edu/sustainability/sites/www.pdx.edu.sustainability/files/BluesReport_online.pdf.

Oregon Solutions. 2013. *Oregon forest collaboratives statewide inventory*. http://orsolutions.org/wp-content/uploads/2011/08/OFCSI_Draft_February_20131.pdf.

Reed, M. G., and S. Bruyneel. 2010. Rescaling environmental governance, rethinking the state: A three-dimensional review. *Progress in Human Geography* 35: 646–653.

USDA (US Department of Agriculture, Forest Service). 2012. National Forest System land management planning. Final rule and record of decision. 36 CFR Part 219. *Federal Register* 77:21162–21267. http://www.fs.usda.gov/Internet/FSE_DOCUMENTS/stelprdb5362536.pdf.

——2016. Eastside restoration. http://www.fs.usda.gov/detail/r6/landmanagement/resourcemanagement/?cid=stelprdb5423597.

USDA and USDI (US Department of Agriculture and US Department of the Interior). 1994. *Record of decision for amendments to Forest Service and Bureau of Land Management planning documents within the range of the northern spotted owl* [plus Attachment A: Standards and Guides]. [Place of publication unknown]: US Department of Agriculture and US Department of the Interior. http://www.reo.gov/library/reports/newroda.pdf.

——. 2015. *20 year reports for the Northwest Forest Plan*. Interagency Regional Monitoring Program. http://www.reo.gov/monitoring/reports/20yr-report/.

Wondolleck, J. M., and S. L. Yaffee. 2003. Collaborative ecosystem planning processes in the United States: Evolution and challenges. *Environments* 31(2):59.

SECTION 3

Science-Based Management: How Has New Science Shaped Our Thinking?

The scientific underpinnings of management in Northwest moist forests have changed over the last several decades. There has been an increased appreciation for how heterogeneity of forest characteristics and processes plays out at the landscape scale under intentional management. Our understanding of how disturbances, wildlife habitat and biodiversity values, stream conditions, biomass production, carbon storage, and other attributes change across forests and through time tempers an earlier focus on wood commodity production. Key lessons learned are highlighted in this section.

Chapter 10 explores how silvicultural practices have changed and assesses their compatibility with this dynamic landscape and desired ecosystem services. Scientific advances in the studies of long-term productivity (chapter 11) and carbon sequestration (chapter 12) are synthesized, with both bearing on larger metrics for ecosystem health. Chapter 13 describes the relationship between forest dynamics and biodiversity concerns of regional natural-resource managers and the public as informed by scientific advances in our understanding of native Northwest moist-forest species and their ecological functions. Chapter 14 describes the functions of aquatic and riparian ecosystems in moist forests. Chapter 15 explores forest-management challenges of integrating watershed-, landscape-, and disturbance-ecology frameworks. These play out over time at a range of scales from stream reaches to stands, to watersheds and landscapes.

A

B

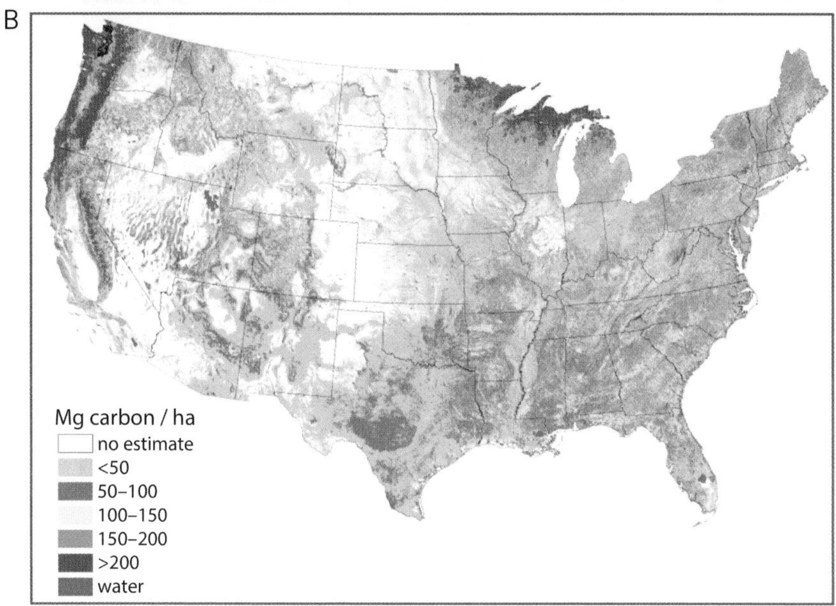

PLATE I. (chap. 1). Pacific Northwest moist coniferous forests (A) are known for their productivity, here (B) reflected by the carbon density of the total forest ecosystem, derived from forest inventory plots for the conterminous United States, 2000–2009. This metric includes carbon from above- and below-ground live trees, standing dead trees, dead and down wood, the understory, the forest floor, and mineral soil organic matter, and above- and below-ground pools. [Credits: (A) Tom Iraci, USFS; (B) Wilson et al. 2013]

A

B

PLATE 2. (chap. 2). (A) Old-growth moist forests in the Pacific Northwest are notable for their structural complexity, massiveness, and the dominance of evergreen conifers, characteristics that differentiate them from most other moist temperate forest regions of the world, which are dominated by deciduous and evergreen hardwoods. (B) The pre-forest stage, the period between a major stand-replacing disturbance and reestablishment of a closed forest stand, is notable for diverse plant life-forms (herbs, shrubs, and scattered trees) and for the fact that it is not dominated by the tree life-form. This stage has high biodiversity with many habitat specialists (not weeds), including songbirds, butterflies, and large grazing animals such as deer and elk. Under natural conditions, the pre-forest stage, with its significant structural legacies of large dead trees (snags) and down logs—important habitat for many animal species—may persist for several decades. [Credits: (A) Tom Iraci, USFS; (B) Jerry F. Franklin]

Plate 3. (chap. 3). The diversity of life-forms in moist forests—fungi, mosses, lichens, and plants—creates a many-layered structure from subsoil to canopy. (A) Lichens drape an old-growth tree in the Washington Cascade Range. (B) Salmonberry in the Oregon Coast Range. (C) American matsutake mushroom. (D) Trillium and deer ferns. [Credits: (A) Bruce G. Marcot; (B) and (C) Doni McKay, USFS; (D) Kathryn Ronnenberg, USFS]

A

B

PLATE 4. (chap. 3). (A) Siletz basketry cap (hazel sticks, split spruce root, beargrass, and maidenhair fern). (B) Sawtooth huckleberry fields in southwest Washington. Controlled burning by crews from the US Forest Service Gifford Pinchot National Forest and the Yakama Nation stimulate growth of forest understory plants. [Credits: (A) Hallie Ford Museum of Art, Willamette University, accession no. 2001.059.002, D. Peterson; (B) Jon Nakae, USFS]

A

B

PLATE 5. (chap. 4). (A) Hull-Oakes Lumber Company—a family-owned, vertically integrated company that manages its own timberland—operates one of the few mills in Oregon still capable of milling large-diameter timber. (B) Logging forwarder at work on a second-growth timber sale. As compared with older equipment, forwarders reduce ground disturbance during harvesting but can't move large logs. Modern mechanized logging requires much smaller crews than once were employed in harvest operations, resulting in fewer available jobs in forest-based communities. [Credits: (A) Hull-Oakes Lumber Company; (B) Robert Deal]

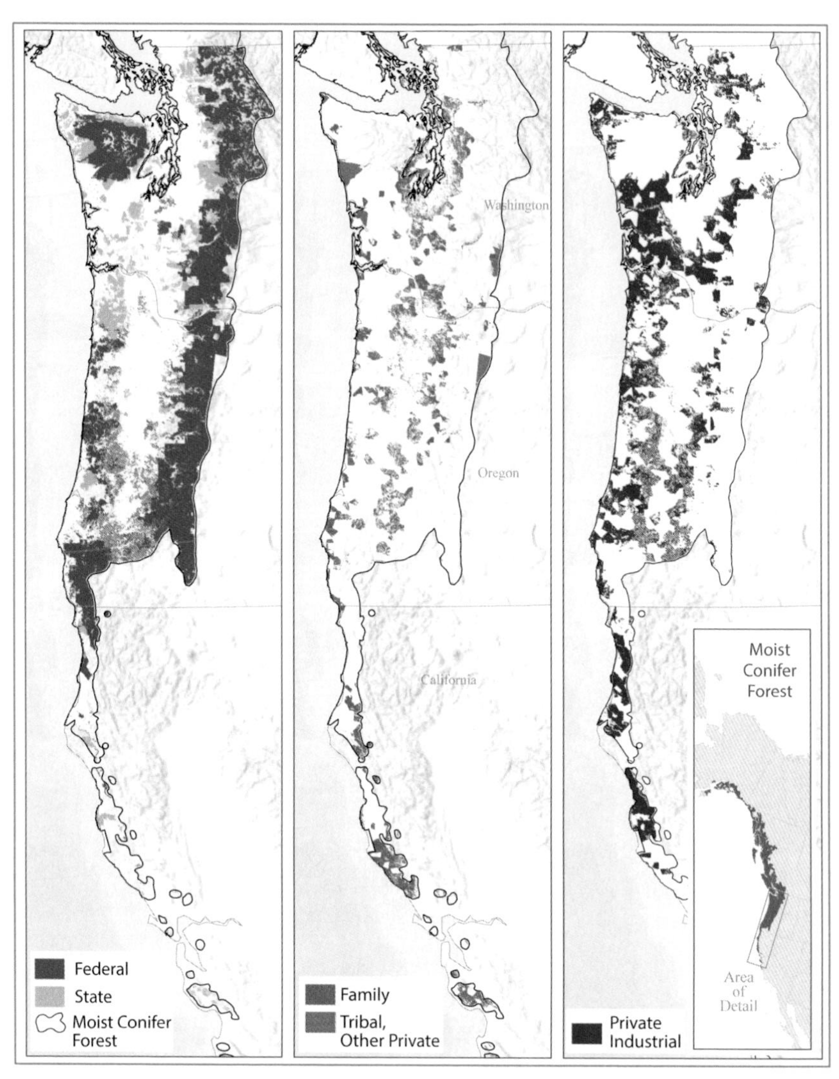

Plate 6. (chap. 6). Landownerships in the moist coniferous forests of the conterminous US Pacific Northwest. [Credit: Kelly Christiansen, USFS]

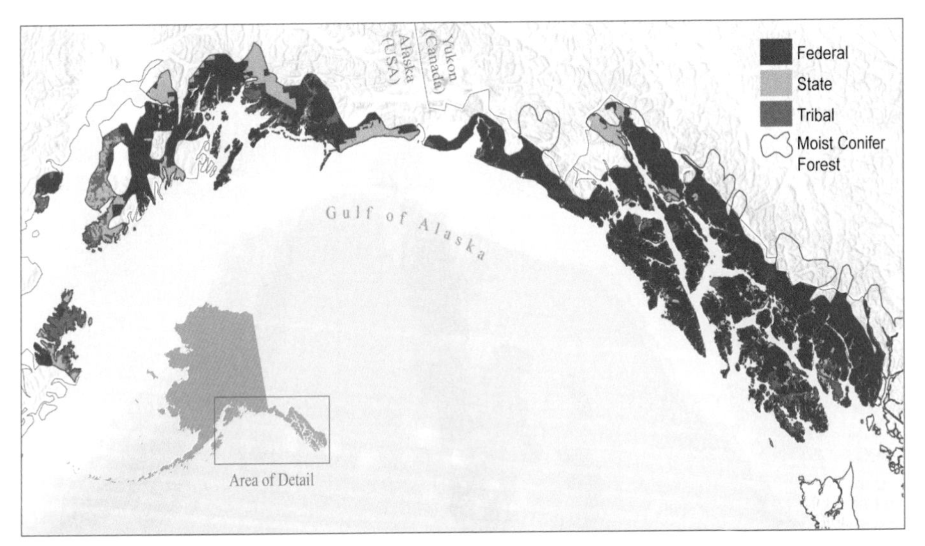

PLATE 7. (chap. 6). Landownerships in the moist coniferous forests of Southeast Alaska. [Credit: Kelly Christiansen, USFS]

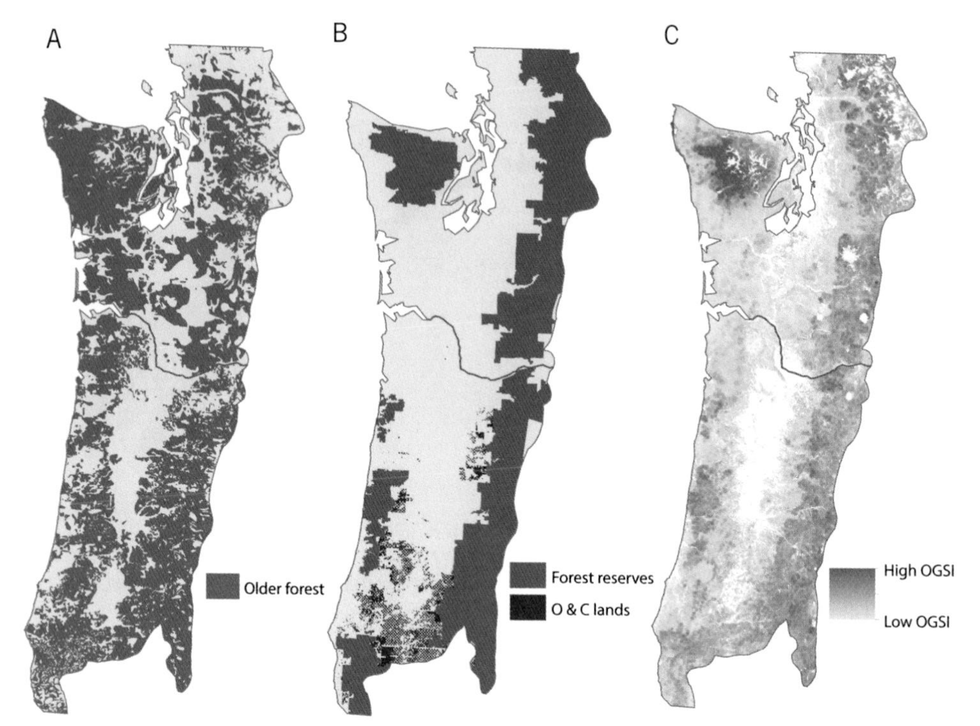

PLATE 8. (chap. 7). Patterns of older forests at the turn of the twentieth century (A) were largely shaped by natural disturbances such as infrequent, large, stand-replacing wildfires. Societal decisions to reserve forests and place them under federal management (B) resulted in distinctly different patterns of older forests by 2012, as represented by (C) the old-growth structural index (OGSI). Higher OGSI values are associated with older forests, lower OGSI values with younger forests. [Credits: (A) Plummer et al. 1902 and OSBF 1914; (C) Davis et al. 2015]

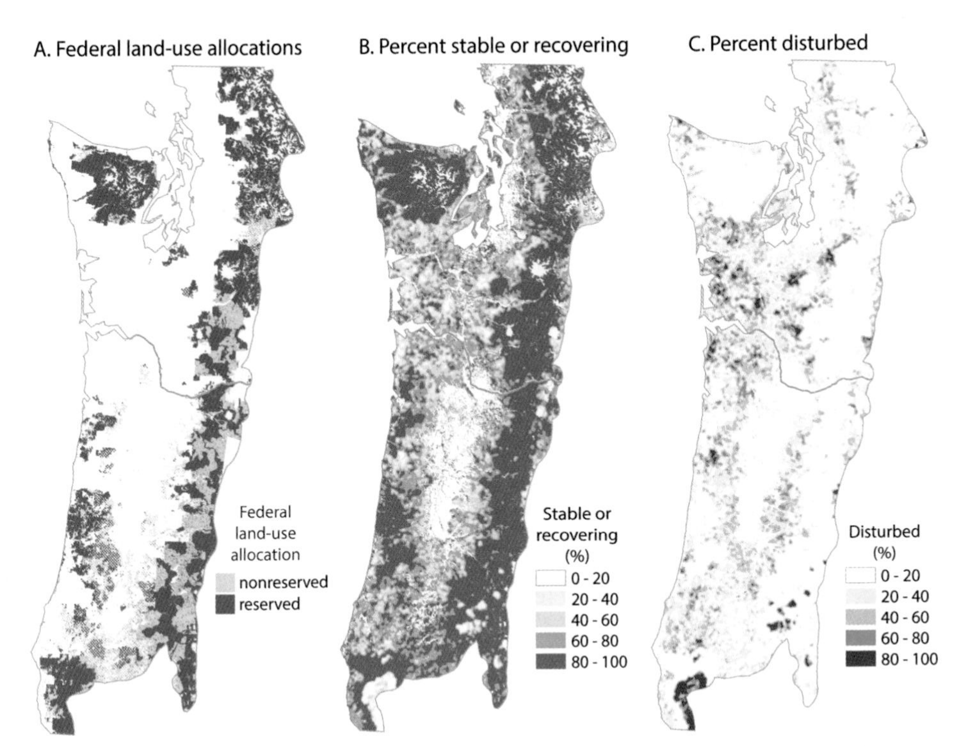

PLATE 9. (chap. 7). Federal land use allocations (A) as implemented under the Northwest Forest Plan in 1994, compared with (B) patterns of forest recovery and (C) disturbance dynamics between 1990 and 2012. [Credit: Data from Davis et al. 2015]

PLATE 10. (chap. 10). Thinning to encourage the growth of larger trees creates a more open structure within the stand that may permit understory development (A) but doesn't leave large open areas of new growth. Regeneration of shade-tolerant tree species (B) creates a multilayered stand structure in thinned forests. [Credit: Paul Anderson, USFS]

PLATE 11. (chap. 13). Fauna of Pacific Northwest moist forests include (A) northern spotted owl; (B) marbled murrelet nesting; (C) northern flying squirrel feeding on a truffle; (D) Oregon slender salamanders guarding eggs in log; (E) Pacific marten; (F) coho salmon. [Credits: (A) Al Levno, USFS; (B) www.HamerEnvironmental.com; (C) Jim W. Grace, USFS; (D) Loretta Ellenburg, USFS; (E) Katie Moriarty, USFS; (F) Christine Hirsch, USFS]

PLATE 12. (chaps. 14 and 15). Headwater streams may make up over 70% of the stream network in Pacific Northwest moist coniferous forests, significantly influencing downstream conditions. [Credit: Raymond J. Davis, USFS]

A

B

PLATE 13. (chap. 13). (A) Across the Northwest moist coniferous forest landscape, wildfire varies in frequency, intensity, and area of coverage. High-severity wildfires often are accompanied by other types of disturbances, such as landslides. (B) In contrast, timber harvest creates a very different landscape pattern of small patches of more frequent disturbance. [Credits: (A) Joe Benjamin, USFS; (B) Bruce G. Marcot]

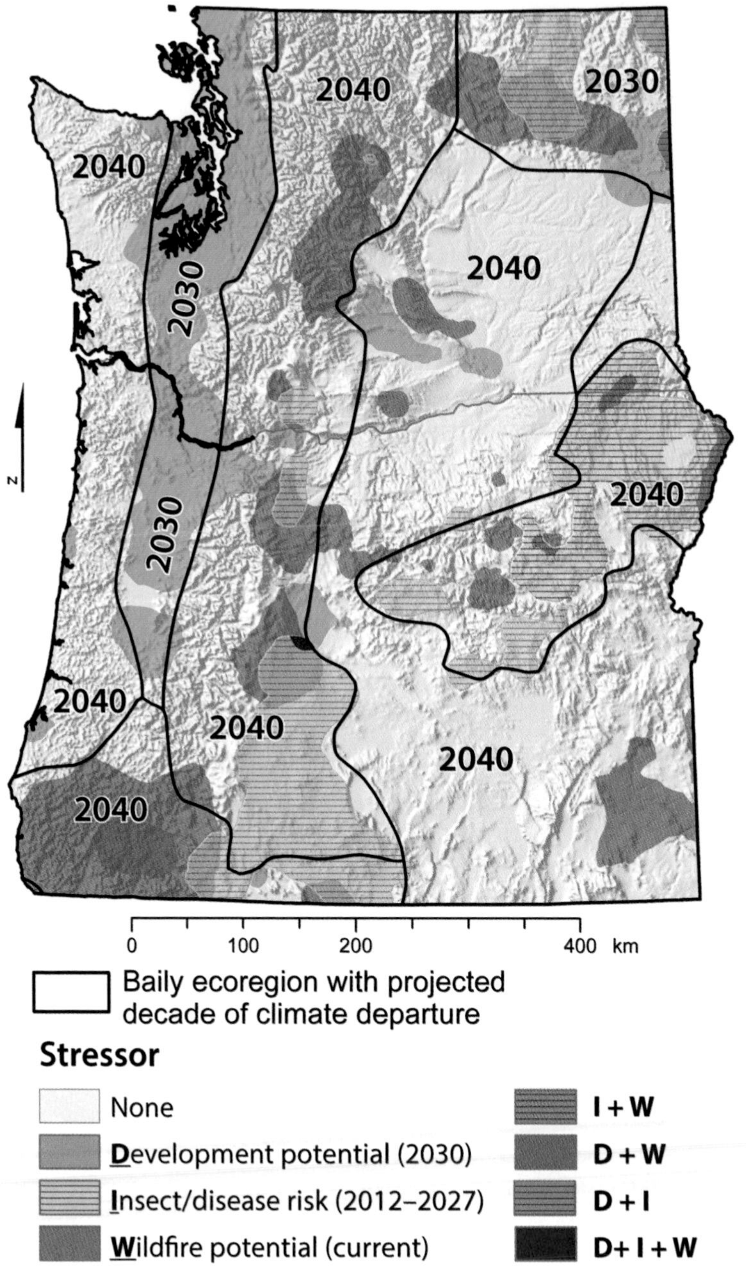

PLATE 14. (chap. 16). Exposure to multiple environmental stressors in relation to decade of climate departure—the decade at which the mean annual temperature is projected to permanently exceed the historical (1895–2012) range of variability (calculated from NASA NEX-DCP30 downscaled climate data set for RCP8.5 climate change scenario). [Credit: Adapted from Kerns et al. 2016]

PLATE 15. (chap. 17). Example of mass timber construction using engineered wood products, including cross-laminated timber (CLT) and glue-laminated beams (glulam). [Credit: Lever Architecture]

PLATE 16. (chap. 20). Managed forest landscape in western Oregon, including (A) federal ownership square in a checkerboard landownership pattern, with (1) stream buffers aligned toward corners, (2) dispersed and aggregated green-tree retention aiding habitat connectivity, and (3) gaps (white circles) potentially providing for early-seral species; (B) emerging mosaic of conditions across federal and nonfederal lands.

Chapter 10

Silviculture for Diverse Objectives

Paul D. Anderson and Klaus J. Puettmann

Several factors have influenced silvicultural practices over the last three decades in the moist coniferous forests of the US Pacific Northwest. Since the peak of timber harvesting during the 1980s, attitudes, expectations, and understanding have shifted dramatically among segments of the public, management, and science communities regarding management objectives. On many landownerships, the historical focus on timber and employment in rural communities has been replaced by societal expectations for sustainability of numerous ecosystem services, including habitat for species of conservation concern or endemic old-growth-associated species, recreation, and other cultural values (chap. 5).

This shift in priorities occurred more profoundly on public lands, especially federal forests. On private and industrial forests, priorities typically remain wood production and maximum economic return (chaps. 4, 6). In contrast, increased scrutiny of forest management operations by environmental organizations and the public has broadened management objectives on public lands. Forest management proposals have become topics of debate, and management decisions have been challenged, some becoming embroiled in the judicial system. Consequently, foresters lost the option to use several common silvicultural practices. For example, clearcutting and application of herbicides to encourage quick tree regeneration were two practices conventionally applied in these forests. Both treatments are now highly restricted on federal land, forcing use of other options to achieve management goals, such as manual weed control or partial-harvesting operations.

Many of the social pressures leading to restrictions on silvicultural tools and practices arose from an improved understanding of the ecological implications of such practices. Recognition of the effects of large-scale clearcut harvesting operations and associated landscape fragmentation on northern spotted owl habitat (Meyer et al. 1998) is the most prominent source of challenges to forest management in the region. Concurrently, shrinking budgets and reduced staffing of federal agencies mean that silvicultural practices are applied to a smaller portion of the forest land base. For example, lower staffing levels limit how many harvesting operations agencies can plan, implement, and monitor.

Meanwhile, foresters have recognized the limitations of their conventional planning procedures and practices, which assumed constant forest ecological and socioeconomic conditions and predictable trends in timber production and markets. The last few decades showed that economic trends and conditions can change very quickly through trade agreements or shifting consumer preferences. As a result, the Northwest forest industry has undergone dramatic restructuring toward more "nimble" timber-management organizations (TIMOs) and real estate investment trusts (REITs) that manage forests for maximum profits, which come from asset appreciation and timber sales. Consequently, there is less integration of decisions regarding land management and the production of wood products by forest companies (Collins et al. 2008).

Similarly, people eager to express their opinions about forest management decisions are no longer limited to local communities but increasingly represent a broader set of regional, national, or international interests. Ecological conditions are also becoming more unpredictable. For example, the potential for species invasions increases with increased global travel and trade, suggesting that foresters take into account risks that new insects or diseases pose to local tree species (Waring and O'Hara 2005). Finally, climate change is projected to lead to novel forest conditions that have no historical equivalent (Hobbs et al. 2006). Although general trends may be predictable, how the changing climate will play out on the landscape is uncertain, especially as extreme weather events will likely drive major shifts (chap. 16).

These recent trends suggest that conventional tools and practices aimed at efficient provision of wood through homogenizing forests may not be the best way to achieve current and future management objectives, especially in a more variable and unpredictable future. Instead, foresters need access to a broader array of silvicultural approaches and practices (Puettmann et al. 2009).

Evolving Silvicultural Paradigms

The increasing diversity of objectives and outcomes expected from managing forests is leading to more sophisticated silvicultural paradigms across ownerships. On private lands managed for commodities, evolving best management practices and state forest-practices regulations have modified silvicultural regimes to mitigate some ecological and societal concerns, especially related to water quality (Loehle et al. 2014). In contrast to a deterministic goal focused on optimal production of a narrow suite of benefits, typically leading to simplified forest structures and composition, silviculture on public lands is increasingly drawing on a more holistic ecological foundation to guide forest development toward attainment of multiple compatible benefits, but often with less certain outcomes (Holling and Meffe 1996).

Forests as Complex Adaptive Systems

Paralleling these evolving silviculture paradigms is a conceptual recognition of forests as complex adaptive systems (Puettmann et al. 2009). Forest ecosystems comprise many functionally diverse components and processes arrayed with varying degrees of heterogeneity, which interact across scales of space and time. For example, small-scale processes such as seed germination may influence larger-scale stand and landscape attributes such as species composition, habitat quality, and ecosystem productivity, which in turn influence the fine-scale germination processes. Such interactions among forest components and processes result in dynamic ecosystems in which stability is the exception rather than the rule. Instead, ecosystems can switch among alternate states, self-organizing in response to localized interactions among components and disturbances. The resultant ecosystem states and trajectories are emergent properties that cannot necessarily be controlled or predicted.

The concept of forest ecosystems as complex adaptive systems provides a robust foundation for explicitly linking ecological principles to the practice of silviculture. Traditionally, local experiences and observations were a key basis for silvicultural decisions. These experiences are less useful in a world with novel and quickly changing social, economic, and ecological conditions. Although local historical knowledge can provide important insights into forest conditions and processes, it is often difficult to translate these insights directly into management decisions, making it more important to recognize and acknowledge future uncertainties in silvicultural prescriptions. We may accomplish this by assessing whether a range of possible outcomes is acceptable, rather than focusing on the most likely or most desirable outcomes. Silviculture then becomes a less "prescriptive" undertaking: active guidance at neighborhood, stand, and landscape scales to encourage forest developmental trajectories can provide the desired ecosystem goods and services, even in the event of unexpected perturbations. To accomplish this, silvicultural practices must embrace the diversity of these forests and acknowledge uncertainty.

This can mean that foresters may accept some deviation from the norm in individual stands, such as stocking success after planting, if the overall stocking goals are achieved at watershed or ownership scales. In many instances, simply switching assessment scales—for example, using criteria defining successful provision of ecosystem services measured at larger spatial and time scales—can accommodate a diversity of initial conditions and associated variability in ecosystem development at smaller scales (fig. 10.1).

Another implication of acknowledging future uncertainties is the inclusion of ecosystem resilience and adaptive capacity. Examples include an assessment of whether certain treatments increase or decrease the likelihood that the future forests will provide desired ecosystem goods and services if climate change results in increased temperatures and more frequent drought events (Neill and Puettmann 2013).

Silviculture as a Multiscale Endeavor

The recent interest in diversifying stand structure at various spatial scales pushes foresters—particularly those managing public lands for more diverse ecological, social, or landscape objectives—to abandon their traditional management focus on *stands* (uniform operational units). Instead, foresters are acknowledging the importance of processes that act at and across various spatial scales (fig. 10.1). An emphasis on within-stand variability emerges from recognition that many silviculturally relevant processes act at multiple spatial scales. For example, competition among trees does not occur at stand scales but rather is a local process influenced by resource and environmental conditions in the immediate neighborhood of the trees. At

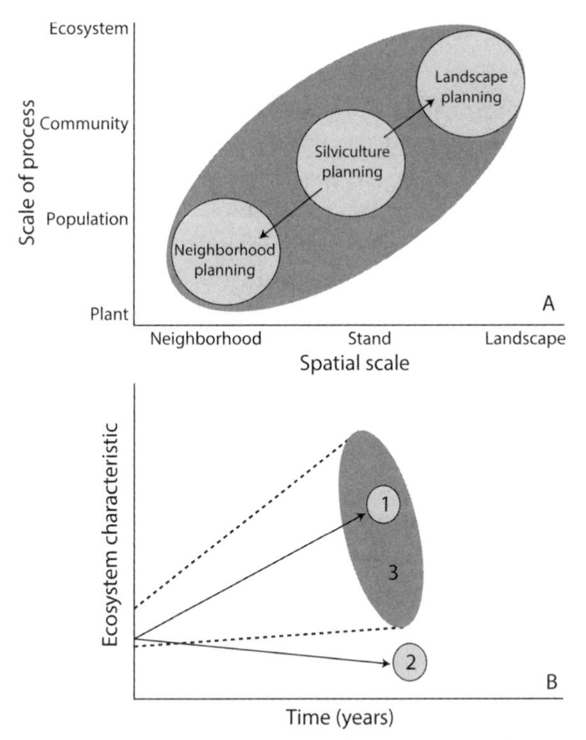

FIGURE 10.1. Silviculture in the context of forests as complex adaptive systems. (A) Ecological and social processes influencing forest management act at several spatial and process scales. Silviculture framed by complex adaptive systems crosses scales, taking interactions of ecosystem structures and processes into account at neighborhood, stand (typically used for silvicultural prescriptions), and landscape scales. (B) Silviculture based on the deterministic command-and-control approach typically focuses on managing stands for a narrow range of stand structures and ecosystem characteristics (two possible sets shown in circles 1 and 2—e.g., even-aged monocultures and balanced uneven-aged stands). Managing forests as complex systems and with scale interactions (A) may mean that a larger range of stand structure is acceptable for individual stands (ellipse 3), as long as landscape goals are achieved. This allows foresters to acknowledge the variability of existing conditions and uncertainties in ecosystem dynamics, let emergent properties play out, and integrate these aspects into management prescriptions. From Puettmann et al. (2009).

the same time, success for many new management objectives—such as provision of clean water, wildlife habitat, or recreational values—is measured at larger watershed or landscape scales. These scales do not act in isolation; the surrounding landscape can influence responses to silvicultural prescriptions at the stand or smaller scales, such as seed production of nearby trees or local fire behavior. Conversely, stand-level silvicultural manipulations, when distributed across a landscape, can influence interactions among ecosystem processes and disturbances, such as the propagation and spread of wildfire, insects, and pathogens over much larger areas. Thus not only are stand-level silvicultural decisions framed by the larger landscape context, they may also be the initiation point for dynamics that uniquely define each landscape as an often unpredictable emergent product of many smaller-scale ecological and human processes.

Advances in Silvicultural Practices for Moist Temperate Forests

With societal rejection of continued old-growth harvest in the 1980s, 35- to 80-year-old second-growth Douglas-fir–western hemlock forests became the predominant forest type available for harvest on federal lands in western Washington and Oregon. Initially established and managed for commercial timber production, these stands were commonly dense (often >500 trees/ha [~200 trees/ac]) and composed of a single age cohort, with simplified structures and little tree regeneration or vegetation in understories. How long these stands would persist in this "stem-exclusion" successional phase (Franklin et al. 2002) was uncertain, although experts suggested several decades. Thus an objective for land managers was to develop silvicultural approaches for accelerating the development of these younger stands toward older-forest conditions with complex ages, structures, and understories (e.g., Bauhaus et al. 2009).

Historically, old-growth stands often developed at stand densities substantially lower than the current typical second-growth forest. For example, growth rates through the first century of stand development in coastal old-growth forests were similar to the rates observed in experimental young-growth stands developing at 100 to 120 trees/ha (~40 to 50 trees/ac) (Tappeiner et al. 1997) and much faster than the growth rates of typical high-density young stands. Further, the wide range of tree ages in old-growth stands demonstrates protracted or recurrent recruitment over periods of 100 to 420 years, in contrast to the 5- to 10-year age range common to the current second-growth stands of plantation or natural seed origins. Second-growth stands of Douglas-fir grown for decades at high density experience substantial senescence of lower limbs and therefore lack the deep crowns typical of old-growth Douglas-fir (Van Pelt and Sillett 2008). Such observations spurred the use of thinning as a means to accelerate the development of old-growth characteristics in younger, second-growth stands (plate 10). The following discussion highlights changes arising from the implementation of the US federal Northwest Forest Plan (USDA and USDI 1994).

A New Approach to Density Management

In contrast to conventional density-management practices directed toward optimizing the growing space for commercially important crop trees, density management has become a means for diversifying forest structure for landowners who target compatible ecological and economic objectives. Increasingly, thinning specifications retain varying densities and a broadened range of conifer diameters, as well as a proportion of minor conifer and deciduous species to diversify tree composition. Often the intent of density reductions also includes encouraging development of understory vegetation and a multilayer canopy in spatially heterogeneous patterns (plate 10). The goal is to create compositional and structural heterogeneity similar to old-forest conditions that provides species habitat and other ecosystem services.

Contemporary ecological and economic objectives may require repeated thinning over time. Increased availability of site resources (water, nutrients, light) following a thinning is temporary, as tree growth and rapid canopy closure (commonly 2% to 4% per year) can result in rapid reoccupation of growing space. Although conventional density-management regimes promote stand growth and vigor by optimizing site occupancy while limiting density-dependent mortality, the desire for diverse understory vegetation, the retention of advanced tree regeneration, and the recruitment of snags and down wood imply a need for greater spatial variation in overstory density and site occupancy. Different ecological and economic benefits accrue at different stages in stand development, yielding varying mixes of benefits and trade-offs over space and time. At a landscape scale, increased continuity of tree harvesting can translate to a more consistent supply of raw materials to support forest industries and employment.

Ongoing studies have characterized the developmental trajectory of recently thinned stands relative to the untreated second-growth condition. However, it remains difficult to assess exactly how much thinning accelerated the development of these young stands toward the structural and compositional diversity associated with old-forest reference stands.

Likely, the answer will vary based on initial stand and site conditions and by the specific structural element, such as tree diameter versus crown structures.

New Role for Forest Legacies

Historically, the silvicultural importance of forest legacies after harvests—live trees, down wood, established seedlings and saplings—focused on promoting crop tree regeneration, for example, by leaving seed or shelterwood trees. We now recognize that legacies can facilitate a broader array of ecological functions and processes, including specific wildlife habitat components or visual qualities. New knowledge of the distinct animal and plant communities supported by canopies of older trees, for instance, suggests that these live trees can serve as "lifeboats." Further, these legacies can serve as sources of inoculum for restoring the old-growth-associated species (e.g., fungi, bryophytes, lichens), communities, or ecosystem processes in the developing stand or the adjacent landscape (Evans et al. 2012).

The influence of legacies on associated species may depend on their distribution (dispersed or aggregated) and abundance. For example, arboreal bryophytes can be vulnerable to microclimate alterations arising from reduced canopy density or along forest edges. Similarly, habitat quality for arboreal rodents may depend on forest structural attributes related to cover from predation, nesting opportunities in bole cavities or on branch platforms, and crown-to-crown movement through a stand. Aggregated clumps of reserve trees may need to be ≥1 ha (≥2.5 ac) to provide a core area where habitat conditions are comparable to intact older forests and free of edge effects (Aubry et al. 2009). Individual or aggregated reserve trees dispersed throughout a stand or landscape may facilitate movement of organisms in the short term (e.g., home range definition) and population migrations in the longer term (e.g., shifting species distributions).

Key silvicultural considerations for effective legacy management are reserve tree life span, spatial distribution, response to silvicultural treatments, and vulnerability to disturbance. Indicators of topographic vulnerability, wind firmness, and general vigor (crown density) are useful criteria for selecting persistent leave-trees. Marking guidelines can emphasize retaining trees with crown damage or stem defects if those characteristics contribute to important biodiversity functions. Retention of advanced regeneration (established seedlings and saplings) may require specific operations planning, particularly in stands undergoing repeated harvest entries. In these stands, advanced regeneration or developing understory flora is often concentrated in previously disturbed areas such as yarding corridors, skid trails, and landings. Advances in geographic information systems (GIS) and global positioning systems (GPS) technology provide greater fine-scale precision in layout of harvest operations, aiding preservation of legacy features.

Morticulture

Dead trees are a striking feature differentiating old-growth from younger managed-forest structure. The low abundance of snags and down wood typical in managed second-growth forests results from earlier practices that removed such wood during harvest operations. Without disturbances, mortality rates in second-growth stands are typically low and consist of smaller trees that die from competition (self-thinning). Such dead trees are too small to provide many desired habitat features.

Recent silvicultural practices are enhancing snag and down-wood abundance and functionality in managed forests. Harvest prescriptions often specify retention of existing snags and live trees designated as future snags or down wood. Snags are created from live trees by girdling or herbicide injection. Pathogens injected into stems of live trees can kill tops or promote rot and cavity formation. Managers may consider damage to residual trees during harvest operations as beneficial to achieving snag and down-wood objectives. Snags can be retained in dispersed or clumped patterns or along forest edges to provide varying species habitats. Trees can be felled and left on-site if down-wood levels are depauperate. Trees felled directionally into the stream channel can contribute to aquatic habitat and channel geomorphology (chap. 14).

Questions persist about the abundance, sizes, and spatial distribution of snags and down wood needed to meet ecological objectives in terrestrial and aquatic ecosystems. Greater abundance and size increase the number of species that use snags or down wood as habitat. Although snag and down-wood size and abundance are generally smaller for managed stands, amounts vary substantially for both managed and unmanaged stands. Decision-support tools can assist managers in establishing snag and down-wood targets based on reference distributions and cumulative species-use relationships (Mellen et al. 2002). However, managers face the dilemma of balancing short- and long-term needs, such as leaving live trees for future recruitment of larger snags versus creation of snags and down wood from smaller trees to address an immediate scarcity, and decisions about snag retention for habitat and snag removal to ensure the safety of forest workers during subsequent management operations. Manipulation of spatial patterns of legacy retention (dispersed, clumped, along edges) can provide varying species-habitat suitability and possibly mitigate risks to forest workers.

New Emphasis on Small-Scale, Within-Stand Variation

The US federal Northwest Forest Plan aimed to integrate ecological and economic demands by spatial separation of management objectives through land-use allocations. Management of reserve areas (in late-successional reserves) focused on achieving ecological goals but could include some harvesting in younger stands for restoration priorities. Forests in other allocations, such as matrix and adaptive-management areas, were open to timber harvests to accommodate economic goals, but with ecological constraints, such as the Aquatic Conservation Strategy. Over time, it became obvious that spatial separation of management goals was not effective, and foresters were challenged to develop silvicultural regimes that accommodated multiple goals or sets of objectives on the same piece of ground. Evidence of the increased emphasis on managing nontree vegetation such as shrubs, herbs, and grasses can be seen in management prescriptions. Federal foresters are expected to provide early-seral plant communities and habitat, even though social pressures make it difficult to create larger canopy openings.

Foresters have been responding to these needs by emphasizing spatial heterogeneity in their stand prescriptions. Creation of gaps (patch openings) by removing clusters of trees can make thinning more profitable and the operations more efficient. Such gaps can provide for higher within-stand diversity in vegetation and, depending on size, may promote early-seral plants and forage without some of the adverse effects associated with clearcutting. Similarly, unharvested leave-islands (intact skips) further encourage spatial variability in otherwise homogeneous stands, as they provide habitat for a different suite of species. Strategically placed leave-islands can be designed as corridors for the movement of wildlife across the landscape.

The functionality of both gaps and leave-islands varies with size. Gaps or leave-islands <1 ha (<2.5 ac) foster transitional environments and vegetation responses that differ from both open and closed forest conditions (Fahey and Puettmann 2007). Hence, to retain core areas of closed-forest environment, leave-islands likely need to be ≥1 ha (Heithecker and Halpern 2007). Conversely, development of early-seral vegetation likely requires patch openings of at least 1 ha.

Vigor, Acclimation, and Adaptive Capacity in an Uncertain Future World

Despite abundant annual rainfall, dry summer conditions may limit the growing season of trees in Northwest moist forests. Climate projections predict increased temperatures and more variable precipitation regimes that collectively will result in altered seasonality of weather and attendant increased exposure to stressors and disturbances (chap. 16). Seasonally warmer temperatures may directly affect trees by increasing the likelihood of weather events misaligned with important life-cycle events. For example, trees may be insufficiently hardened to autumnal or vernal frosts. Successive years of low precipitation may lead to summer moisture deficits that stress trees. Warmer temperatures coupled with changing precipitation regimes may accelerate the life cycles of insects and pathogens (e.g., Douglas-fir bark beetle, Swiss needle cast).

The age structures of older Douglas-fir forests reflect a variety of stand initiation and developmental pathways resulting from hundreds of years of climate and disturbance interactions (Poage et al. 2009). Fire frequency is variable over this geographically large and ecologically diverse biome. In a warmer, more variable future climate, these moist forests may experience periods of increased potential for fire activity that might seem extraordinary compared with recent decades, but which in fact have historical precedent.

Climate change poses different risks and therefore warrants different management considerations for different locations in the landscape (ridges, north- versus south-facing slopes, riparian areas, climate refuge areas) and for ownerships with different objectives. General trends in regional climate are locally modified by topography and interactions with vegetation. Vulnerability to desiccation and windthrow is greater for trees on exposed ridges and forest edges. Riparian areas are cold-air drainages and are often less influenced by increases in temperature, but they are not immune to influences of droughts and decreases in canopy cover that locally reduce shade and increase radiation loads for streams.

The planning horizon for different ownerships and land allocations is also an important consideration when implementing silvicultural treat-

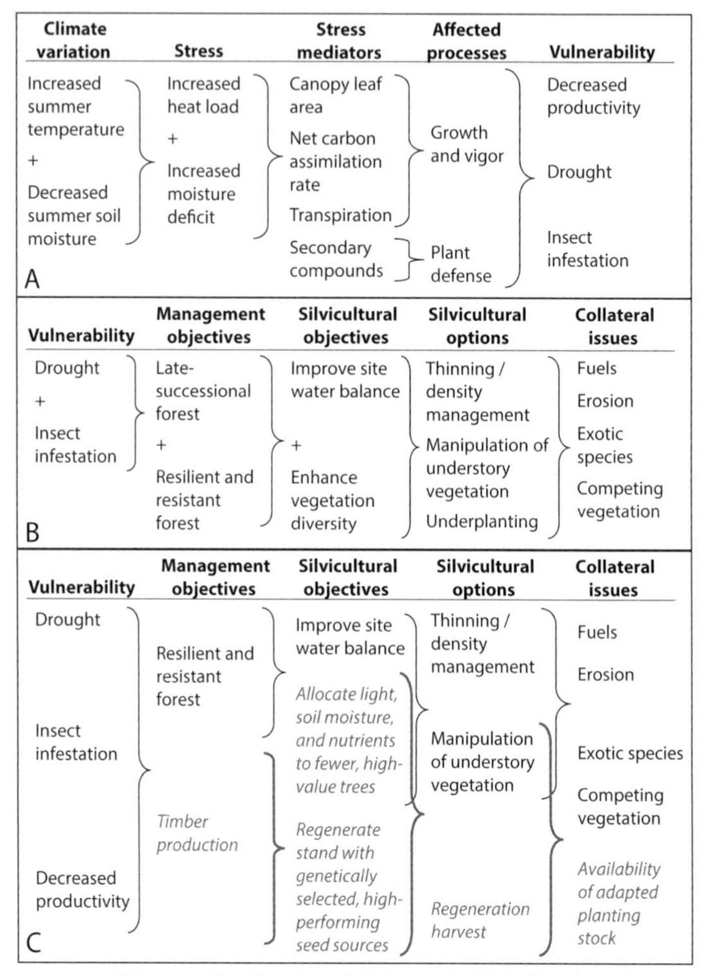

FIGURE 10.2. Climate-related vulnerabilities and silvicultural options vary by development stage and management objectives. Using the stem-exclusion phase as an example, logic models can be assembled to show (A) how climate variations generate environmental stresses, mediated by tree physiological processes to result in various vulnerabilities; (B) how these vulnerabilities can be addressed when management objectives include forest resilience and development of late-successional forest attributes, which translate into silvicultural objectives that are then addressed through various silvicultural treatment options, with recognition of an associated array of collateral issues; and (C) how stand conditions and vulnerabilities, with a different set of management objectives (e.g., timber production), may warrant different silvicultural treatments and associated collateral issues. Italics show differences between B and C.

ments. The development of old-forest structures may take a long time, perhaps a century or more. Therefore species or genotypes regenerating in the near future require the capacity to acclimate to stressful environments over this extended period of older-forest development. On the other hand, the shorter planning horizons on industrial forests (35- to 50-year rotations) provide more frequent opportunities for regeneration of better-adapted species or genotypes.

Vulnerabilities change with stage of forest development. Seedlings have little water-storage capacity; hence forests in the regeneration phase can be particularly vulnerable to seasonal drought and high evaporative demand. Alterations of weather patterns may inhibit natural regeneration from seed by disrupting the coordinated phenological processes of pollen dispersal and flower receptivity. Established stands may experience vulnerabilities arising at population and community levels, such as competition for resources intensified or altered by climate dynamics and associated stresses, such as insect or disease infestations (fig. 10.2).

Vulnerabilities of older stands may be more subtle, as tree sizes and stand structures can buffer the influences of climate extremes, and will vary by species and individual characteristics (e.g., wind events may disproportionately topple taller trees with dense crowns or shallow root systems). The increased frequency and severity of disturbances due to climate change can potentially simplify the species composition in older, mixed-species forests. Subsequently, many other elements of the diverse biological communities that make up distinct old-growth canopy communities may be vulnerable, such as arthropods and bryophytes.

In the near term, relatively intensive density reductions to decrease intertree competition may be most useful for increasing tree vigor and forest resistance to climate-related stressors. Foresters are familiar with the effects of density on tree and stand productivity and for staying below thresholds for bark beetle epidemics. Assuming climate stresses manifest themselves as the decreased availability of water or other site resources and lower thresholds for competition-related mortality, maintaining stands well below maximum stand density would buffer these effects.

Management of overstory density alone may be insufficient to address climate change and may have unintended consequences. For example, increased light and moisture resources resulting from overstory thinning can provide an opportunity for tree regeneration and growth of understory vegetation that places demands on site resources. This may partially offset the benefits of thinning to the vigor of overstory trees.

Increasing the adaptive capacity of forests at stand and landscape scales may require regenerating a wider variety of species consistent with the anticipated range of climate variability. At lower elevations where the frequency of summer water deficits is predicted to increase, planting a mixture of drought-tolerant species such as ponderosa pine, incense cedar, or Oregon white oak may provide a greater breadth of adaptive characteristics than a monoculture of Douglas-fir. Partial overstory removals or forest gaps created in variable-density thinning may provide mixed-species regeneration opportunities.

Recent studies provide useful information on survival, growth, and productivity of various conifer and hardwood species planted under the residual overstory of thinned Douglas-fir forests (e.g., Chan et al. 2006). Understanding forests as complex adaptive systems and accepting variability and uncertainty may also result in viewing disturbances as opportunities for management. For example, silviculturists may take advantage of root rot pockets and windthrow gaps as local, small-scale opportunities to manage for multiple species, especially when there is relatively low risk to near-term stand productivity.

Another approach to enhancing diversity and adaptability is planting seedlings comprising a mixture of local seed sources with seed sources adapted to projected climate stresses. For example, to match current populations of Douglas-fir to the prospective climate of 2100, seedlings could be planted from sources that are 450 to 600 meters (~1,480 to 1,970 ft) lower in elevation, or from a distance of 1.8 to 2.5 degrees latitude to the south. For the 20 or so Northwest species that have defined seed zones (areas within which plant materials can be transferred with little risk of being poorly adapted to their new location), the general transfer guidelines are similar—from lower to higher elevation or from southern to northern latitudes. Although Douglas-fir demonstrates a relatively high degree of localized population adaptation, other widely distributed species such as western white pine are broadly adapted. Initially, seed transfers for species with strong local adaptation might be restricted to adjacent seed zones, or to shifts between distant locations within a seed zone. There is always some risk that introduced seed sources or species may be less fit in the short term than local seed sources or current species. Silvicultural practices that provide suitable microsites throughout the regeneration and maturation stages will be a major consideration when introducing genotypes to target a future environment.

Applying principles of complex adaptive systems to silviculture should provide for more diverse and potentially climate-adapted landscapes. Although silvicultural prescriptions may continue to be written at stand and finer scales, doing so with explicit awareness of the larger temporal and spatial contexts and applying different silvicultural approaches at different spatial and temporal scales can produce unique, diverse landscapes as emergent properties. Planting of more diverse species and genotypes that are better adapted to future climates to create adaptive capacity on the landscape need not occur everywhere; distribution should be strategic. Once introduced, they will disperse through the interactions among climate, disturbance, and natural regeneration processes.

Literature Cited

Aubry, K. B., C. B. Halpern, and C. E. Peterson. 2009. Variable-retention harvests in the Pacific Northwest: A review of short-term findings from the DEMO study. *Forest Ecology and Management* 258:398–408.

Bauhaus, J., K. J. Puettmann, and C. Messier. 2009. Silviculture for old-growth attributes. *Forest Ecology and Management* 258:525–537.

Chan, S. S., D. J. Larson, W. H. Emmingham, K. G. Maas-Hebner, S. Johnston, and D. Mikowski. 2006. Overstory and understory development in thinned and underplanted Oregon Coast Range Douglas-fir stands. *Canadian Journal of Forest Research* 36:2696–2711.

Collins, S., D. Darr, D. Wear, and H. Brown. 2008. Global markets and the health of America's forests: A Forest Service perspective. *Journal of Forestry* 106:47–52.

Evans, S. A., C. B. Halpern, and D. McKenzie. 2012. The contributions of forest structure and substrate to bryophyte diversity and abundance in mature coniferous forests of the Pacific Northwest. *Bryologist* 115:278–294.

Fahey, R. T., and K. J. Puettmann. 2007. Ground-layer disturbance and initial conditions influence gap partitioning of understorey vegetation. *Journal of Ecology* 95:1098–1109.

Franklin, J. F., T. A. Spies, R. Van Pelt, and 9 coauthors. 2002. Disturbances and structural development of natural forest ecosystems with silvicultural implications, using Douglas-fir as an example. *Forest Ecology and Management* 155:399–423.

Heithecker, T. D., and C. B. Halpern. 2007. Edge-related gradients in microclimate in forest aggregates following structural retention harvests in western Washington. *Forest Ecology and Management* 248:163–173.

Hobbs, R. J., S. Arico, J. Aronson, and 15 coauthors. 2006. Novel ecosystems: Theoretical management aspects of the new ecological world order. *Global Ecology and Biogeography* 15:1–7.

Holling, C. S., and G. K. Meffe. 1996. Command and control and the pathology of natural resource management. *Conservation Biology* 10:328–337.

Loehle, C., T. B. Wigley Jr., A. Lucier Jr., E. Schilling, and R. J. Danehy. 2014. Toward improved water quality in forestry: Opportunities and challenges in a changing regulatory environment. *Journal of Forestry* 112:41–50.

Mellen, K., B. G. Marcot, J. L. Ohmann, K. L. Waddell, E. A. Willhite, B. B. Hostetler, S. A. Livingston, and C. Ogden. 2002. DecAID: A decaying wood advisory model for Oregon and Washington. Pp. 527–534 in *Proceedings of the symposium on the ecology and management of dead wood in western forests, 1999 November 2–4, Reno, NV.* Technical coordination by W. F. Laudenslayer Jr., B. Valentine, C. P. Weatherspoon, and T. E. Lisle. General Technical Report PSW-GTR-181. Albany, CA: USDA Forest Service, Pacific Southwest Research Station.

Meyer, J. S., L. L. Irwin, and M. S. Boyce. 1998. Influence of habitat fragmentation on northern spotted owls in western Oregon. *Wildlife Monographs* 139:3–51.

Neill, A. R., and K. J. Puettmann. 2013. Managing for adaptive capacity: Thinning improves food availability under climate change conditions. *Canadian Journal of Forest Research* 43:428–440.

Poage, N. J., P. J. Weisberg, P. C. Impara, J. C. Tappeiner, and T. S. Sensinig. 2009. Influence of climate, fire and topography on contemporary age structure patterns of Douglas-fir at 205 old forest sites in western Oregon. *Canadian Journal of Forest Research* 39:1518–1530.

Puettmann, K. J., D. K. Coates, and C. Messier. 2009. *A critique of silviculture: Managing for complexity*. Washington, DC: Island Press.

Tappeiner, J. C., D. Huffman, D. Marshall, T. A. Spies, and J. D. Bailey. 1997. Density, ages and growth rates in old-growth and young-growth forests in coastal Oregon. *Canadian Journal of Forest Research* 27:638–648.

USDA and USDI (US Department of Agriculture and US Department of the Interior). 1994. *Standards and guidelines for management of habitat for late-successional and old-growth forest related species within the range of the northern spotted owl. Attachment A to the record of decision for amendments to Forest Service and Bureau of Land Management planning documents within the range of the northern spotted owl.* [Place of publication unknown]: US Department of Agriculture and US Department of the Interior. http://www.reo.gov/library/reports/newsandga.pdf .

Van Pelt, R., and S. C. Sillett. 2008. Crown development of coastal *Pseudotsuga menziesii*, including a conceptual model for tall conifers. *Ecological Monographs* 78:283–311.

Waring, K. M., and K. L. O'Hara. 2005. Silvicultural strategies in forest ecosystems affected by introduced pests. *Forest Ecology and Management* 209:27–41.

Chapter 11

Long-Term Forest Productivity

Bernard T. Bormann, Steven S. Perakis, Robyn L. Darbyshire, and Jeff Hatten

Planning for forest sustainability has been a hallmark of US national resource management, beginning with the work of several visionaries of the previous century, including Gifford Pinchot and US presidents Grover Cleveland and Theodore Roosevelt. Their efforts created the US national forests in 1905 to address concerns about sustainable, long-term supplies of both water and timber. Congress subsequently passed the Multiple-Use Sustained Yield Act of 1960 to fulfill needs beyond water and timber resources. The National Forest Management Act of 1976 better assured sustainably by defining it as "the achievement and maintenance in perpetuity of a high-level annual or regular periodic output of the various renewable resources of the national forests without impairment of the productivity of the land."

Potential impediments to forest sustainability gained attention in the 1970s and 1980s, when cumulative effects of intensive management became easily discernible to the public, as airline passengers bore witness to the effects of widespread clearcut harvesting. By this time, nearly 50% of the area of national forests in western Oregon and Washington had been converted to even-aged stands (chap. 7), under management designed to maximize mean annual volume increment, with rotations of around 80 years on the most productive forests. On state and private lands, previously unmanaged stands became quite rare; rotation length (time between clearcut logging operations) on private lands dropped to as short as 25 years on highly productive sites. These diverse issues refocused on the question of planning for forest sustainability and the science and management of long-term productivity.

Because long-term forest or soil productivity effects are difficult to study, research has generally focused on shorter-term surrogate measures and indicators. By about 1990, broad concerns about long-term forest productivity included (1) repeated nutrient removals in harvests and suggestions from Australia, Europe, and New Zealand of second-growth declines; (2) ecosystem nutrient losses and erosion associated with slash burning, herbicides, and aggressive site preparation; (3) soil compaction and disturbance from logging equipment; (4) losses of organic matter in mineral soils, reducing water- and nutrient-holding capacities, aggregation, and supporting soil organisms; and (5) reduced early-seral vegetation and losses of woody debris.

In 1989, the US Forest Service Pacific Northwest Research Station worked with many partners to develop a regional program consisting of a network of long-term experiments focused on soils. Its purpose was to tease apart the ecosystem processes underlying ecosystem productivity changes related to harvesting, woody debris, and different kinds of vegetation, including early-seral stages (Little et al. 2000). However, the scaling back of even-aged rotations driven by the northern spotted owl lawsuits and the development of the US federal Northwest Forest Plan (USDA and USDI 1994) diverted attention from productivity. As the Plan was implemented, concerns about long-term productivity on federal forests in this portion of the Northwest took a backseat, addressed mainly by meeting standards for temporary soil disturbance and coarse woody debris. The implicit assumption was that if these standards were met, long-term productivity would not decline. Since Plan implementation in 1994, the science of forest productivity has developed considerably. Herein, we review major scientific developments that continue to reshape the long-term productivity debate across landownerships and harvest practices, and summarize future directions for research and management.

New Science

Scientific knowledge regarding the factors that influence forest productivity has expanded significantly since the Northwest Forest Plan went into effect.

Soil Diversity

Franklin and Dyrness (1973) documented the remarkable array and diversity of forest vegetation in the Pacific Northwest region (chap. 2) and developed a detailed ecological classification system that set the stage for site-specific land management. Although parallel work on soils did not follow, the complex spatial diversity of soils is similarly remarkable and almost always higher in wet than in drier forest types. Diversity in soil morphology and function comes from variation in bedrock, topography, and degree of soil development, driven by clashing tectonic plates and resulting volcanism and ocean-floor uplift. Additionally, tectonics provide deposits of odd geologic fragments (e.g., the Klamath Mountains, and many islands of British Columbia and Southeast Alaska). Dynamic forces such as erosion, landslides, and especially windthrow create soil diversity at scales as small as a few meters. This soil diversity has now been well documented, with detailed surveys by the Natural Resource Conservation Service across most of the coastal Northwest (NRCS 2015).

Forest Growth

Temperate forest trees worldwide have been growing faster for the last few decades (Graumlich et al. 1989; Boisvenue and Running 2006; McMahon et al. 2010). Improved soil productive capacity is not clearly linked to this trend; the most likely cause is carbon dioxide (CO_2) fertilization and longer growing seasons, but some portion of this pattern may be attributable to advances in forest-management practices and higher nitrogen in rainfall in some areas. Increased atmospheric CO_2 concentration speeds photosynthesis and allows plants to close their stomates more frequently, conserving water, and perhaps more than overcomes the increasing respiratory cost of higher temperatures (Kirschbaum 2000). On managed, conifer-dominated federal forests in the US Northwest, aboveground growth of stands planted after 1984 also could have benefited from changing practices, which included a move away from broadcast burning for site preparation, use of larger planted seedlings, improved tree spacing from both planting and precommercial thinning approaches, and more intensive monitoring of regeneration success (as per the National Forest Management Act of 1976). Because net primary production declines in older forests as autotrophic respiration increases (Ryan et al. 1997), age demographics also likely have been important.

Aboveground growth of conifers is not the same as total solar energy capture or gross primary production, which are the actual bases for determining net primary production (NPP). Losses or gains in wood volume produced affect provision of wood commodities and popular measures of carbon sequestration (chap. 12), but losses or gains in NPP have real consequences for nearly all ecosystem services. Net primary production sets the upper limit on secondary production (herbivore and detritivore food chains) and influences many other provisioning (water yield, stand structures, and habitat) and regulating (hydrological flow and nutrient retention) services. Long-term ecosystem studies in moist conifer forests that compare different management strategies are helping us to better understand NPP during early succession (e.g., Little et al. 2000).

Nonconiferous vegetation influences productivity of Northwest forests, both directly and indirectly. Long-term experiments have demonstrated that early-seral strategies favoring hardwood trees and shrubs as well as conifers produced twice the aboveground biomass increment compared with those favoring young conifers alone during the first 15 years after treatment (Bormann et al. 2015). In this study, mixed stands also differed in the amounts and kinds of forage and litter and had increased plant biodiversity. Douglas-fir also grew well, perhaps in part because it was able to obtain nutrients from recalcitrant organo-mineral complexes in a process called priming (Kuzyakov et al. 2000), resulting in a temporary decline of about a fifth of the soil carbon in the upper mineral (B) horizons.

Soil Compaction

A network of nationwide studies initiated since 1990 found that experimentally compacting mineral soils by nearly 10% uniformly across 0.4-ha (1-ac) plots reduced plot-total 10-year aboveground biomass increment by about 7% (Powers et al. 2005). A more in-depth analysis revealed that productivity increased 40% on sandy soils but declined 50% on clay soils, opening the possibility that harvest systems could be adapted for different soil types to optimize both long-term productivity and economic return. More broadly, these findings infer that less-than-complete compaction in typical operations would have smaller effects and reinforce the long-held adage that forest management needs to be site specific. Many techniques have been developed to minimize compaction, including use of designated skid trails to reduce the area affected by dragging logs, use of suspension harvest techniques, bedding trails with slash beforehand, and ripping (deep plowing) afterward. We know more about soil types and compaction, but the positive or negative effects of compaction and amelioration over a stand rotation have not been well studied.

Forest Fires

Wild and prescribed fires have long been suspected of affecting long-term productivity, though their effects are mostly indirectly related to erosion and other losses of nutrients and soil organic matter that support productivity. Slash burning has been reduced to a minor issue on US federal lands in the Northwest as its application has declined since 1994, but the forested area consumed by wildfire in Oregon and Washington accelerated after 1985 (Bormann et al. 2006; chap. 7). Wildfire regimes in most moist conifer forests are described as low frequency and high severity—with large fuel buildups between fires, they burn at high intensity (>300°C [570°F]). These wildfires have long been thought to harm soils, but available studies lacking extensive replication must be interpreted cautiously when using only a single nearby unburned stand as the comparison. This uncertainty fuels public debate, leading to a range of perspectives from "wildfire sterilizes the soil for centuries" to "it has little more effect than a prescribed fire."

In the single study contrasting a range of management regimes, with quantitative soil data before and after intense fire, wildfire effects appeared more dramatic than previously realized, with soil-surface temperature exceeding 600°C (1,100°F), accompanied by losses of nearly all decomposing logs and 23 Mg C/ha from surface organic O-horizons and mineral soil, with nearly 60% coming from mineral soil. (Bormann et al. 2008). In this study, much of the mineral soil carbon loss was explained by transport of mineral-soil fine substrates into the aerial smoke plume, leaving behind a surface accumulation of coarse fragments. Although short-term before-after studies do a poor job of foreshadowing long-term effects on forest productivity, well-replicated comparisons of adjacent sites with different long-term wildfire histories show that nitrogen availability and soil carbon and nitrogen pools in the forest floor can take centuries to recover following intense fire, whereas changes in mineral-soil carbon and nitrogen are often smaller than in organic horizons (Giesen et al. 2008; Perakis et al. 2015).

Prefire stand conditions can affect postfire resilience. When Homann et al. (2014) measured soil carbon 8 years after fire had burned young regenerating forest, thinned stands, and uncut controls, the young forest showed no sign of recovery in soil carbon, whereas thinned stands recovered some, and uncut controls recovered all the soil carbon lost in the fire. The uncut forest burned at lower intensity because of lower fuel mass and a slow-burning hardwood understory, whereas thinned stands burned hotter because of higher fuels but gained carbon from a massive needle-fall after the fire.

Role of Active Management

Like farmers, most private- and industrial-forest landowners recognize the soil as the capital from which their wealth flows. Upon learning of potential negative effects of tree harvest on soil structure or function, such as nutrient and soil organic-matter losses and soil compaction, they have used active remediation techniques such as fertilization and soil loosening to sustain yields (Fox 2000). In a recent example, studies in Scandinavia confirmed a growth decline of 4% to 13% over 20 years in pine and spruce stands following whole-tree harvesting compared with bole-only harvesting (Helmisaari et al. 2011). However, the effect of nutrient losses, mainly in needles, was reversible with fertilizer applications. Remediation strategies after intense wildfire have yet to be developed or tested.

Future Research Directions

Maintaining forest productivity is fundamental to sustaining myriad ecosystem services and the human-forest ecosystem in general. New perspectives on how to maintain or increase long-term productivity and thereby support sustainability have been emerging in various public and scientific circles, but all require avoiding soil and nutrient losses. Reliable production, whether it represents sustainable yield, even flow, or resilience, will thus remain a concern in all forests, whether or not commodities are a primary focus.

Long-term commodity supply is a foundational goal for forests contributing to industrial manufacturing products that depend on capital investment. The challenge for forest planners is to calculate potential sustainable supply while considering both uncertainties and dynamic processes, including outright surprises. One approach could be to present production capacity as a range, based on an understanding of the uncertainties.

Here we highlight several research areas that are helping to narrow the range of predictions of the effects of management on productivity and hence ecosystem services. Science in support of ecosystem sustainability will eventually combine all these perspectives to provide metrics to track ecological and social well-being and to better understand the processes that determine them.

How Productivity Changes with Time

Applied long-term productivity research is needed regarding the strategic analysis of contemporary forestry activities (chap. 10). For example, how does feedback of active management in support of ecological and community well-being affect the capacity of the ecosystem over a decade or longer to (1) capture solar energy; (2) convert the energy, as carbon flows back to the atmosphere, into food chains, habitats, long-lasting commodities, and bioenergy production; and (3) sustain desired ecosystem services?

Soil carbon and productivity. Increased capture of solar energy by moist forests and soils is one of many approaches that can be used to combat rising atmospheric CO_2 concentrations (Noormets et al. 2015; chap. 12). How to achieve an increase, especially in soil storage, has yet to emerge. Where nitrogen limits net primary production (NPP), adding nitrogen fertilizer increases carbon storage in tree biomass; however, synthetic nitrogen fertilizer is produced with fossil-fuel carbon. On moist sites that are deficient in mineral-derived nutrients (phosphorus, calcium, potassium, and micronutrients), management of plants and soils to increase access to nutrients by deep rooting, priming of recalcitrant soil organic matter, and weathering of primary minerals seems a promising but complicated way to increase NPP. In cool, moist forests of Southeast Alaska, priming is focused on the carbon- and nutrient-rich upper mineral soil (Bh horizon), helping to create the bleached (E) horizons for which podzols (soils typical of conifer or boreal forests) are famous. The result, however, is an increasingly dense lower Bh horizon that resists root penetration (Bormann et al. 1995). When this happens, nutrient uptake declines from both organic and mineral layers. Mixing of soil layers by windthrow can reverse this situation but also increases decomposition. The trade-off is between decreasing NPP and increasing soil-carbon storage without disturbance, versus increasing NPP and decreasing soil-carbon storage with disturbance. The short- and long-term implications have yet to be worked out.

Woody debris. On federal Northwest forestlands since 1994, down-wood regulations have addressed the dual purposes of late-seral habitat and long-term productivity. Yet progress on understanding the effects of wood on long-term productivity has been slow. Most conifers are adept at accumulating wood carbon in boles. Tree mortality has great potential to add carbon to soils, but boles go though many stages before residues can be incorporated into mineral soils, including fungal and bacterial decomposition and insect consumption long before a snag falls to the ground. Leaf and branch litterfall and root turnover are more constant, but relative effects on soils are not well known. Intense fire and microbial attack on down wood (some driven by tree-associated mycorrhizae) further short-circuits the wood-to-soil conversion. Intense wildfire consumes nearly all but the most recently added woody debris (Bormann et al. 2008), and mineral soils are heated more under burning debris (Hebel et al. 2009). Nevertheless, soil development often is more apparent under logs (Spears et al. 2003). Results from long-term trials will be available in the next decade to assess the effects of woody additions on soil carbon, net primary productivity, and other ecosystem processes.

Hardwood trees and shrubs. Many hardwood trees and shrubs are known for high productivity at early ages, more rapid and plentiful return of litter to the soil surface, faster initial decomposition and movement into mineral soils, and deeper rooting compared with many conifers (Bormann et al. 2015). With climate change driving temperatures up, soil moisture retention becomes a critical variable in determining fire susceptibility, species occurrences and migrations, and summertime low flows in rivers and streams, as well as ecosystem productivity. Moisture retention is largely controlled by evapotranspiration and mineral-soil organic-matter content, so any means to avoid soil-carbon losses or build reserves would be valuable. Landscape management that focuses on creating and maintaining early-seral vegetation, especially by extending occupation by hardwoods, would bring back critical processes that are truncated by conversion of hardwoods to conifers. A recent study (Naudts et al. 2016) suggests that the lower albedo (absorption of solar radiation) of hardwoods relative to conifers can reduce above-canopy air temperatures and affect local to continental climate.

Bioenergy and nutrients. Managers considering short-rotation harvests for biomass, bioenergy, biofuels, wood pellets, or new long-fiber boards will need to track effects of removing nutrients on subsequent rotations and replacing them through silvicultural practices. Across much of the region where nitrogen limits growth, losses of fertility due to nitrogen removal in biomass (Himes et al. 2014) can be ameliorated by direct fertilization or by encouraging nitrogen-fixing red alder or *Ceanothus*, which provide both biomass and improved soil fertility (Bormann et al. 1994). In highly productive Coast Range forests where nitrogen is naturally abundant in soil, however, calcium or phosphorus may be deficient (Mainwaring et al. 2014). Calcium in particular may be highly depleted from nitrogen-rich acidified soils, regardless of underlying bedrock, with atmospheric inputs setting a limit to calcium removals in harvest (Perakis et al. 2006; Hynicka et al. 2016). In comparison, phosphorus is less mobile and less readily depleted than calcium (Hynicka et al. 2016) and may limit short-term plant demands even though abundant storage in soil organic matter should sustain long-term productivity (Perakis et al. 2013). On less weathered soils, how early-seral hardwoods or conifers and their mycorrhizal symbionts thrive by obtaining needed nutrients through weathering, deep rooting, or both needs to be better understood. While deep soil may also contain abundant nutrient reserves, we don't know whether these nutrients are available to trees, particularly in conifers harvested on short rotations that favor aboveground growth (Callesen et al. 2015). New sustainable management approaches could be used to test how different tree species shape soil carbon and different nutrients (Bormann et al. 1994; Cross and Perakis 2011) while concurrently increasing biodiversity.

Soil biota. The biodiversity of soil microbes and insects contributing to general soil diversity patterns is not yet fully understood, especially relative to ecological processes associated with forest productivity. However, next-generation DNA-sequencing methods have recently opened up a host of new possibilities in this regard, especially for microbiota. Initial research needs include characterizing soil biota in response to a host of natural and anthropogenic disturbances, including the various forest-management practices in current use across the moist-forest landscapes of the Northwest. Experimental treatments designed to examine the role of this biota in forest ecological functions and processes are the next, more crucial, step for research.

Critical zone integration. Between the tops of trees and the bottom of the groundwater is a living, breathing, boundary layer where rock, soil, water, air, and living organisms interact. These complex interactions regulate the natural habitat; determine the productivity of life-sustaining resources, including food and water; and affect climate itself. The Critical

Zone integrative research framework (Brantley et al. 2007), adopted by the National Science Foundation in 2006, is being applied worldwide and is likely to advance knowledge of longer-term ecosystem processes as described above.

Measuring the Effects of Management on Productivity

Measuring forest-management effects on decadal- to century-scale changes in soils as well as aboveground vegetation is central to assuring long-term ecosystem productivity, economic and ecological sustainability, long-term ecological and social well-being, and provision of ecosystem services. Measurement is most efficiently accomplished via field studies that contrast a range of management practices. This reality, recognized in the Multiple-Use Sustained Yield Act of 1960, deserves to be reaffirmed in light of our greater appreciation of dynamic processes and opportunities for structured learning.

Literature Cited

Boisvenue, C., and S. W. Running. 2006. Impacts of climate change on natural forest productivity: Evidence since the middle of the 20th century. *Global Change Biology* 12:862–882.

Bormann, B. T., K. Cromack Jr., and W. O. Russell III. 1994. Influences of red alder on soils and long term ecosystem productivity. Pp. 47–56 in *The biology and management of red alder*. Edited by D. E. Hibbs, D. S. DeBell, and R. F. Tarrant. Corvallis: Oregon State University Press.

Bormann, B. T., R. L. Darbyshire, P. S. Homann, B. A. Morrissette, and S. N. Little. 2015. Managing early succession for biodiversity and long-term productivity of conifer forests in southwestern Oregon. *Forest Ecology and Management* 340:114–125.

Bormann, B. T., P. S. Homann, R. L. Darbyshire, and B. A. Morrissette. 2008. Intense forest wildfire sharply reduces mineral soil C and N: The first direct evidence. *Canadian Journal of Forest Research* 38:2771–2783.

Bormann, B. T., D. C. Lee, A. R. Kiester, T. A. Spies, R. W. Haynes, G. H. Reeves, and M. G. Raphael. 2006. Synthesis: Interpreting the Northwest Forest Plan as more than the sum of its parts. Pp. 23–48 in *Northwest Forest Plan—the first ten years (1994–2003): Synthesis of monitoring and research results*. Edited by R. W. Haynes, B. T. Bormann, and J. R. Martin. General Technical Report PNW-GTR-651. Portland, OR: USDA Forest Service, Pacific Northwest Research Station.

Bormann, B. T., H. Spaltenstein, M. H. McClellan, F. C. Ugolini, K. Cromack Jr., and S. M. Nay. 1995. Rapid soil development after windthrow disturbance in pristine forests. *Journal of Ecology* 1:747–757.

Brantley, S. L., M. Goldhaber, and K. V. Ragnarsdottir. 2007. Crossing disciplines and scales to understand the critical zone. *Elements* 3:307–314.

Callesen, I., R. Harrison, I. Stupak, J. Hatten, K. Raulund-Rasmussen, J. Boyle, N. Clarke, and D. Zabowski. 2015. Carbon storage and nutrient mobilization from soil minerals by deep roots and rhizospheres. *Forest Ecology and Management*. doi:10.1016/j.foreco.2015.08.019.

Cross, A., and S. S. Perakis. 2011. Complementary models of tree species–soil relationships in old-growth temperate forests. *Ecosystems* 14:248–260.

Fox, T. R. 2000. Sustained productivity in intensively managed forest plantations. *Forest Ecology and Management* 138:187–202.

Franklin, J. F., and C. T. Dyrness. 1973. *Natural vegetation of Oregon and Washington*. General Technical Report PNW-8. Portland, OR: USDA Forest Service, Pacific Northwest Forest and Range Experiment Station.

Giesen, T. W., S. S. Perakis, and K. Cromack Jr. 2008. Four centuries of soil carbon and nitrogen change after stand-replacing fire in a forest landscape in the western Cascade Range of Oregon. *Canadian Journal of Forest Research* 38: 2455–2464.

Graumlich, L. J., L. B. Brubaker, and C. C. Grier. 1989. Long-term trends in forest net primary productivity: Cascade Mountains, Washington. *Ecology* 70: 405–410.

Hebel, C., J. A. Smith, and K. Cromack Jr. 2009. Invasive plant species and soil microbial response to wildfire burn severity in the Cascade Range of Oregon. *Applied Soil Ecology* 42:150–159.

Helmisaari, H.-S., K. H. Hanssen, S. Jacobson, M. Kukkola, J. Luiro, A. Saarsalmi, P. Tamminen, and B. Tveite. 2011. Logging residue removal after thinning in Nordic boreal forests: Long-term impact on tree growth. *Forest Ecology and Management* 261:1919–1927.

Himes, A. J., E. C. Turnblom, R. B. Harrison, K. M. Littke, W. D. Devine, D. Zabowski, and D. G. Briggs. 2014. Predicting risk of long-term nitrogen depletion under whole-tree harvesting in the coastal Pacific Northwest. *Forest Science* 60:382–390.

Homann, P. S., B. T. Bormann, B. A. Morrissette, and R. L. Darbyshire. 2014. Postwildfire soil trajectory linked to prefire ecosystem structure in Douglas-fir forest. *Ecosystems* 18:260–273.

Hynicka, J. D., J. C. Pett-Ridge, and S. S. Perakis. 2016. Biological nitrogen enrichment regulates calcium sources in forests. *Global Change Biology*. doi:10.1111/gcb.13335.

Kirschbaum, M. 2000. Will changes in soil organic carbon act as a positive or negative feedback on global warming? *Biogeochemistry* 48:21–51.

Kuzyakov, Y., J. K. Friedel, and K. Stahr. 2000. Review of mechanisms and quantification of priming effects. *Soil Biology and Biochemistry* 32:1485–1498.

Little, S. N., B. T. Bormann, L. Bednar, and 10 coauthors. 2000. Integrated site research plan: Long-term ecosystem productivity [LTEP] program. http://www.fsl.orst.edu/LTEP.

Mainwaring, D. B., D. A. Maguire, and S. S. Perakis. 2014. Three-year growth response of young Douglas-fir to nitrogen, calcium, phosphorus, and blended fertilization treatments in Oregon and Washington. *Forest Ecology and Management* 327:178–188.

McMahon, S. M., G. G. Parker, and D. R. Miller. 2010. Evidence for a recent increase in forest growth. *Proceedings of the National Academy of Sciences of the United States of America* 107:3611–3615.

NRCS (National Resource Conservation Service). 2015. http://websoilsurvey.sc.egov.usda.gov/DataAvailability/SoilDataAvailabilityMap.pdf.

Naudts, K., Y. Chen, M. J. McGrath, J. Ryder, A. Valade, J. Otto, S. Luyssaert. 2016. Europe's forest management did not mitigate climate warming. *Science* 351: 597–600.

Noormets, A., D. Epron, J. C. Domec, S. G. McNulty, T. Fox, G. Sun, and J. S. King. 2015. Effects of forest management on productivity and carbon sequestration: A review and hypothesis. *Forest Ecology and Management* 355:124–140.

Perakis, S. S., D. A. Maguire, T. D. Bullen, K. Cromack Jr., R. H. Waring, and J. R. Boyle. 2006. Coupled nitrogen and calcium cycling in forests of the Oregon Coast Range. *Ecosystems* 9:63–74.

Perakis, S. S., E. R. Sinkhorn, C. E. Catricala, T. D. Bullen, J. Fitzpatrick, J. D. Hynicka, and K. Cromack Jr. 2013. Forest calcium depletion and biotic retention along a soil nitrogen gradient. *Ecological Applications* 23:1947–1961.

Perakis, S. S., A. Tepley, and J. E. Compton. 2015. Disturbance and topography shape nitrogen availability and $\delta^{15}N$ over long-term forest succession. *Ecosystems* 18:573–588.

Powers, R. F., D. A. Scott, F. G. Sanchez, R. A. Voldseth, D. Page-Dumroese, J. D. Elioff, and D. M. Stone. 2005. The North American long-term soil productivity experiment: Findings from the first decade of research. *Forest Ecology and Management* 220:31–50.

Ryan, M. G., M. B. Lavigne, and S. T. Gower. 1997. Annual carbon cost of autotrophic respiration in boreal forest ecosystems in relation to species and climate. *Journal of Geophysical Research: Atmospheres* 102:28871–28883.

Spears, J. D., S. M. Holub, M. E. Harmon, and K. Lajtha. 2003. The influence of decomposing logs on soil biology and nutrient cycling in an old-growth mixed coniferous forest in Oregon, USA. *Canadian Journal of Forest Research* 33:2193–2201.

USDA and USDI (US Department of Agriculture and US Department of the Interior). 1994. *Record of decision for amendments to Forest Service and Bureau of Land Management planning documents within the range of the northern spotted owl* [plus Attachment A: Standards and Guides]. [Place of publication unknown]: US Department of Agriculture and US Department of the Interior. http://www.reo.gov/library/reports/newroda.pdf.

Chapter 12

Managing Carbon in the Forest Sector

Mark E. Harmon and John L. Campbell

There is an increasing need to manage forest carbon (Mackey et al. 2013). Forests store much of the globe's terrestrial carbon (US forest carbon density, plate 1B) and could play a significant role in climate mitigation by decreasing the amount of atmospheric carbon (Smith et al. 2014). However, forests respond to natural disturbances and management, which means that forests could be subject to future carbon losses if climate change increases the occurrence and severity of natural disturbances altering forests (Law and Waring 2015). These possible losses explain why forests are a focus of climate adaptation strategies designed to increase the resilience of forests to natural disturbance, even when such efforts result in lower carbon stocks at present.

Our intent here is not to support the case of mitigation over adaptation or vice versa. Determining this balance involves policy debates hinging around more than carbon and involving an assessment of the future risks to many ecosystem services in a changing climate. Nor is our intent to make specific recommendations; that, too, will require a wider conversation. Rather we provide a scientific overview of how carbon behaves in Pacific Northwest forests and consider how future management may influence this carbon. We conclude by addressing trade-offs between carbon and other management objectives.

Carbon in Moist Forests of the Pacific Northwest

Of all the forests on Earth, those in the moist regions of the Pacific Northwest have some of the highest potential to store carbon. Studies of old-growth forests show that this region not only stores a great deal of carbon, but that a large share (~75%) is contained in live and dead wood relative to the soil compared with other regions (~50%; Smithwick et al. 2002). This difference exists not because the amount of soil carbon being stored in Northwest moist forests is unusually low compared with other temperate forests, but because the amount of live and dead wood is so large. The very high productivity of these forests does not explain the relative proportions of live, dead, and soil carbon; likely, moist Northwest forests have particularly long-lived components such as live and dead trees that potentially store large amounts of carbon. Trees in the Northwest generally have the longest life spans in their genera (some over 1,000 years) and can grow very large. When these trees die, parts such as large-diameter stems can take centuries to fully decompose. Also, although major disturbances in this region can be large in area, they are naturally infrequent. Hence the era of fire suppression, which has altered many (but not all) dry forests, has had relatively little impact here.

Although Northwest forest management has greatly improved forest productivity (yield at harvest), these forests are inherently slow to develop relative to other regions (e.g., the southeastern United States; Waring and Franklin 1979). The primary species planted, Douglas-fir, is relatively slow to develop initially, but is long-lived, continues moderate growth for at least 150 years, and provides superior wood for construction. Although the inherent productivity of these forests (net primary production, NPP) is not the highest globally, it is very high compared with those in many boreal and temperate regions, and this forest occurs over an extensive land base—all factors that have made it an attractive timber-producing region. Management, particularly clearcutting, has reduced the amount of carbon stored in many of these forests (Hudiburg et al. 2009) but also has resulted in the accumulation of stores in buildings and landfills that offsets some of these forest losses.

Combined with natural disturbances and human-caused fires, past forest management has created four unique starting points (initial conditions) for future carbon management: (1) former heavily forested land that is now under sparse forests, shrubs, or agriculture that could be returned to a closed-forest condition; (2) forest lands that have been converted to plantations and are managed primarily for timber production; (3) young to mature forests, many on US federal lands, that are being managed for biological conservation with some timber harvest; and (4) mature to old-growth forests that are not being harvested to any significant degree. Although one prescription for carbon management could be used for all, a more nuanced approach would acknowledge the differences in initial conditions.

Dynamics of Carbon

The forest sector can be thought of as having three parts: the forest ecosystem; products resulting from forest harvesting (both in use and after disposal); and substitutions of materials and energy sources that may reduce the use of fossil carbon. Here we refer to these as the *forest, products,* and *substitutions* carbon pools. At a general level, these three parts (and the forest sector as a whole) act like a leaky bucket, with carbon flowing in as well as out. This represents a conceptual challenge; envisioning a system in which there is only inward flow or only outward flow is less complex. Many explanations of the forest sector invoke simpler concepts by describing the *live* part of the forest, products, and substitutions as having inputs without losses and the *dead* part of the ecosystem as having losses without inputs. Although such simplifications may seem adequate to describe the effect of certain management actions, they fail to capture critical aspects of the forest carbon cycle as a whole. For instance, although systems with simultaneous inputs and outputs may reach a steady state (inputs equal outputs) in which carbon stores do not appear to be changing, they often act in "unexpected" ways: two steady-state systems with the same input and output may store different amounts of carbon; a system thought to be generally losing carbon (e.g., the dead part of ecosystems) can accumulate carbon, and one thought to generally gain carbon (e.g., the live part of ecosystems) can lose it; or the time to reach steady state might be largely controlled by the size of the leaks and not the rate of input.

Forest

Carbon enters forests via photosynthesis and exits via multiple pathways, including respiration, combustion, harvest, and export to aquatic ecosystems. Forest carbon is stored in various pools, each having specific functions but all conforming to dynamic input-output systems. At a fundamental level, we may divide forest carbon into that stored in living plants (live), dead plant material (dead), and soil (highly decomposed, relatively stable).

The live pool is unusual in that it controls its own inputs via photosynthesis, affording at times a positive feedback between the amount of leaves and fine roots and the rate at which they accumulate carbon. This is particularly important for very young, rapidly aggrading forests. However, the live pool is similar to the others because it can lose carbon, specifically by plant mortality, combustion, and harvest. Live carbon reaches high levels in Pacific Northwest moist forests because trees live so long—they have a very small mortality leak compared with other forests. It is also why live carbon can accumulate in these forests for centuries. Although many believe that older, larger trees stop growing, as individuals they can add much more carbon each year than younger, smaller trees (Stephenson et al. 2014). Live carbon accumulation slows down as forests age, not because individuals cease to grow, but because mortality takes a larger and larger share of the production.

The dead pool is linked to the live part because its input is plant mortality, related to death of plant parts (e.g., leaves) occurring in individual plants or to entire stands and landscapes influenced by disturbances. Losses from the dead pool take several forms: respiration associated with decomposers; combustion; and the formation of relatively stable material that is considered soil. Although soil can be viewed in different ways, in terms of carbon dynamics it can be considered a pool of relatively decomposed dead organic carbon that is stabilized chemically, physically, or environmentally, thus protecting it from further losses. Like the dead pool, soil is subjected to respiration and combustion losses, but also to losses via erosion and leaching (water-dissolved carbon).

The soil pool can contain a large store of carbon, but the input to this pool is low compared with photosynthesis. Instead, the large store results from a small proportion being lost each year, and though sometimes considered a "permanent" carbon store, soil carbon remains only as long as inputs remain. Therefore soil carbon has greater potential to change than is commonly assumed. In general, soil carbon is the slowest-changing part of the forest carbon system, and when primary succession occurs on fresh volcanic rock, landslides, and glacial and river deposits, carbon can accumulate in soils for a millennium or more. When partially combusted, the live, dead, and soil pools can produce charcoal that, once incorporated into the soil, can be stored for an exceptionally long period.

The way these pools act depends on the "level" of examination, and unless one is aware of this, conclusions of various studies can seem contra-

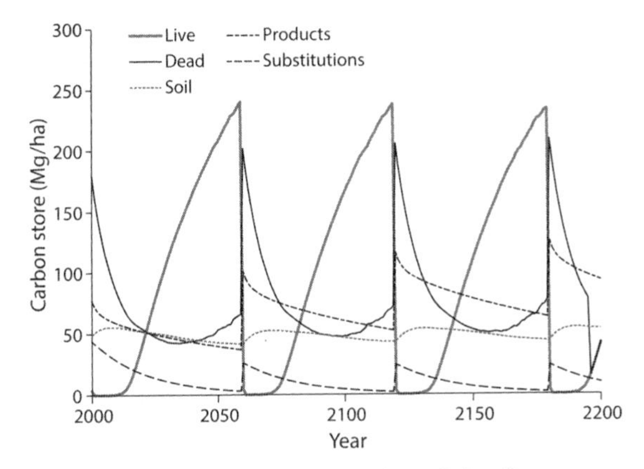

FIGURE 12.1. Hypothetical change in the size of the forest-sector carbon pool (store) for a stand that was first harvested in 1950 and placed on a 60-year clearcut harvest rotation. At each harvest, roughly 50% of the live carbon store is removed, and the other half is added to the dead pool. The majority (75%) of manufactured products were assumed to be buildings that are eventually disposed in landfills. Substitution losses are related to building longevity and leakage to other sectors.

dictory and confusing (Harmon 2001). In forests, individual live trees gain carbon, often at increasing rates as they become larger (Stephenson et al. 2014), and dead trees lose carbon. This is because live trees have minimal losses (although processes such as wind damage and heart rots are exceptions), and once a dead tree is created it has no way to add to itself. However, at the next higher level—a collection of either live or dead trees—both can either gain or lose carbon depending on the relative sizes of the inputs or losses. At the stand level, live carbon generally decreases quickly and increases slowly (fig. 12.1). In contrast, the dead-tree component appears to gain carbon quickly as trees die from fire, wind, or insect outbreaks and to lose it slowly through decomposition. At a landscape or regional level, all forest carbon pools act much like soil at the stand level, changing slowly over time in response to long-term environmental, disturbance, and management regimes.

Products

The products pool is created from harvested carbon used in many different ways (paper, posts, shipping pallets, buildings, bioenergy) that have different life spans and eventually have a range of fates (recycled, landfills, incineration) that influence how this part of the forest sector stores carbon. As products are manufactured, some carbon is lost to the atmosphere through decomposition and combustion. When energy is captured in the latter, a substitution is possible (see below). Products are subject to losses in use, including decomposition of buildings, accidental combustion, and disposal. Some disposed products may be recycled, reducing the net losses from these pools; some are placed in landfills, which can have extremely small rates of loss (comparable to soil); others are incinerated, and although this adds carbon to the atmosphere, if energy is captured it may result in a substitution with potentially positive carbon impacts.

The dynamics of products in use most resemble that of the dead part of the forest in temporal pattern and timing: at the individual level (e.g., a house), they tend to decrease over time; for products produced at the stand level they have sudden increases and slow decreases; and for products produced at the landscape level they appear to steadily change. Although salvage is often proposed as a way to store more carbon, the dead-tree pool and the forest-products pool have similar dynamics. Products disposed in landfills most resemble soil in terms of temporal dynamics but differ in that soil carbon has had many centuries if not millennia to accumulate, whereas landfills are a relatively recent invention and are in an accumulation phase worldwide. Incinerated carbon has no direct store, but may indirectly if it is part of an energy substitution that "stores" fossil carbon.

Substitutions

Use of harvest carbon can lead to substitutions that may have consequences on fossil carbon use and result in a form of "virtual" carbon store. There are two general types of substitution: product substitution, in which wood products replace building materials with higher embodied energy (such as concrete and steel); and the use of forest biomass for energy generation, which we will refer to as *bioenergy* that replaces fossil carbon. Although both forms have been promoted as forest-related climate-mitigation strategies, several factors need to be kept in mind in evaluating these findings. Use of wood products often involves less energy for manufacturing than some other building materials, but it is not always clear how much of this energy is fossil based; whether the amount of carbon involved in fossil energy is constant (use of coal versus natural gas changes the carbon intensity by a factor of two); and whether the amount of energy involved is constant over time. Some of these factors would likely lower the initial displacement of fossil carbon and reduce the long-term benefits. Moreover, product substitutions are usually modeled without losses, but as with all other pools they are clearly subject to losses: fossil carbon unused by the building sector could be used by other sectors (leakage), and the longevity of product substitution displacements is tied to the life span of buildings. Hence it is likely that product substitution benefits have been overestimated in some studies (Law and Harmon 2011).

Bioenergy substitutions are subject to some of the same limitations: the form of fossil energy being substituted is an important consideration. Moreover, fossil carbon unused by the energy sector could be used by another sector. It must be borne in mind that, on average, biomass carbon contains at best approximately 90% of the energy of coal and 45% of the energy of natural gas, implying a maximum displacement factor of 0.45 to 0.90 of fossil carbon per unit of biogenic carbon. Using energy to create the forest-derived biofuel (in feedstock processing and drying) could lower the amount of fossil carbon displaced even further. Thus consuming a ton of biofuel carbon necessarily displaces something less than a ton of fossil carbon, and since doing so necessarily reduces the other forest-sector stocks, use of bioenergy can sometimes add more carbon to the atmosphere than simply using fossil fuel carbon.

Effects of Natural Disturbances

Natural disturbances in the Pacific Northwest such as wildfires, windstorms, and insect outbreaks kill many trees and appear to dramatically reduce carbon stores, but their actual impact is less significant than one might think, provided that forests regenerate. The notion that disturbed forests lose all or even most of their carbon is simply incorrect. Although severe disturbances can kill most if not all the trees, much of the large dead wood remains on the landscape for decades to centuries. At the stand level, this dead carbon store will decline over time, whereas the live carbon store rebuilds as long as the forest regenerates in a suitable time frame (several decades).

Stand-replacing fires in the US Northwest forests west of the Cascade Range have been infrequent but can be exceedingly large relative to the entire ecoregion (e.g., the Nestucca, Tillamook, and Biscuit fires of ca. 1840, 1930, and 2002, respectively). Combined with a multicentury successional trajectory, this means that a single disturbance event can impose region-wide disequilibria in live and dead carbon stores lasting much longer than in forests having either lower biomass or more frequent stand-replacing disturbances.

Wildfire can remove dead and soil carbon, but this is quite limited because fires of all sorts, including high-intensity fires, tend to mostly combust only the smaller surface materials and not larger or buried ones (Campbell et al. 2012). The result is that, although carbon stores decline in disturbed forests, this is a temporary and small decline relative to the effects of clearcut harvest, which can remove much of the live carbon (Mitchell et al. 2012).

The key concern related to climate adaptation (box 12.1) is not that forests will be disturbed—that will happen regardless—but that forests will be permanently replaced by less "carbon-dense" ecosystems and will not regenerate. Increases in disturbance frequency and severity, like those predicted to result from climate change, will lead to a decrease in carbon stores, but to some degree this is self-limiting because less carbon stored means there is less carbon to lose. In contrast, a change from forest to nonforest can lead to a less-carbon-stored-in-vegetation and hence a more-carbon-in-the-atmosphere scenario. While the exact effects of climate change on forest composition remain difficult to predict, it is reasonable to believe that forests on the dry end of their distribution, those bordering nonforest ecological communities, would be most vulnerable to regeneration failure. With the exception of its narrow southern border with chaparral vegetation, the moist forests of the Pacific Northwest are surrounded by other forest communities, and although regeneration rates can vary widely, natural permanent conversion of forest to nonforest vegetation has not yet been documented across the region.

Effects of Management

Management can have different effects on forest-sector carbon: (1) increased harvesting, thinning, and salvage lower the forest carbon store, but increase the store in products and may increase the virtual store associated with substitutions; (2) increased use of prescribed fire with no harvest decreases the forest carbon store but does not affect products and substitutions; and (3) increased use of silvicultural practices related to improved growing stock and fertilization can potentially lead to an increase in all parts of the forest carbon sector. As with natural disturbance, increasing the intensity and frequency of harvest, thinning, and salvage make the forest leakier and hence less able to store carbon. The same actions lead to more input into products

BOX 12.1. TO SEQUESTER OR TO STORE: THAT IS THE QUESTION

To mitigate climate change, do we maximize atmospheric carbon sequestration or maximize carbon stores? It may seem that directly taking carbon from the atmosphere is optimal. However, a forest may remove atmospheric carbon at a high rate but not increase carbon storage (Mackey et al. 2013). Also, it may not remove atmospheric carbon on a net basis (zero net sequestration) but store a great deal of carbon. For example, if carbon is harvested in a forested landscape, the forest can be kept in a permanent sequestration state, but it can also store very little carbon relative to an unharvested landscape. Converting the unharvested landscape to one with harvest would achieve increases in carbon removal from the atmosphere (sequestration) by decreasing the forest carbon store, since younger harvested forests contain less live carbon than older ones. Harvesting carbon can result in a store of products, but since the dynamics of this pool are similar to the forest's dead and soil carbon, this does not necessarily lead to a net gain in atmospheric carbon removed.

Time is another consideration. A forest stand may initially lose carbon and eventually regain it, with stored carbon levels being relatively constant from one harvest to the next. This is also true across stands of similar treatment but in different stages of "recovery." While this suggests that carbon in these systems is sustainable and renewable, it does not mean the system is carbon neutral. The key to understanding whether a practice will have positive or negative climate mitigation effects is to compare the landscape average carbon stores before and after treatments. Practices that increase long-term carbon stores have positive mitigation effects; those that decrease stores have negative effects; and those resulting in no change can be deemed carbon neutral.

There are limits to how much carbon can be stored in forests. Globally, maximizing forest coverage and biomass would reduce atmospheric carbon concentration only 40 to 70 ppm (Mackey et al. 2013). What range of carbon storage is possible for Northwest moist forests? They are thought to contain two-thirds of the carbon that they could if we ceased all harvest, suppressed all fire, and allowed every stand to mature into old growth (Hudiburg et al. 2009; Krankina et al. 2012). In contrast, if all forests were managed on a 60-year clearcut harvest interval they would store 13% to 23% less carbon (Krankina et al. 2012).

and substitutions, although increasing intensity too far can affect soil fertility and lower input to the forest sector, and harvesting below certain stand ages can decrease inputs to the live and products pools. In many cases, the increases in wood products and substitution stores caused by increases in harvest intensity and frequency do not offset the carbon losses incurred in

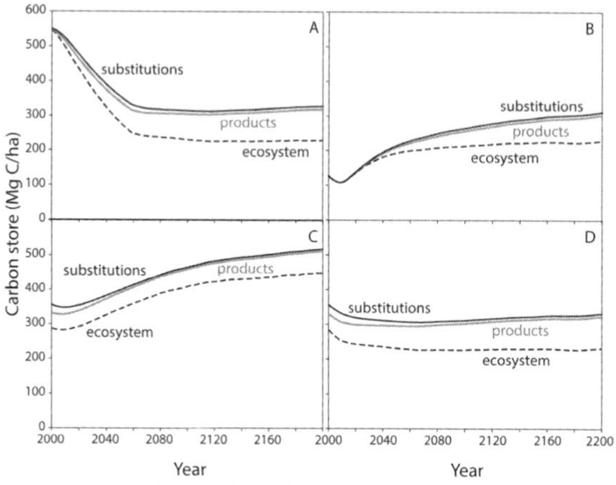

FIGURE 12.2. Landscape change in carbon stores for various management systems: (A) conversion of an old-growth moist coniferous forest to a 60-year clearcut harvest; (B) conversion of an agricultural system to a 60-year clearcut harvest; (C) conversion of a 60-year clearcut harvest to one with a 150-year harvest interval; and (D) maintaining a 60-year clearcut harvest system. The ecosystem includes the stores of live, dead, and soil carbon; products include those in use and disposed; and substitutions include virtual stores related to products and biomass energy. Note that the values for any time are cumulative—that is, the uppermost line is the sum of ecosystem, products, and substitutions.

the forest. This is because the effects of harvesting practices depend on the initial condition of the land (fig. 12.2).

Creating an intensive short-rotation plantation system leads to gain of forest-sector carbon if an agricultural field is the initial condition, but the same system leads to a loss of carbon to the atmosphere if an old-growth forest is the initial condition.

Given the likelihood that a changing climate will raise temperatures and increase water loss from soil and vegetation, thus increasing the severity and frequency of wildfire (chap. 16), there has been interest in thinning forests, including those in the moist region of the Pacific Northwest, to reduce water demand and fuel levels. Thinning forests may reduce water demand temporarily, but unless thinning is frequent, surviving trees and newly established ones will soon use this available water. If thinning is too intense, allowing more light to reach the forest floor, understory plants such as herbs, grasses, and shrubs may use much of this newly available water.

The same temporary effect occurs for fuel reduction: eventually the fuel level will recover unless the treatment is repeated. One difference between water-use reduction and fuel reduction is that while water is used on all the forestland each year, and in drought years can limit vegetation growth, survival, and reproduction throughout the landscape, the area burned by severe fire is relatively small even over a decade. This means that far more area needs to be treated, with subsequent treatments over time, than will be burned very severely—a trade-off that may lead to more carbon being lost via the fuel treatment than by fire. Another consideration regarding fuel reduction is that while fuel treatments can reduce the severity and size of fires (but not prevent fires), the amount of carbon lost to the atmosphere by fire is not as variable as many suppose. Reducing a fire from very severe to moderate prevents some carbon being added to the atmosphere, but the former releases perhaps 25% to 35% of the forest carbon, whereas the latter releases 10% to 15%. The real benefit to carbon stores from fuel-reduction treatments may be related to decreasing fire severity to assure reestablishment of forest after fires via residual trees.

Because it decreases inputs and increases forest leakiness, thinning generally leads to less carbon being stored in forests than not thinning (Zeide 2001). There are several reasons for this: (1) if it involves harvest or combustion of stems, thinning is a form of removal, and hence it makes the forest more leaky; (2) by killing trees, thinning at least temporarily decreases input into the forest (net primary production, or NPP) until the leaves and fine roots needed to capture the released resources grow back; and (3) thinning does not increase the total amount of resources available to trees but reallocates the same total resources to fewer individuals. The latter two points deserve elaboration, given that thinning is seen as increasing tree growth. There is no doubt that thinning generally increases individual tree growth, but by its very nature it leads to fewer trees, so at the stand level, thinning does not generally increase total NPP without the addition of fertilizers or irrigation. Moreover, the recovery of NPP after thinning is not instantaneous, but takes time, during which NPP is depressed.

Salvage of trees killed by disturbance can lead to an increase in products and substitutions, but one must also account for losses incurred in the forest. If salvaged carbon is made into very long-lived products or landfilled, and if manufacturing efficiency is very high, it is possible that increases in these pools could mostly offset losses in forest carbon. Currently it is likely that increases in product stores are equal to or less than forest carbon losses. Because salvaged wood is of lower quality than green wood, it may be combusted for bioenergy-related substitution. In the short term, burning salvaged wood can release more carbon to the atmosphere than allowing it to decompose, because combustion is instantaneous, whereas decomposition can take decades to centuries, over which time the disturbance-generated dead material stores carbon (TerMikaelian et al. 2015).

Although management directed at increasing substitutions has potential for mitigating climate change, the impacts of these parts of the forest sector depend on initial conditions. To be most effective, bioenergy substitutions need to involve minimal reductions in forest carbon stores and low amounts of additional energy related to transport and processing, and must substitute for high-carbon fossil fuels such as coal, and not low-carbon fuels such as methane (natural gas). Product substitution would be more effective if harvests did not reduce forest carbon greatly and if wood were used in buildings that either have a longer life span than at present, or are nonresidential. (Wood is the primary material in current residential buildings, and therefore replacing these buildings does not increase product stores.) For both substitutions to be most effective, a policy is needed that reduces the rate at which other sectors use fossil fuels "stored" by the products and energy provided by the forest sector.

Trade-Offs with Management of Other Resources

Carbon is not the only forest resource to be managed or subject to future change. If removing carbon from the atmosphere were the only objective, the best strategy is to protect these moist forests of the Pacific Northwest from harvest and disturbance and allow them to grow as old as possible. However, there are other objectives to forest management, including protecting water quality; supplying timber, fuel, or habitat; providing a recreational resource; and a host of other values. Also, forest management has to consider the present but plan for the future, and when this future includes climate change it involves considerable uncertainty and risk. Attaining some management objectives may come with a carbon cost. The key point is that while some carbon costs may be unavoidable to meet management objectives, they may be minimized by being more aware of how the forest sector responds to management and disturbance.

Literature Cited

Campbell, J. L., M. E. Harmon, and S. R. Mitchell. 2012. Can fuel reduction treatments really increase forest carbon sequestration by reducing future fire emissions? *Frontiers in Ecology and the Environment* 10:83–90.

Harmon, M. E. 2001. Carbon sequestration in forests: Addressing the scale question. *Journal of Forestry* 99:24–29.

Hudiburg, T. M., B. E. Law, D. P. Turner, J. Campbell, D. Donato, and M. Duane. 2009. Carbon dynamics of Oregon and Northern California forests and potential land-based carbon storage. *Ecological Applications* 19:163–180.

Krankina, O. N., M. E. Harmon, F. Schnekenburger, and C. A. Sierra. 2012. Carbon balance on federal forest lands of western Oregon and Washington. *Forest Ecology and Management* 286:171–182.

Law, B. E., and M. E. Harmon. 2011. Forest sector carbon management, measurement and verification, and discussion of policy related to climate change. *Carbon Management* 2:73–84.

Law, B. E., and R. H. Waring. 2015. Carbon implications of current and future effects of drought, fire and management on Pacific Northwest forests. *Forest Ecology and Management* 355:4–14.

Mackey, B., I. C. Prentice, W. Steffen, J. I. House, D. Lindenmayer, H. Keith, and S. Berry. 2013. Untangling the confusion around land carbon science and climate change mitigation policy. *Nature Climate Change* 3:552–557.

Mitchell, S. R., M. E. Harmon, and K. E. B. O'Connell. 2012. Carbon debt and carbon sequestration parity in forest bioenergy production. *Global Change Biology, Bioenergy* 4:818–827.

Smith, P., M. Bustamante, H. Ahammad, and 46 coauthors. 2014. Agriculture, forestry and other land use (AFOLU). Chap. 11 in *Climate Change 2014: Mitigation of climate change. Contribution of Working Group III to the Fifth Assessment Report of the Intergovernmental Panel on Climate Change*. Edited by O. Edenhofer, R. Pichs-Madruga, Y. Sokona, and 13 co-editors. Cambridge, UK: Cambridge University Press. http://www.ipcc.ch/pdf/assessment-report/ar5/wg3/ipcc_wg3_ar5_chapter11.pdf.

Smithwick, E. A. H., M. E. Harmon, S. M. Remillard, S. A. Acker, and J. F. Franklin. 2002. Potential upper bounds of carbon stores in forests of the Pacific Northwest. *Ecological Applications* 12:1303–1317.

Stephenson, N. L., A. J. Das, R. Condit, and 34 coauthors. 2014. Rate of tree carbon accumulation increases continuously with tree size. *Nature* 507:90–93.

TerMikaelian, M. T., S. J. Colombo, and J. Chen. 2015. The burning question: Does forest bioenergy reduce carbon emissions? A review of common misconceptions about forest carbon accounting. *Journal of Forestry* 113:57–68.

Waring, R. H., and J. F. Franklin. 1979. Evergreen coniferous forests of the Pacific Northwest. *Science* 204:1380–1386.

Zeide, B. 2001. Thinning and growth: A full turnaround. *Journal of Forestry* 99:20–25.

Chapter 13

Biodiversity

Deanna H. Olson, Brooke E. Penaluna, Bruce G. Marcot, Martin G. Raphael, and Keith B. Aubry

Sustainability of a human-forest ecosystem refers to its continuing capacity to maintain characteristic species, processes, and functions; to be resistant or resilient to most perturbations; and to provide commodities, uses, and other public benefits in the face of changing environmental, social, economic, and cultural circumstances. The nature of these societal benefits is expected to change during the twenty-first century, and this may be especially true for biodiversity values.

Biodiversity is a broad term that refers to the composition and abundance of life, spanning levels of organization that include genes, traits, populations, species, communities, and ecosystems. In addition to concerns for threatened or endangered species, over the last few decades there has been an expanded focus on sustaining biodiversity of forests, especially once-overlooked endemic taxa and species assemblages that are often rare, little known, and vulnerable to change (e.g., Raphael and Molina 2007). Simultaneously, the pace of forest landscape change from disturbance has increased. Forest management, wildfire, pests and pathogens, floods and landslides, windstorms, invasive species, chemical contamination, drought, and other disturbances—and their complex interactions—collectively affect biodiversity. Novel ecological communities may emerge as native biota reassemble after major disturbances and as species ranges shift in response to climate change. Consequently, our concept of human-forest ecosystem sustainability is also changing.

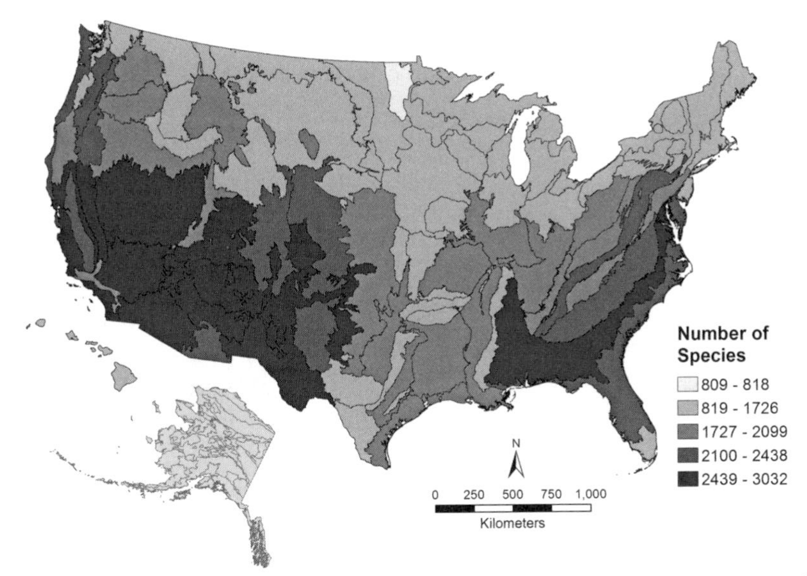

FIGURE 13.1. Number of species of vascular plants and vertebrates that are associated with forests in the United States, stratified by Omernik Level III Ecoregions (Omernik 1987), showing significant biodiversity in the moist forests of the Pacific Northwest. From Oswalt and Smith (2014).

Herein we provide some lessons learned during the last several decades of research and management of a broad array of species in the Pacific Northwest that inhabit moist coniferous forests. The high biomass of Northwest forests (Krankina et al. 2014) is associated with a rich and diverse biota (fig. 13.1), largely because of diverse abiotic and biotic conditions across this extensive region. Terrestrial vertebrates alone number ~317 species for western forests of Oregon and Washington, including 29 amphibians, 20 reptiles, 164 birds, and 104 mammals (Olson et al. 2001). Other species-rich groups include taxa that are not yet fully described, especially fungi, mollusks, and soil invertebrates.

In this chapter, we aim to provide new insights about how biodiversity management and research priorities are changing. We draw from selected examples across taxa to show that providing for a diversity of forest conditions in the future will increasingly require consideration of (1) biodiversity from genes to ecosystems, inclusive of their key ecological functions; (2) all lands, with a balance of coarse- and fine-filter approaches; and (3) all potential stressors.

Context of Biodiversity Conservation in US Northwest Moist Forests

The conservation of the northern spotted owl (plate 11A) became a major public policy concern in the US Northwest during the late 1980s. Its numbers were diminishing and it was considered for listing under the US Endangered Species Act. The old-growth forests that provide optimal habitat for this species had declined precipitously from decades of timber harvesting (chap. 7). This triggered collaborative efforts to manage for owls and numerous other species associated with late-successional and old-growth (LSOG) forest conditions, while also prioritizing socioeconomic goals across the region—especially on federally managed forests, which encompass ~10 million ha (24 million ac). Pacific salmon and trout quickly moved to the forefront of these discussions, along with the marbled murrelet (plates 11B and 11F). Alternatives for a regional ecosystem-management plan on federal lands (which would eventually become the Northwest Forest Plan) were developed in 1993 (FEMAT 1993), each with a coarse-scale set of upland forest and riparian reserves to maintain or restore LSOG habitats for owls, murrelets, salmon, and their associated communities.

Owing to the diversity of LSOG-associated taxa, concerns were raised about the efficacy of proposed coarse-scale habitat set-asides. Over 1,000 LSOG-associated taxa were identified and screened for viability under 9 alternative federal forest-management scenarios. These taxa included fungi (n = 527 species); lichens (157); bryophytes (106); vascular plants (124); mollusks (102); arthropods (15 species groups); amphibians (18 species); birds other than spotted owls and murrelets (36); bats (11); and mammals other than bats (15). Many of these species were rare or little known, with low mobility and restricted ranges. During the expert panel assessments that informed the writing of the Northwest Forest Plan (where species experts convened to evaluate both published literature and unpublished information), many species appeared poorly protected by the proposed coarse-filter approach of establishing an extensive system of federally reserved lands (Molina et al. 2006). This concern led scientists and managers to design a parallel fine-filter approach for a subset of species whose primary habitat conditions or occupancy would be assessed at local sites in each project area as forest-management activities were being planned. Surveys would be conducted for these species and, if present, measures would be implemented to protect the site. This became the Northwest Forest Plan's Survey and Manage program.

Before the Plan was implemented in 1994, forest-management priorities were largely restricted to threatened and endangered birds and mammals. The Survey and Manage program developed an array of inventory and monitoring approaches for taxa that are little known and not easily detected, including pre-disturbance surveys, strategic surveys, purposive surveys, equivalent effort surveys, and opportunistic surveys. These taxa included fungi that may or may not produce above-ground fruiting bodies, arboreal mammals, and woodland salamanders (plate 11D) that occupy below-ground habitats. Among these approaches, those most effective for informing conservation of rare or little-known species were (1) historical data compilation; (2) probability sampling; (3) genetic evaluation; and (4) habitat modeling (Olson et al. 2007).

Although the Survey and Manage program initially cost >US$10 million per year, which was financially unsustainable, it yielded significant new information on many key species. In 2016, ~300 species continued to be managed under the Survey and Manage program across 9.7 million ha (24 million ac; USDI 2013), and this program had substantially advanced our knowledge of a wide array of native species in old-forest ecosystems. Examples of little-known species addressed include taildropper slugs and jumping slugs, which feed on fungi or facilitate the decomposition of forest-floor debris; arboreal mammals such as the red tree vole, key prey for northern spotted owls; and previously unknown species, such as the Scott Bar salamander, which was discovered during surveys for other species.

The Survey and Manage program enabled consideration of a broader array of taxonomic groups within the other federal sensitive species programs in the region. In 2016 the Interagency Special Status and Sensitive Species Program (ISSSSP), managed jointly by the US Forest Service and Bureau of Land Management (BLM), considered ~800 native species of concern in Washington and Oregon (USDA and USDI 2014) during federal land-management planning.

In 2015, federal natural resource managers in the US Northwest were asked to identify their highest biodiversity priorities. The resulting top 10% included 200 different species, and working groups then were assembled to address associated knowledge gaps (USDA and USDI 2015). In addition, conservation assessments were being developed to synthesize biological information, identify potential threats, and develop management alternatives (USDA and USDI 2014), and new studies were being conducted on several taxa that occur in moist coniferous forests, including fungi, bryophytes, bumblebees, caddisflies, aquatic snails, salamanders, and bats (USDA and USDI 2015).

Fine-filter approaches to species management can be expensive, but are far less costly than allowing species to become threatened or endangered. Together, the Survey and Manage and ISSSSP programs cost ~US$1 million per year as of 2016 and make significant contributions to the conservation of ~1,000 species that were largely overlooked prior to these efforts. When these rare or little-known species are detected where management projects are proposed, there is often a focus on habitat management to support continued persistence at forest-stand scales, greatly reducing the need to monitor populations at broader spatial scales. We do not yet know whether these approaches will sustain native biodiversity over the long term, but they forestall short-term species losses, especially for taxa with limited mobility that function at small scales.

In contrast, the combined federal effectiveness monitoring programs for owls, murrelets, late-successional and old-growth forests, aquatic-riparian environments, and socioeconomic issues in the Northwest Forest Plan currently cost ~US$4.5 million per year. Programs for monitoring northern spotted owls and marbled murrelets have assessed population trends and habitat in the region (owl, Davis et al. 2016, Dugger et al. 2016; murrelet, Falxa and Raphael 2016). In a summary of northern spotted owl habitat and demographic trends, Davis et al. (2016) acknowledged that there is a significant challenge to achieving the ultimate objective of effectiveness monitoring, to develop a model-driven, habitat-based approach for owl population assessment, due to uncertainties in mapping habitat and the increased presence of barred owls that compete for similar habitat. Falxa and Raphael (2016) reported no trend for murrelet populations at the regional scale but found a declining trend at smaller spatial scales; in addition, they estimated that ~2% of higher-suitability nesting habitat was lost on federally managed lands since 1994 from fire, harvest, and windthrow. It was expected to take several decades to see increasing trends for these species owing to extensive habitat loss within their ranges (chap. 7).

Similarly, ~1.5% of owl habitat was lost on federally managed lands between 1993 and 2012, but these losses were partially offset by the development of new owl habitat. During the same period, a new threat emerged for northern spotted owls when barred owls, a competitor of spotted owls, expanded their range into the region (Wiens et al. 2014). Thus, although dramatic losses may have been forestalled for many other late-successional species, the efficacy of the Northwest Forest Plan to conserve the northern spotted owl may ultimately be rendered moot by an unanticipated risk factor. Although time will be needed to evaluate the long-term effectiveness of the combined coarse- and fine-filter approaches developed in the northwestern United States for species management on federal lands, the suite of protective measures aims to provide important benefits for a broad range of species (Raphael and Molina 2007). Accordingly, these approaches have been incorporated into the new Forest Service planning rule (USDA 2012).

New Challenges of Old-Growth Forest Management

Since the 1990s, the management of moist coniferous forests has focused on species associated with late-successional forest conditions, although the need to actively manage for early-seral species has only been broadly recognized in the last decade. Historically, the primary source for early-seral habitats was catastrophic wildfires (chap. 2). With the implementation of the Northwest Forest Plan in 1994, clearcut harvests ended and second-growth forests flourished on federal lands, reducing the prevalence of early-seral conditions.

Early-seral vegetation is characterized by a distinct vascular plant community that provides the pollen, fruits, and seeds that benefit many wildlife species. Some bryophytes also exhibit early-seral associations, with higher diversity tied to the variety of substrates found in early-seral conditions compared with later conditions (Swanson 2012). Other taxa occur in early-seral conditions but are not obligates: these include 78 vertebrates associated with nonconiferous vegetation in Northwest forests (Hagar 2007). As the extent of early-seral conditions has waned in this region, natural resource specialists have become increasingly concerned about the population status of several fruit- and seed-eating birds and a suite of pollinators, including butterflies, moths, and hummingbirds. Forest-management activities can provide benefits for both late- and early-seral forest species through harvest practices that retain key biological legacies (Franklin and Johnson 2012) and potentially through restoration practices that create gaps within a thinned matrix to accelerate the development of multilayered forest canopies.

Genes, Communities, Key Ecological Functions, and Ecosystem Services

Biodiversity-management objectives in moist coniferous forests include multiple scales of biological organization that may range from genetic attributes to the structuring of communities and functioning of ecosystems. Increased biological complexity may provide long-term stability for communities and ecosystems (Marcot 2007) owing to complex interactions among fine-scale components (e.g., populations, individuals, or genes) that result in more efficient use of resources and increased biomass production. High levels of biodiversity may also decrease the risks of living in a challenging environment by providing local resilience against changing environmental conditions. Several examples are available to demonstrate the role of biological complexity (*biocomplexity*) in moist forests beyond species to consider phenotypic and genetic variation and ecological functions. (Phenotype is the observable characteristics of an organism, which depend on the interaction of genetics and environment.)

Among the Pacific salmon and trout species that occupy streams in moist coniferous forests of the Northwest, distinct traits and genetic characteristics are used to identify *evolutionarily significant units* of concern (e.g., Oregon Coast coho salmon [plate 11F] and Oregon Coast cutthroat trout). Pacific salmon and trout have enhanced morphological, behavioral, life-history, and genetic variability relative to many other bony fishes, enabling them to survive in or adapt to rapidly changing conditions. For example, they may spawn in spring, summer, or autumn in the same or different streams throughout the range of their distribution. Also, cutthroat and rainbow trout or steelhead may or may not move between freshwater and marine environments, with sea-run and freshwater-resident strategies occurring within a single population. Sometimes these strategies vary geographically within watersheds, whereby upstream fish are more likely to be freshwater resident. This biocomplexity has likely contributed to their persistence in highly dynamic environments in the past and could be key to their persistence in the future. Pacific salmon and trout with the lowest amount of biocomplexity, such as hatchery-stocked individuals, are more vulnerable to losses once they are released in the wild (Araki and Schmid 2010).

The biocomplexity that results from multiple food-web interactions within a community may also help to protect forest-floor communities in the Pacific Northwest, where Douglas-fir, mycorrhizal fungi, northern flying squirrels (plate 11C), and northern spotted owls are intimately connected via trophic dynamics that include microbiota, raptors, small mammals, and plants (Carey 2003). Management for such community-level biocomplexity is a proposed restoration goal for managed moist coniferous forests (Carey 2006a, b).

In the mid-1990s, a new approach was developed for evaluating the functionality of forest ecosystems based on key ecological functions, the ecological roles played by organisms that, in turn, affect the resources and habitats of other species (Marcot and Vander Heyden 2001). Examples include primary burrow excavation, seed dispersal, soil digging, chewing of soil organic matter, nutrient turnover, and conifer pathogens that aid in creation of tree cavities and dead wood used by a host of other species (chap. 6; box 6.1). Each of these activities affects or engineers resources and environmental conditions for other species.

Assessing patterns of key ecological functions is useful for understanding the effects of disturbance or management activities. Important patterns include functional diversity (the variety and abundance of species performing functions), functional redundancy (number of species with similar functional roles), and functional attenuation (loss of function categories, Marcot and Vander Heyden 2001). The presence or absence of these functions provides an empirically based estimate of the degree to which ecosystems are fully functional. Previously, the more limited concepts of *indicator* and *keystone* species were used, but they applied only to those species that had the greatest influence on other species or their environment. Those earlier concepts failed to capture the full range of biotic roles and could not reveal changes in the functional integrity of forest ecosystems. Databases of key ecological functions were developed in conjunction with species-habitat relationships data for Washington and Oregon (Johnson and O'Neil 2001). This effort created important new opportunities for evaluating ecosystem functionality (Marcot and Aubry 2003).

Public interest in both the ecological integrity of forest ecosystems and the management of forests for commodities has, in part, resulted in the recognition of ecosystem services that depend on biodiversity. This has added a new dimension to understanding and prioritizing biodiversity values from a management perspective. The public's desire to sustain native species has been a driving force in the expansion of programs designed to inventory and protect potentially sensitive species. Also, the protection of streams and riparian areas to provide clean water has merged with priorities for ensuring the persistence of native salmon and trout, which helps to protect a broad assemblage of aquatic and riparian taxa. For example, in small streams where cutthroat trout are protected from upslope forest harvest by riparian buffers, additional measures to prevent anthropogenic siltation and provide clean water also protect other taxa found in these stream reaches, including endemic species like coastal tailed frogs and coastal giant salamanders, aquatic invertebrates, and streamside riparian communities.

More directly, recreational opportunities such as fishing, wildlife viewing, and natural immersion experiences are creating new opportunities for the public to understand and appreciate the value of biodiversity. One upshot is the involvement of citizen scientists in forest wildlife research and monitoring (box 13.1). Yet ecosystem services can be at odds. For example, the desire for fishing experiences has led to stocking programs of non-native fish or hatchery-raised fish that often have different traits than native fish. The widespread introduction of hatchery-raised rainbow trout to lakes and streams where they were not present historically has led to hybridization and reduced diversity in native lineages of Pacific trout. This example highlights an additional concern: as non-native and invasive species proliferate, novel assemblages result and native species may be lost. Exotic species can outcompete or prey upon native species or may introduce new pests and pathogens. For example, several alternative management actions are currently being implemented to protect northern spotted owls from invasive barred owls and to reduce the ecological impacts of injurious weeds and several aquatic invasive species.

All Lands Considered

Two key lessons learned during the last few decades are that moist coniferous forests in all stages of development are important for native assemblages, and that management at landscape scales will be needed to protect native biodiversity. A good example is the intrinsic potential models used to identify streams or portions of a watershed that provide essential habitat for Pacific salmon based on regional, ecological, and life-history differences (Burnett et al. 2007). These models assume that suitability of aquatic habitat is strongly influenced by stable attributes, such as geomorphology. Identifying streams and portions of watersheds that have high intrinsic potential is useful in prioritizing conservation actions for Pacific salmon. Resulting maps demonstrate the need to address multiple landownerships and expand the geographic scope of management activities for salmon. For example, some of the most productive habitat for coho salmon is located on private land, such that widespread recovery of coho salmon is unlikely unless habitat can be improved in reaches of high intrinsic potential on private lands (Burnett et al. 2007; Reeves et al. 2016; chap. 15).

Marbled murrelet life history also demonstrates the importance of considering the broader landscape. Marbled murrelets nest on the limbs

BOX 13.1. CITIZEN SCIENTISTS AND FOREST CARNIVORE RESEARCH

Citizen scientists are volunteers who collect data for research, including rigorous experiments and studies contributing new information on species occurrences. One broadly applied citizen-science project is the National Audubon Society's Christmas Bird Count, initiated in 1900 and resulting in almost 350 scientific papers and important insights about the changing status of North American bird populations (Silverton 2009). Citizen scientists bring unique skills and perspectives to research and allow researchers to expand the scope of their projects at little or no cost. Direct involvement is one of the best ways for the public to gain a better understanding and appreciation of science contributions to society.

Until recently, most distribution records for rare and elusive forest carnivores in moist coniferous forests (e.g., Canada lynx, wolverine, fisher, montane red fox, gray wolf, grizzly bear) were obtained by public agencies or environmental organizations in the form of sighting reports (anecdotal observations) submitted by outdoor recreationists or amateur biologists. However, because of species misidentifications, such data have been of limited value to science, especially for rare and elusive species for which reliance on such data can actually impede conservation efforts (McKelvey et al. 2008). The recent development of simple, effective, and relatively inexpensive devices for obtaining reliable occurrence data noninvasively (digital remote cameras, also known as camera traps; hair-snagging devices to obtain DNA samples) has revolutionized the compilation of occurrence data for potentially threatened forest carnivores. These devices produce verifiable evidence of species occurrence, much like museum specimens, and require little expertise to operate. Similarly, automated sound recording has advanced bird surveys (Brandes 2008), and the emerging field of environmental DNA (eDNA) enables aquatic inventories by collecting water samples and screening the samples for the DNA of resident organisms (Biggs et al. 2015). These methods are expanding the efficacy of citizen-science efforts.

Since 2001, hundreds of volunteers each year have helped to document the current distributions of rare and elusive forest carnivores as part of the Citizen Wildlife Monitoring Project (CWMP) directed by Conservation Northwest in the Pacific Northwest. Recently, the CWMP has made important discoveries for several key species, including documenting in 2008 the first wild gray wolf pups to be born in Washington in nearly 80 years, and determining the southernmost extent of the recovery area for wolverines in the northern Washington Cascade Range (http://www.conservationnw.org/what-we-do/wildlife-habitat/wildlife-monitoring).

of older coniferous trees, especially those with epiphytic plants that help provide a cup for the bird's single egg (plate 11B). Suitable nesting habitat depends not only on the presence of large trees, but also on their pattern in the landscape, because murrelets select relatively large patches of contiguous forest for nesting. The Northwest Forest Plan has been successful in conserving much of the remaining suitable habitat on federal lands, but >40% of current suitable habitat occurs on nonfederal lands and is decreasing in extent owing to continuing timber harvest (Raphael et al. 2016). Those losses are resulting in gaps in the distribution of murrelet nesting habitat, such as in southern Washington and northern Oregon where there are no federal lands, and these gaps may in turn result in the isolation of populations and the potential loss of genetic diversity. In addition, the marbled murrelet forages in marine waters adjacent to its nesting habitat. Viability of murrelet populations depends not only on the amount and pattern of suitable inland nesting habitat, but also on ocean conditions that foster production of the small fishes and krill upon which murrelets feed (Falxa and Raphael 2016). The challenge for conserving and restoring murrelet populations is to find a way to manage across all landownerships to enhance their ecological resilience.

Recent research has also revealed the spatial scale at which some species use habitats and has identified landscape-scale correlates of connectivity. For example, some amphibians that occupy forested headwaters most often occur and move within 15 m (~50 ft) of streams (plethodontid salamanders, Kluber et al. 2008; Olson and Kluber 2014). Gene flow, a measure of connectedness of populations, can be influenced by factors such as sun exposure (Spear et al. 2012), high heat-load, and forest cover (Emel and Storfer 2015).

Broad-ranging forest carnivores such as fishers and martens (plate 11E) use relatively contiguous forest stands with moderate to dense levels of canopy cover, complex forest structure, and elements of older forests such as logs, snags, and decadent trees to provide denning and resting sites. Understanding the spatial and temporal scale of species movements is particularly important when habitat management plans encompass multiple watersheds or landownerships over time. Maintaining biodiversity in landscapes that have been fragmented into habitat *islands* by timber harvest and other large-scale disturbances (chap. 7), or in areas where available habitat is naturally fragmented, such as in the island archipelagos of Southeast Alaska (fig 1.1), will require a multiscale approach across landownerships.

All Disturbances Considered

Historically, the conservation of fish and wildlife in moist coniferous forests focused primarily on mitigating the effects of forest-management practices. However, it is important to consider all potential stressors. Disturbances that affect species are not restricted to timber harvest and related road construction and maintenance activities, but also include altered biotic communities (changes in predator-prey relationships; invasive species), chemical contaminants, diseases and pathogens, fire, and altered climate niches (chap. 16). Two avian examples include interactions between barred owls and northern spotted owls, and between marbled murrelets and corvids (crows and jays). Barred owls have expanded into the range of northern spotted owls, resulting in the displacement and loss of spotted owls. Predation by corvids on murrelet eggs or nestlings directly reduces murrelet recruitment and, ultimately, population size. Such stressors were not recognized during the development of the Northwest Forest Plan, and have only come to light in the course of 20 years of intensive population monitoring.

Pests and pathogens are other biotic interactions that influence biodiversity. In Northwest moist coniferous forests, there is ongoing scrutiny for signs of disease in fish (whirling disease); amphibians (fungal, viral, bacterial pathogens, trematodes); snakes (snake fungal disease); bats (white-nose syndrome); deer and elk (chronic wasting disease); birds (avian influenza and West Nile virus); and coniferous trees (Swiss needle cast). Researchers are examining methods for managing such outbreaks and are seeking better understanding of synergisms with other disturbances. For example, drought conditions may exacerbate bark beetle infestations, and fungal diseases in animals may signal altered skin microbiota or immune responses caused by other stressors.

Several forest taxa can be used to recognize ecosystems with degraded environmental conditions. Mosses and lichens are used to identify potential air-quality issues (Root et al. 2015; Gatziolis et al. 2016). The *index of biotic integrity* of aquatic systems, based on macroinvertebrate assemblages, has been applied broadly to evaluate water quality of streams and has been extended to assess fish (Wan et al. 2010), amphibians (Stapanian et al. 2015), birds (Bryce et al. 2002), and terrestrial conditions (Karr and Kimberling 2003). Amphibians have emerged as bio-indicators, as they respond to a host of stressors in both aquatic and terrestrial habitats and, owing to their central position in food webs, to changes in predator or prey communities.

Tools Developed to Aid Biodiversity Management across Spatial Scales

Since 1994, threat or risk assessments have been used in a variety of applications, including evaluating species-listing decisions, determining effects of past and future management activities, and advising conservation and recovery planning. Assessments have been conducted for northern spotted owls, fishers, marbled murrelets, western snowy plovers, and other species, typically using structured, rigorous expert panels and by combining information from a variety of sources. In addition, taxonomic assemblages and communities have been evaluated in bioregional assessments, such as the expert-panel viability assessments conducted on hundreds of vertebrate species for appraising alternatives that led to the Northwest Forest Plan. Similar multispecies assessments have been part of the federal forest-planning and project development processes, especially for species considered sensitive to forest management.

Sophisticated methods for managing risk and uncertainty using a decision-science framework also have been developed during the past two to three decades. Previously, simpler decision-making tools, such as decision trees and linear programming optimization, were used in forest conservation and management, but these tools proved to be too limited to address the broadening set of social needs and multiple-use interests in forest resources. To help managers make complex decisions involving multiple resources and stakeholders, the Ecosystem Management Decision Support (EMDS) system, in particular, was developed recently to integrate fuzzy-logic models with geographic information systems (GIS) as a risk assessment tool (Reynolds et al. 2014).

The Science of Forest Biodiversity Unfolds

Although we have learned much about biodiversity in moist coniferous forests during the last couple of decades, we still lack a clear understanding of the ecological functions and processes many species provide, or how native taxa contribute to forest ecosystem resilience. Research attention to food-web dynamics can help to unravel many complex species interactions and the intermediate or supporting ecosystem services for many forest species. To address the maintenance of biodiversity in moist coniferous forests of the Pacific Northwest, we have described a number of approaches that were developed since 1994, each linking scientific advances to management. These approaches include (1) quantification of genetic diversity and identification of habitat-management strategies that preserve it; (2) coarse- and fine-filter forest management to address taxa of concern; and (3) describing and using biocomplexity, key ecological functions, suites of species types, and environmental indicators to manage ecological communities.

By integrating these approaches, many more organisms and the potential ecological roles they play can be evaluated for their contributions to the sustainability of forest ecosystems, resource production, and social values. Research is now focused on how multiple disturbances, both intentional and natural, may interact across a heterogeneous landscape of species assemblages. This reinforces the main lesson learned about forest biodiversity, which is that all species, all lands, and all stressors matter, forming a complex multistate system. This is an enormous challenge, yet we now understand that managing to sustain forest biodiversity will be a dynamic process in both space and time, and that all components of the ecosystem contribute to the process and its outcome.

Literature Cited

Araki, H. and C. Schmid. 2010. Is hatchery stocking a help or harm? Evidence, limitations and future directions in ecological and genetic surveys. *Aquaculture* 308:S2–S11.

Biggs, J., N. Ewald, A. Valentini, and 8 coauthors. 2015. Using eDNA to develop a national citizen science-based monitoring programme for the great crested newt (*Triturus cristatus*). *Biological Conservation* 183:19–28.

Brandes, T. S. 2008. Automated sound recording and analysis techniques for bird surveys and conservation. *Bird Conservation International* 18:S163–S173.

Bryce, S. A., R. M. Hughes, and P. R. Kaufmann. 2002. Development of a bird integrity index: Using bird assemblages as indicators of riparian condition. *Environmental Management* 30:294–310.

Burnett, K. M., G. H. Reeves, D. J. Miller, S. Clarke, K. Vance-Borland, and K. Christiansen. 2007. Distribution of salmon-habitat potential relative to landscape characteristics and implications for conservation. *Ecological Applications* 17:66–80.

Carey, A. B. 2003. Biocomplexity and restoration of biodiversity in temperate coniferous forest: Inducing spatial heterogeneity with variable density thinning. *Forestry* 76:127–136.

Carey, A. B. 2006a. Active and passive forest management for multiple values. *Northwestern Naturalist* 87:18–30.

Carey, A. B. 2006b. AIMing to restore forests: Evaluation with SER criteria. *Northwestern Naturalist* 87:31–42.

Davis, R. J., B. Hollen, J. Hobson, J. E. Gower, and D. Keenum. 2016. *Northwest Forest Plan—the first 20 years (1994–2013): Status and trends of the northern spotted owl habitats*. General Technical Report PNW-GTR-929. Portland, OR: USDA Forest Service, Pacific Northwest Research Station.

Dugger, K. M., E. D. Forsman, A. B. Franklin, and 35 coauthors. 2016. The effects of habitat, climate, and barred owls on long-term demography of northern spotted owls. *The Condor: Ornithological Applications* 118:57–116.

Emel, S. L., and A. Storfer. 2015. Landscape genetics and genetic structure of the southern torrent salamander, *Rhyacotriton variegatus*. *Conservation Genetics* 16:209–221.

Falxa, G.A., and M. G. Raphael, tech. coords. 2016. *Northwest Forest Plan—the first 20 years (1994–2013): Status and trend of marbled murrelet populations and nesting habitat*. General Technical Report PNW-GTR-927. Portland, OR: USDA Forest Service, Pacific Northwest Research Station.

FEMAT (Forest Ecosystem Management Assessment Team). 1993. *Forest ecosystem management: An ecological, economic, and social assessment. Report of the Forest Ecosystem Management Assessment Team*. Portland, OR: US Department of Agriculture; US Department of the Interior [and others].

Franklin, J. F., and K. N. Johnson. 2012. A restoration framework for federal forests in the Pacific Northwest. *Journal of Forestry* 110:429–439

Gatziolis, D., S. Jovan, G. Donovan, M. Amacher, and V. Monleon. 2016. *Elemental atmospheric pollution assessment via moss-based measurements in Portland, Oregon*. General Technical Report PNW-GTR-938. Portland, OR: USDA Forest Service, Pacific Northwest Research Station.

Hagar, J. 2007. Wildlife species associated with non-coniferous vegetation in Pacific Northwest conifer forests: A review. *Forest Ecology and Management* 246: 108–122.

Johnson, D. H., and T. A. O'Neil, manag. dirs. 2001. *Wildlife-habitat relationships in Oregon and Washington*. Corvallis: Oregon State University Press.

Karr, J. R., and D. N. Kimberling. 2003. A terrestrial arthropod index of biological integrity for shrub-steppe landscapes. *Northwest Science* 77:202–213.

Kluber, M. R., D. H. Olson, and K. J. Puettmann. 2008. Amphibian distributions in riparian and upslope areas and their habitat associations on managed forest landscapes of the Oregon Coast Range. *Forest Ecology and Management* 256:529–535.

Krankina, O. N., D. A. DellaSala, J. Leonar, and M. Yatskov. 2014. High-biomass forests of the Pacific Northwest: Who manages them and how much is protected? *Environmental Management* 54:112–121.

Marcot, B. G. 2007. Biodiversity and the lexicon zoo. *Forest Ecology and Management* 246:4–13.

Marcot, B. G., and K. B. Aubry. 2003. The functional diversity of mammals in coniferous forests of western North America. Pp. 631–664 in *Mammal com-*

munity dynamics: Management and conservation in the coniferous forests of western North America. Edited by C. J. Zabel and R. G. Anthony. Cambridge, UK: Cambridge University Press.

Marcot, B. G., and M. Vander Heyden. 2001. Key ecological functions of wildlife species. Chap. 6. Pp. 168–186 in *Wildlife-habitat relationships in Oregon and Washington*. Managing direction by D. H. Johnson and T. A. O'Neil. Corvallis: Oregon State University Press.

McKelvey, K. S., K. B. Aubry, and M. K. Schwartz. 2008. Using anecdotal occurrence data for rare or elusive species: The illusion of reality and a call for evidentiary standards. *BioScience* 58:549–555.

Molina, R., B. G. Marcot, and R. Lesher. 2006. Protecting rare, old-growth forest associated species under the survey and manage guidelines of the Northwest Forest Plan. *Conservation Biology* 20:306–318.

Olson, D. H., J. C. Hagar, A. B. Carey, J. H. Cissel, and F. J. Swanson. 2001. Wildlife of westside and high montane forests. Pp. 187–212 in *Wildlife-habitat relationships in Oregon and Washington*. Managing direction by D. H. Johnson and T. A. O'Neil. Corvallis: Oregon State University Press.

Olson, D. H., and M. R. Kluber. 2014. Plethodontid salamander distributions in managed forest headwaters in western Oregon. *Herpetological Conservation and Biology* 9:76–96.

Olson, D. H., K. J. Van Norman, and R. D. Huff. 2007. *The utility of strategic surveys for rare and little-known species under the Northwest Forest Plan*. General Technical Report PNW-GTR-708. Portland, OR: USDA Forest Service, Pacific Northwest Research Station.

Omernik, J. M. 1987. Ecoregions of the conterminous United States. Map (scale 1:7,500,000). *Annals of the Association of American Geographers* 77:118–125.

Oswalt, S. N., and W. B. Smith, eds. 2014. *U.S. forest resource facts and historical trends*. FS-1035. USDA Forest Service.

Raphael, M. G., G. A. Falxa, D. Lynch, S. K. Nelson, S. F. Pearson, A. J. Shirk, and R. D. Young. 2016. Status and trend of nesting habitat for the marbled murrelet under the Northwest Forest Plan. Chap. 2. Pp. 37–94 in *Northwest Forest Plan—the first 20 years (1994–2013): Status and trend of marbled murrelet populations and nesting habitat*. Technical coordination by G. A. Falxa and M. G. Raphael. General Technical Report PNW-GTR-927. Portland, OR: USDA Forest Service, Pacific Northwest Research Station.

Raphael, M. G., and R. Molina, eds. 2007. *Conservation of rare or little-known species: Biological, social, and economic considerations*. Washington, DC: Island Press.

Reeves, G. H., B. R. Pickard, and K. N. Johnson. 2016. *An initial evaluation of potential options for managing riparian reserves of the Aquatic Conservation Strategy of the Northwest Forest Plan*. General Technical Report PNW-GTR-937. Portland, OR: USDA Forest Service, Pacific Northwest Research Station.

Reynolds, K. M., P. F. Hessburg, and P. S. Bourgeron, eds. 2014. *Making transparent environmental management decisions*. Berlin and Heidelberg: Springer.

Root, H. T., L. H. Geiser, S. Jovan, and P. Neitlich. 2015. Epiphytic macrolichen indication of air quality and climate in interior forested mountains of the Pacific Northwest, USA. *Ecological Indicators* 53:95–105.

Silverton, J. 2009. A new dawn for citizen science. *Trends in Ecology and Evolution* 24:467–471.

Spear, S. F., C. M. Crisafulli, and A. Storfer. 2012. Genetic structure among coastal tailed frog populations at Mount St. Helens is moderated by post-disturbance management. *Ecological Applications* 22:856–869.

Stapanian, M. A., M. Micacchion, and J. V. Adams. 2015. Wetland habitat disturbance best predicts metrics of an amphibian index of biotic integrity. *Ecological Indicators* 56:237–242.

Swanson, M. E. 2012. *Early seral forests in the Pacific Northwest: A literature review and synthesis of current science*. US Forest Service, Central Cascades Adaptive Management Partnership. https://ncfp.files.wordpress.com/2012/06/swanson_20120111.pdf.

USDA (US Department of Agriculture, Forest Service). 2012. National Forest System land management planning. Final rule and record of decision. 36 CFR Part 219. *Federal Register* 77(68):21162–21267. http://www.fs.usda.gov/Internet/FSE_DOCUMENTS/stelprdb5362536.pdf.

USDA (US Department of Agriculture) and USDI (US Department of the Interior) 2014. Interagency Special Status/Sensitive Species Program (ISSSSP). http://www.fs.fed.us/r6/sfpnw/issssp/.

———. 2015. Interagency sensitive and special status species program update—June2015.http://www.fs.fed.us/r6/sfpnw/issssp/documents3/update-2015-06.pdf.

USDI (US Department of the Interior). 2013. Survey and Manage. http://www.blm.gov/or/plans/surveyandmanage/.

Wan, H., C. J. Chizinski, C. L. Dolph, B. Vondracek, and B. N. Wilson. 2010. The impact of rare taxa on a fish index of biotic integrity. *Ecological Indicators* 10:781–788.

Wiens, D. J., R. G. Anthony, and E.D. Forsman. 2014. Competitive interactions and resource partitioning between northern spotted owls and barred owls in western Oregon. *Wildlife Monographs* 185:1–50.

Chapter 14

Aquatic-Riparian Systems

Deanna H. Olson, Sherri L. Johnson, Paul D. Anderson, Brooke E. Penaluna, and Jason B. Dunham

As water plays a defining role for the development of moist coniferous forests, aquatic-riparian systems are defining components of these landscapes. In the Pacific Northwest, aquatic-riparian ecosystems in moist forests are increasingly being recognized as multistate systems, with a complex mix of heterogeneous habitats within and among watersheds and a host of disturbances at play over time (box 14.1). Although a mosaic of conditions has been recognized previously for aquatic systems (e.g., Pringle et al. 1988), there is a renewed focus on the thermal (Torgersen et al. 1999; Armstrong and Schindler 2013) and habitat (Stanford et all 2005) heterogeneity of aquatic-riparian systems and on multifaceted approaches for their management (Rieman et al. 2015).

This chapter highlights new knowledge about aquatic-riparian ecosystems and their management, especially relative to Northwest moist forests over the last few decades. We first focus on advances in basic science of aquatic-riparian ecological structure, functions, processes, and spatiotemporal heterogeneity, addressing the hypothesis of a multistate aquatic-riparian ecosystem in these forests. Second, we focus on experiments examining riparian-management approaches, many of which have been conducted along smaller headwater streams. We link the multistate ecological perspective to the need for a well-integrated multistate management approach. Perspectives of larger watersheds and landscapes are addressed in chapter 15, and interrelated topics in other chapters (chaps. 13, 16).

BOX 14.1. PRIMER ON PACIFIC NORTHWEST AQUATIC-RIPARIAN CONDITIONS

Aquatic-riparian ecosystems in Northwest moist coniferous forests are heterogeneous. Geomorphology (how earth surface processes shape the landscape, including soils and topography) affects the flow of water, resulting in lotic (flowing water) and lentic (pooling water) habitats. Both vary in size, from small seeps and potholes to larger rivers and lakes. Mountains lead to a dominance of stream networks across the Northwest; small headwater streams can account for 70% to 80% of stream-network length in a watershed (Gomi et al. 2002). Precipitation varies seasonally (wet winters, dry summers), producing perennial and ephemeral aquatic habitats with variable microclimates in their adjacent riparian areas. Precipitation (rain or snow) affects runoff, water levels, and peak flows. The *hyporheic zone* (belowground component of streams where areas of the streambed and floodplain are saturated with stream-source water) also varies, influencing stream temperature and nutrient cycling.

Streams have additional connections with their riparian zones and with the broader watersheds through which they flow. For example, forested riparian zones intercept light and provide litterfall and wood for stream-channel structure. The density of riparian forest cover and the type of streamside vegetation influences direct light, leaf litter, and down-wood inputs, and hence stream temperatures and primary production. Topography affects these processes as well through hill shading. Leaf litter and primary production provide energy resources and habitat for multiple taxa (fungi, bacteria, invertebrates, vertebrates). Large down wood provides habitat for upland, riparian, and stream species. Riparian tree roots provide structure for stream-bank stability, and riparian trees and microbes fix nitrogen and influence nutrient content of the water moving to the stream.

Natural and anthropogenic disturbances contribute to aquatic-riparian heterogeneity. Extreme climatic events may result in floods, droughts, and fire, which may lead to peak and low flows, along with debris flows and landslides. Human-caused disturbances include dams, roads and culverts, timber harvest, mining, grazing, urbanization, channelization, chemical pollution, water withdrawals, and species introductions (e.g., game fish, mollusks, bait releases) and removals (e.g., overexploitation of beaver, fish) (Penaluna et al. 2016). Hence, with many contributors to habitat heterogeneity, aquatic-riparian portions of Northwest moist forests are increasingly being recognized as multistate ecosystems—with component parts, functions, and processes varying at different locations within and among watersheds, as well as changing at single locations over time (fig. 14.1).

Stream-Riparian Ecology

Application of disturbance-ecology concepts to stream systems has resulted in a rich body of research and concept development since the late 1980s (Pringle et al. 1988; Resh et al. 1988; Lake 2000; Benda et al. 2004; Beechie et al. 2006; Naiman et al. 2010; Stanley et al. 2010; chap. 15). Stanley et al. (2010) developed a time line of landmark phases in stream disturbance ecology. Before 1986, disturbance concepts were rarely applied to streams. Webster et al.'s (1983) concepts of stream community resistance and resilience contributed to the subsequent development of landscape ecology and the framework of dynamic and stochastic processes. A "disturbance renaissance" began in the 1990s, with a plethora of empirical studies that provided support for the general application of the "dynamic equilibrium" hypothesis (Resh et al. 1988) to many stream systems. Stanley et al. (2010) concluded that disturbance is now a broadly recognized driver of stream ecosystems, but as yet there is little knowledge of how natural and anthropogenic disturbances interact, how to apply disturbance concepts to humanized systems, how to examine disturbances at broader spatial scales, and how to apply disturbance as a management tool.

Relative to riparian zones, riparian forest composition and stand densities likewise are naturally variable when they are influenced by frequent and patchy natural disturbances. Riparian disturbances in Northwest moist forests include flooding, landslides, and treefalls (e.g., from self-thinning processes, tree diseases, wind, or ice storms) and, less frequently, fire, all of which can create gaps that are quickly colonized by nitrogen-fixing red alder or other rapidly growing species (e.g., thimbleberry and salmonberry). As these colonizers fill in the disturbed areas, they can create a diverse matrix of early- and late-seral plants. This floral diversity contributes to the multistate nature of aquatic-riparian ecosystems in moist coniferous forests and has led to support for different management strategies for near-stream riparian areas as compared with upland forests (Pabst and Spies 1998).

Aquatic-Land Interactions

Several processes link riparian and stream systems; for instance, water flowing through the riparian zone and into the stream can influence nutrient concentrations and stream temperatures. Riparian and hyporheic (beneath a stream channel or floodplain) zones can be sources or sinks for nitrogen and carbon to the stream, depending on conditions that determine whether biogeochemical transformations increase or decrease concentrations (e.g., carbon flux in Alaska coastal rain forests, D'Amore et al. 2015). Hyporheic exchange creates both upwelling and downwelling flow patches in the streambed (e.g., in Southeast Alaska, Wondzell et al. 2009), which may contribute to creating local streambed areas with unique biochemical and thermal environments. Increased stream temperatures caused by timber harvest and climate change affect cold-adapted salmonids, amphibians, and other species in Northwest forest streams. Deep groundwater in hyporheic zones has a fairly constant temperature, and upwelling can cool streams in summer, creating thermal refuges for organisms. In other stream reaches, however, high air temperatures and low streamflow conditions may be drivers of warming stream temperatures. Statistical models have developed significantly over the last decade to assess the stream "thermalscape" based on a variety of local conditions and predicted climate futures (Isaak et al. 2010, 2015; USDA 2015).

Past studies have focused on the influence of riparian zones on streams, and more recent research has begun to demonstrate how streams in turn influence their surroundings (Baxter et al. 2005). High streamflows deposit materials onto floodplains, providing a reverse subsidy of sediment and organic matter back to the riparian zone, whereas hyporheic flows can transport solutes away from streams (Wondzell 2011). Carbon sequestration occurs in the floodplain as a result of burial of large wood in alluvial sediments. Riparian microclimates, including air temperature and relative humidity, are influenced by the presence of water nearby, particularly evident within 15 to 20 m (~50 to 65 ft) of the stream (Anderson et al. 2007). Moisture-dependent organisms occur in these near-stream cool-moist areas, including hydrophilic plants and animals. Organisms also move from streams to terrestrial zones, including insects and amphibians that use the streams for egg deposition and larval development and move into terrestrial habitats after metamorphosis: they transport aquatic energy subsidies to uplands, some of which return as terrestrial subsidies to the stream during subsequent breeding cycles. There are also unidirectional flows of energy into uplands, such as salmon carrying subsidies of nutrients from aquatic ecosystems to terrestrial ecosystems.

Spatial and Temporal Considerations

Riparian-stream interactions change spatially, from headwaters to downstream, and also over time in fixed locations. Headwater drainages and their streams in Northwest moist forests (plate 12) are becoming recognized for several biophysical contributions to larger downstream networks and diverse watershed conditions, including episodic pulses of instream sediment and down wood from landslides (May and Gresswell 2004; Miller and Burnett 2008; Reeves et al. 2016); invertebrate prey as food for downstream fish (Wipfli and Gregovich 2002); and their unique biotic assemblages (e.g., amphibians, Olson and Weaver 2007). Headwater streams are strongly influenced by the riparian forest, and as the stream widens and shading is reduced downstream, solar energy inputs directly to streams become more of a factor. Stream volume also increases downstream, and species adapted to larger rivers become more prevalent. These changes in interactions among physical, chemical, and biological components of streams from headwaters to downstream river were described in the River Continuum Concept (Vannote et al. 1980).

Recently, the River Continuum Concept has been augmented by several complexities, such as the roles of landscape-scale conditions for aquatic species habitats (Burnett et al. 2007) and occasional disturbances that drive ecological river-network conditions, functions, and processes. For example, May and Gresswell (2004) found a relationship between drainage basin area and the chance of a landslide occurring in the basin with a storm event; they also found an association between this landslide occurrence rate, valley width of the mainstem river, and instream sediment-storage patterns. In a follow-up study, May et al. (2013) found associations between historical landslides, drainage area, stream valley width, and productive fish habitat. Hence, adding a temporal component of landslide probability at a watershed scale has improved our understanding of variable instream abiotic and biotic conditions. Landslide-potential models can inform land-use planning decisions about riparian management in debris-flow-prone areas at both stream-reach and watershed scales, especially relative to sediment delivery and down-wood recruitment, factors that contribute to instream organism habitat suitability (e.g., "NetMap" models of slope stability, Benda et al. 2015a; Reeves et al. 2016; Terrain Works 2016).

Aquatic-Riparian Management

The key ecosystem services to protect or restore in aquatic-riparian systems include water quality, fisheries, and endemic species (Penaluna et al. 2016). These services depend in turn upon more extensive assemblages of organisms in the aquatic and riparian food webs and upon the ecological processes contributing to cold-water and species habitat conditions. We have described a multistate aquatic and riparian system. However, although knowledge of historical variability of conditions and processes can provide a template for forest management (e.g., from fire history, Cissel et al. 1998), we are still developing our understanding of diverse aquatic-riparian ecosystem states that may have historically existed across most forested watersheds. Furthermore, today's aquatic-riparian conditions may show signatures of past anthropogenic activities, obscuring natural variation. Significant challenges to managing these resources include uncertainties around how to integrate a disturbance-ecology framework into riparian forest management, and how to retain system complexity (Sibley et al. 2012). Additionally, there are questions regarding how to balance short- versus long-term effects of actions and how to consider trade-offs between protective riparian management to aid some attributes while accelerating active management for restoration of others.

Some current stream-riparian standards and guidelines may be inconsistent with multistate management because fixed widths are prescribed for streamside buffers for particular stream-reach types (e.g., fish-bearing or non-fish-bearing streams). Such fixed-width buffers may be considered a "controlling paradigm" of natural resource management and could result in a reduction in the range of natural variation and lost ecosystem resilience (Holling and Meffe 2002). Management for diverse aquatic-riparian conditions is a contemporary pivot point in Northwest moist coniferous forests. With recognition of the importance of a variety of naturally occurring aquatic-riparian conditions, a broader array of approaches could help to address site-specific conditions to sustain habitat heterogeneity. Such mixed-approach management prescriptions can promote multistate conditions, enhancing aquatic-riparian ecological resilience and forestalling potential losses of system components (fig 14.1).

What are the key components for designing management approaches for aquatic-riparian systems? Penaluna et al. (2016) described a management process of conceptually reverse-engineering key services to the ecological conditions that will provide them. For biodiversity-related services, many aquatic-dependent species of Northwest moist forests are cold adapted, so attention has focused on key factors that influence riparian and water temperatures, such as overstory shading, groundwater sources, geomorphic conditions, and other factors contributing to low water flows and high air temperatures. Also, many decades of timber harvest resulted in sources of disturbance such as splash dams (chap. 6; box 6.2) that scoured stream channels to bedrock, and road networks with stream-crossing culverts prone to

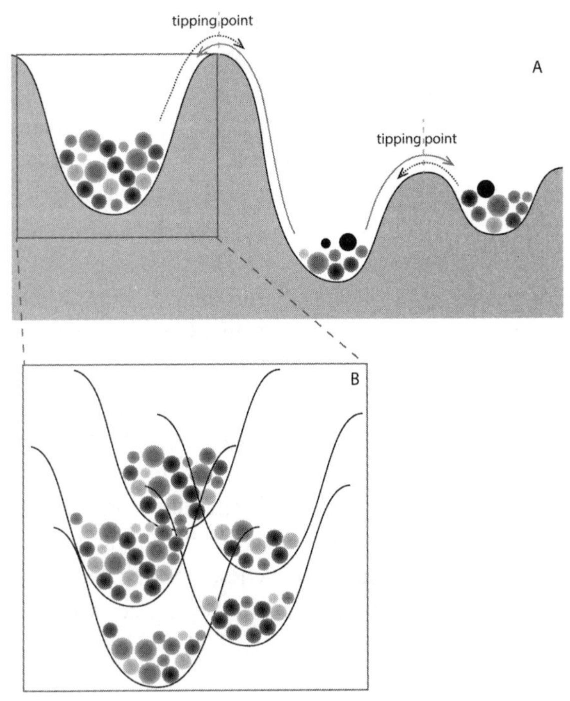

FIGURE 14.1. Schematic representation of the multistate and multidomain concept of aquatic-riparian ecosystems in forests (modified from Penaluna et al. 2016). (A) Cups represent aquatic-riparian ecosystem domains, with multiple states possible per domain. (B) Balls in cups represent the variety of conditions (species, habitats) over space and time, defining the multistate capacity of domains. Arrows represent disturbances or restoration efforts that may move the ecosystem among states and domains. For aquatic-riparian systems, a growing body of literature supports their potential lack of resilience (ability to retain a set of conditions and states over space and time) to cumulative contemporary disturbances (Penaluna et al. 2016), potentially leading to domain shifts.

erosion of fine sediments. The importance of habitat attributes provided by large down wood and variable stream-substrate conditions is increasingly recognized. Providing for large-wood recruitment (e.g., Meleason et al. 2003; Benda et al. 2015b) and managing landslide-prone areas for delivery of both wood and substrates to streams (e.g., Miller and Burnett 2008) are key considerations for forest management (Reeves et al. 2016).

Habitat for salmonid fishes is of special concern owing to their sensitive status and direct ties to commercial fisheries, recreation, native tribes, and other societal benefits. Habitat suitable for Pacific salmonids depends on attributes of the entire riverine network, from the estuary upstream through fish-bearing reaches to the small streams of the forested headwaters that deliver food, sediment, and wood to downstream fish habitats. The components of particular concern in management of Northwest forested aquatic systems include (1) the varying scales or types of aquatic habitat that are implied by fish presence or stream size (representative of diverse ecological conditions; balls, fig. 14.1); (2) critical habitats for aquatic-riparian species of concern; (3) the main habitat attributes linked to aquatic species and affected by disturbances (temperature, large down wood, sediments); and (4) probability of disturbances such as fire, debris flows, and climate change effects. In addition to fish, habitats and dispersal corridors critical for low-mobility riparian species are a concern (see tables 1 and 2 in Olson and Burnett 2013). Restoration of minority tree species in riparian forests that were replaced by plantings of Douglas-fir after harvest can support substantial biodiversity. As these components are discovered and described for stream reaches, watersheds, and landscapes, management priorities for sustaining or improving aquatic-riparian values could be fine-tuned, with an underlying objective of creating heterogeneous conditions.

Today, delineation of streamside buffer zones is a primary management tool for maintaining salmonid habitat and stream-water quality (Richardson et al. 2012). Riparian vegetative buffers contribute to a number of management objectives, including stabilization of stream banks and upland slopes to control erosion and sedimentation; nutrient and chemical filtration and cycling; litter and wood inputs; and provision of shade, insulation, and other habitat features that support biodiversity composition and function. Important questions related to riparian buffers are How wide do buffers need to be to provide these functions effectively? Can effective widths vary over time and space? What types of management actions in riparian areas, including inside buffers, can retain or enhance ecological conditions and functions? Are there actions that pose short-term risks but long-term gains, or vice versa? Although initial science syntheses described distances from streams from which various habitat attributes derived (e.g., FEMAT 1993), a variety of buffer widths are currently implemented across the region (Olson et al. 2007), and lessons learned from several subsequent stream-reach to watershed-scale studies in western Oregon and Washington have addressed these interrelated questions.

Important contributions have come from retrospective studies, several of which have reported negative effects of historical timber-harvest practices (clearcutting) on amphibians. For example: (1) 11 of 20 cut-over streams had 1 or none of 4 species present, as compared with 2 of 23 uncut streams having less than 3 of 4 species (Corn and Bury 1989); (2) amphibian abundances were highest in old-forest sites and lowest in young sites with no buffers, and Van Dyke's salamanders were not found in any logged site without a buffer (Raphael et al. 2002); (3) coastal tailed frogs were associated with stands >105 years old, and coastal giant and torrent salamanders and tailed frogs were positively associated with buffers (Stoddard and Hayes 2005); and (4) tailed frog and torrent salamander densities were 2 to 7 times lower in managed forest compared with old-forest stands and were less abundant in unbuffered clearcut streams than in buffered clearcut streams (Pollett et al. 2010).

Before-after control-treatment management experiments provide powerful evidence of harvest effects because reference treatments and before-treatment data reduce the confounding effects of spatial or temporal variability. Some important timber-management experiments in Northwest moist coniferous forests include the Riparian Ecosystem Management Study of western Washington (Bisson et al. 2013; Raphael and Wilk 2013); the Hinkle Creek Study (e.g., Leuthold et al. 2011; Som et al. 2012; Kibler et al. 2013); the RipStream Project (Groom et al. 2011); the Density Management and Riparian Buffer Study of western Oregon (box 14.2); and two ongoing efforts—the Type N Study in western Washington and the Trask Watershed Study in western Oregon. However, emerging results vary across these studies. At first glance, our hypothesis of aquatic-riparian ecosystems being multistate could be at play here, as the ecological conditions of the sites vary, the historical and recent anthropogenic disturbance contexts are different, and the interactions between site and disturbance possibly vary as well. Certainly these studies are context dependent, spanning different geographies, stream sizes, and stream communities, as well as differing silvicultural treatments, from upland clearcuts to selective harvests and multiple-thinning entries.

Riparian-management prescriptions to preserve or restore key components of stream-riparian systems are influenced by buffer widths, silviculture treatments, and time since treatment. Several studies in the Northwest point to the effectiveness of no-harvest buffers wider than 6 m (20 ft) to preserve water temperatures with clearcutting, and of wider buffers (e.g., 15 m [~50 ft] minimum) to conserve the apparently more-sensitive salamander species with thinning (Anderson et al. 2007; Groom et al. 2011; Leuthold et al. 2011; Olson and Burton 2014). However, results may not be easily transferrable to other places, and fixed-width buffers are not a

BOX 14.2. EXAMPLE EXPERIMENT: STREAM BUFFERS WITH UPLAND THINNING

The Density Management and Riparian Buffer Study of western Oregon examines the effects of upland thinning with different riparian buffer widths on small-stream aquatic-riparian conditions. Pretreatment, 15 aquatic vertebrate species were found in or near streams (Olson and Weaver 2007), and a microclimate *stream effect* was described—a streamside zone ~15 m (50 ft) wide where cool, moist conditions attenuate from streams toward uplands (Anderson et al. 2007; Olson et al. 2007).

Because thinning aimed to accelerate old-growth forest conditions, the first thinning reduced dense overstories to 200 trees per hectare (tph) (80 trees per acre, tpa), with skips and gaps for heterogeneity. Four stream buffer types were examined: (1) the one and two site-potential tree-height buffers of the federal Northwest Forest Plan; (2) a variable-width (~15 m [50 ft] minimum) buffer that varied with site conditions (unique trees, steep slopes, seeps); and (4) a streamside-retention (~6 m, 20 ft) buffer designed to retain stream-bank stability. No-harvest units served as controls. The study was initially implemented at 13 sites from Mount Hood to Coos Bay, Oregon. Today, 8 sites have been followed through a second thinning (~12 years after the first thinning), reducing the largest overstory trees to 85 tph (30 tpa). In this region, the density of the largest trees in late-successional and old-growth forests can be less than 85 tph.

The value of variable-width buffers has been demonstrated for native species (Olson and Burton 2014; Olson et al. 2014). Along 6-m (20-ft) buffer streams, fewer Dunn's salamanders and torrent salamanders were found (Dunn's, 10 years after the first thinning and 1 year after the second thinning; torrent, 1 year after the second thinning). Both taxa had increased counts in streams with the variable-width buffer and the one site-potential tree-height buffer after the second thinning. Woodland salamanders (family Plethodontidae) also were most often found within ~15 m (50 ft) of streams. Instream habitats were generally retained with buffers, and over 80% of instream wood that could be traced to source trees came from within 15 m of streams. Also, growth of riparian trees exhibited an edge effect: greater overstory tree-growth rates extended 15 m into the no-harvest buffer from the adjacent upland thinning. Lastly, thinning had negligible effects on water temperature with variable-width or wider buffers through two harvests, although slightly increased daily maximum water temperatures were found where 0.4-ha (1-ac) patch openings were located directly adjacent to variable-width buffers (Anderson et al. 2007).

panacea, even if prescribed by fish presence or stream size. Heterogeneity of conditions for maintaining ecological resilience could be supported by refining buffer prescriptions and within-buffer management on a case-by-case basis, for an integrated strategy of mixed-width buffers across a watershed or landscape.

Prospectively, understanding of how future disturbances such as climate change will interact with this complex system is incomplete. Building "climate-smart" forests to moderate temperature extremes may translate to retention of more trees and shading, especially along stream channels that have no topographic hill shading (chap. 16; application of thermal loading model, Reeves et al. 2016). Although management of buffer widths to protect salmon and other ecosystem services appears to be moving toward policies that allow tailoring buffer widths and treatments to the site, there is still much learning to be done to adapt buffers to extant and future conditions. Nevertheless, decades of learning are now paying off, with new knowledge providing the basis for refined riparian-buffer mitigations. For example, additional riparian protections are being applied in some circumstances (e.g., to address thermal loading, Oregon State Forest Practices rules). Also, active riparian forest restoration is under consideration to recruit large down wood (e.g., thinning to accelerate riparian tree growth in secondary forest, tree-tipping); establish minority tree species in old plantations; retain little-known species; and manage fuels in fire-prone areas. Scientific support is accruing for forest-management decisions that go beyond protective approaches toward creating complex, heterogeneous riparian forests by considering (1) the probability of unplanned disturbances and successional changes creating dynamic conditions and processes across watersheds over time; (2) resilience to disturbances, such as by both buffering temperature extremes and providing patch openings for productivity; (3) promoting development of future large wood for instream and riparian habitats; and (4) managing for aquatic-terrestrial connectivity. As this century progresses, we see prescriptions for aquatic-riparian protection transitioning to a toolbox of context-driven approaches for both protection and restoration.

Literature Cited

Anderson, P. D., D. J. Larson, and S. S. Chan. 2007. Riparian buffer and density management influences on microclimate of young headwater forests of western Oregon. *Forest Science* 53:254–269.

Armstrong, J. B., and D. E. Schindler. 2013. Going with the flow: Spatial distributions of juvenile coho salmon track an annually shifting mosaic of water temperature. *Ecosystems* 16:1429–1441.

Baxter, C. V., K. D. Fausch, and C. Saunders. 2005. Tangled webs: Reciprocal flows of invertebrate prey link streams and riparian zones. *Freshwater Biology* 50:201–220.

Beechie, T. J., M. Liermann, M. M. Pollock, S. Baker, and J. Davies. 2006. Channel pattern and river-floodplain dynamics in forested mountain river systems. *Geomorphology* 78:124–141.

Benda, L., N. L. Poff, D. Miller, T. Dunne, G. Reeves, G. Pess, and M. Pollock. 2004. The network dynamics hypothesis: How channel networks structure riverine habitats. *BioScience* 54:413–427.

Benda, L. E., S. E. Litschert, G. Reeves, and R. Pabst. 2015b. Thinning and instream wood recruitment in riparian second growth forests in coastal Oregon and the use of buffers and tree tipping as mitigation. *Journal of Forestry Research* 27(4):821. doi:10.1007/s11676-015-0173-2.

Benda, L., D. Miller, J. Barquin, R. McCleary, T Cai, and Y. Ji. 2015a. Building virtual watersheds: A global opportunity to strengthen resource management and conservation. *Journal of Environmental Management* 57:722–739.

Bisson, P. A., S. M. Claeson, S. M. Wondzell, A. D. Foster, and A. Steel. 2013. Evaluating headwater stream buffers: Lessons learned from watershed-scale experiments in southwest Washington. Pp. 169–188 in *Density management in the 21st century: West side story*. Edited by P. D. Anderson and K. L. Ronnenberg. General Technical Report PNW-GTR-880. Portland, OR: USDA Forest Service, Pacific Northwest Research Station.

Burnett, K. M., G. H. Reeves, D. J. Miller, S. Clarke, K. Vance-Borland, and K. Christiansen. 2007. Distribution of salmon-habitat potential relative to landscape characteristics and implications for conservation. *Ecological Applications* 17: 66–80.

Cissel, J. H., F. J. Swanson, G. E. Grant, and 11 coauthors. 1998. *A landscape plan based on historic fire regimes for a managed forest ecosystem: The Augusta Creek study*. General Technical Report PNW-GTR-422. Portland, OR: USDA Forest Service, Pacific Northwest Research Station.

Corn, P. S., and R. B. Bury. 1989. Logging in western Oregon: Responses of headwater habitats and stream amphibians. *Forest Ecology and Management* 29: 39–57.

D'Amore, D. V., R. T. Edwards, P. A. Herendeen, E. Hood, and J. B. Fellman. 2015. Dissolved organic carbon fluxes from hydropedologic units in Alaskan coastal temperate rainforest watersheds. *Soil Science Society of America Journal* 79:378–388.

FEMAT (Forest Ecosystem Management Assessment Team). 1993. *Forest ecosystem management: An ecological, economic, and social assessment. Report of the Forest Ecosystem Management Assessment Team*. Portland, OR: US Department of Agriculture; US Department of the Interior [and others].

Gomi, T., R. C. Sidle, and J. S. Richardson. 2002. Understanding processes and downstream linkages of headwater systems. *BioScience* 52:905–916.

Groom, J. D., L. Dent, L. J. Madsen, and J. Fleuret. 2011. Response of western Oregon (USA) stream temperatures to contemporary forest management. *Forest Ecology and Management* 262:1618–1629.

Holling, C. S., and G. K. Meffe. 2002. Command and control and the pathology of natural resource management. *Conservation Biology* 10:328–337.

Isaak, D. J., C. H. Luce, B. E. Rieman, D. E. Nagel, E. E. Peterson, D. L. Horan, S. Parkes, and G. L. Chandler. 2010. Effects of climate change and wildfire on stream temperatures and salmonid thermal habitat in a mountain river network. *Ecological Applications* 20:1350–1371.

Isaak, D., M. Young, D. Nagel, D. Horan, and M. Groce. 2015. The cold-water climate shield: Delineating refugia for preserving salmonid fishes through the 21st century. *Global Change Biology* 21:2540–2553.

Kibler, K. M., A. Skaugset, L. M. Ganio, and M. M. Huso. 2013. Effect of contemporary forest harvesting practices on headwater stream temperatures: Initial response of the Hinkle Creek catchment, Pacific Northwest, USA. *Forest Ecology and Management* 310:680–691.

Lake, P. S. 2000. Disturbance, patchiness, and diversity in streams. *Journal of the North American Benthological Society* 19:573–592.

Leuthold, N., M. J. Adams, and J. P. Hayes. 2011. Short-term response of *Dicamptodon tenebrosus* larvae to timber management in southwestern Oregon. *Journal of Wildlife Management* 76:28–37.

May, C. L., and R. E. Gresswell. 2004. Spatial and temporal patterns of debris flow deposition in the Oregon Coast Range, USA. *Geomorphology* 57: 135–149.

May, C., J. Roering, L. S. Eaton, and K. M. Burnett. 2013. Controls on valley width in mountainous landscapes: The role of landsliding and implications for salmonid habitat. *Geology* 4:503–506.

Meleason, M. A., S. V. Gregory, and J. P. Bolte. 2003. Implications of riparian management strategies on wood in streams of the Pacific Northwest. *Ecological Applications* 13:1212–1221.

Miller, D. J., and K. M. Burnett. 2008. A probabilistic model of debris-flow delivery to stream channels, demonstrated for the Coast Range of Oregon, USA. *Geomorphology* 94:184–205.

Naiman, R. J., J. S. Bechtold, T. J. Beechie, J. J. Latterell, and R. Van Pelt. 2010. A process-based view of floodplain forest patterns in coastal river valleys of the Pacific Northwest. *Ecoystems* 13:1–31.

Olson, D. H., P. D. Anderson, C. A. Frissell, H. H. Welsh Jr., and D. F. Bradford. 2007. Biodiversity management approaches for stream riparian areas: Perspectives for Pacific Northwest headwater forests, microclimate and amphibians. *Forest Ecology and Management* 246:81–107.

Olson, D. H., and K. M. Burnett. 2013. Geometry of forest landscape connectivity: Pathways for persistence. Pp. 220–238 in *Density management for the*

21st century: West side story. Edited by P. D. Anderson and K. L. Ronnenberg. General Technical Report PNW-880. Portland, OR: USDA Forest Service, Pacific Northwest Research Station.

Olson, D. H., and J. I. Burton. 2014. Near-term effects of repeated thinning with riparian buffers on headwater stream vertebrates and habitats in Oregon, USA. *Forests* 5:2703–2729.

Olson, D. H., J. B. Leirness, P. G. Cunningham, and E. A. Steel. 2014. Riparian buffers and forest thinning: Effects on headwater vertebrates 10 years after thinning. *Forest Ecology and Management* 321:81–93.

Olson, D. H., and G. Weaver. 2007. Vertebrate assemblages associated with headwater hydrology in western Oregon managed forests. *Forest Science* 53:343–355.

Pabst, R. J., and T. A. Spies. 1998. Distribution of herbs and shrubs in relation to landform and canopy cover in riparian forests of coastal Oregon. *Canadian Journal of Botany* 76:298–315.

Penaluna, B. E., D. H. Olson, R. L. Flitcroft, M. Weber, J. R. Bellmore, S. M. Wondzell, J. B. Dunham, S. L. Johnson, and G. H. Reeves. 2016. Aquatic biodiversity in forests: A weak link in ecological and ecosystem service resilience. *Biodiversity and Conservation*. doi:10.1007/s10531-016-1148-0

Pollett, K. L., J. G. MacCracken, and J. A. MacMahon. 2010. Stream buffers ameliorate the effects of timber harvest on amphibians in the Cascade Range of southern Washington, USA. *Forest Ecology and Management* 260:1083–1087.

Pringle, C. M., R. J. Naiman, G. Bretschko, J. R. Karr, M. W. Oswood, J. R. Webster, R. L. Welcomme, and M. J. Winterbourn. 1988. Patch dynamics in lotic systems: The stream as a mosaic. *Journal of the North American Benthological Society* 7:503–524.

Raphael, M. G., P. A. Bisson, L. L. C. Jones, and A. D. Foster. 2002. Effects of streamside forest management on the composition and abundance of stream and riparian fauna of the Olympic Peninsula. Pp. 27–40 in *Congruent management of multiple resources: Proceedings from the wood compatibility initiative workshop*. Edited by A. C. Johnson, R. W. Haynes, and R. A. Monserud. General Technical Report PNW-GTR-563. Portland, OR: USDA Forest Service, Pacific Northwest Research Station.

Raphael, M. G., and R. J. Wilk. 2013. The riparian ecosystem management study: Response of small mammals to streamside buffers in western Washington. Pp. 163–168 in *Density management in the 21st century: West side story*. Edited by P. D. Anderson and K. L. Ronnenberg. General Technical Report PNW-GTR-880. Portland, OR: USDA Forest Service, Pacific Northwest Research Station.

Reeves, G. H., B. R. Pickard, and K. N. Johnson. 2016. *An initial evaluation of potential options for managing riparian reserves of the Aquatic Conservation Strategy of the Northwest Forest Plan*. General Technical Report PNW-

GTR-937. Portland, OR: USDA Forest Service, Pacific Northwest Research Station.

Resh, V. H., A. V. Brown, A. P. Covich, M. E. Gurtz, H. W. Li, G. W. Minshall, S. R. Reice, A. L. Sheldon, J. B. Wallace, and R. C. Wissmar. 1988. The role of disturbance in stream ecology. *Journal of the North American Benthological Society* 7:433-455.

Richardson, J. S., R. J. Naiman, and P. A. Bisson. 2012. How did fixed-width buffers become standard practice for protecting freshwaters and their riparian areas from forest harvest practices? *Freshwater Science* 31:232–238.

Rieman, B. E., C. L. Smith, R. J. Naiman, and 12 coauthors. 2015. A comprehensive approach for habitat restoration in the Columbia Basin. *Fisheries* 40: 124–135.

Sibley, P. K., D. P. Kreutzweiser, B. J. Naylor, J. S. Richardson, and A. M. Gordon. 2012. Emulation of natural disturbance (END) for riparian forest management: Synthesis and recommendations. *Freshwater Science* 31:258–264.

Som, N. A., N. P. Zégre, L. M. Ganio, and A. E. Skaugset. 2012. Corrected prediction intervals for change detection in paired watershed studies. *Hydrological Sciences Journal* 57:134–143.

Stanford, J. A., M. S. Lorang, and F. R. Hauer. 2005. The shifting habitat mosaic of river ecosystems. *Verhandlungen der Internationalen Vereinigung für Theoretische und Angewandte Limnologie* (*Proceedings of the International Association of Theoretical and Applied Limnology*) 29:123–136.

Stanley, E. H., S. M. Powers, and N. R. Lottig. 2010. The evolving legacy of disturbance in stream ecology: Concepts, contributions, and coming challenges. *North American Benthological Society* 29:67–83.

Stoddard, M. A., and J. P. Hayes. 2005. The influence of forest management on headwater stream amphibians at multiple spatial scales. *Ecological Applications* 15:811–823.

Terrain Works. 2016. Terrain Works: Discover and map the hidden essentials of the Earth for improved management, restoration & conservation. http://www.terrainworks.com/.

Torgersen, C. E., D. M. Price, H. W. Li, and B. A. McIntosh. 1999. Multiscale thermal refugia and stream habitat associations of Chinook salmon in northeastern Oregon. *Ecological Applications* 9:301–319.

USDA (US Department of Agriculture, Forest Service). 2016. NorWeST Stream Temp: Regional database and modeled stream temperatures. http://www.fs.fed.us/rm/boise/AWAE/projects/NorWeST.html.

Vannote, R. L., G. W. Minshall, K. W. Cummins, J. R. Sedell, and C. E. Cushing. 1980. The river continuum concept. *Canadian Journal of Fisheries and Aquatic Sciences* 37:130–137.

Webster, J. R., M. E. Gurtz, J. J. Haines, J. L. Meyer, W. T. Swank, J. B. Waide, and J. B. Wallace. 1983. Stability of stream ecosystems. Pp. 355–395 in *Stream ecology: Application and testing of general ecological theory*. Edited by J. R. Barnes and G. W. Minshall. New York: Plenum Press.

Wipfli, M. S., and D. P. Gregovich. 2002. Export of invertebrates and detritus from fishless headwater streams in Southeast Alaska: Implications for downstream salmonid production. *Freshwater Ecology* 47:957–969.

Wondzell, S. M. 2011. The role of the hyporheic zone across stream networks. *Hydrological Processes* 25:3525–3532.

Wondzell, S. M., J. LaNier, R. Haggerty, R. D. Woodsmith, and R. T. Edwards. 2009. Changes in hyporheic exchange flow following experimental wood removal in a small low-gradient stream. *Water Resources Research* 45:W05406.

Chapter 15

Watersheds and Landscapes

Gordon H. Reeves and Thomas A. Spies

Our understanding of what constitutes the freshwater ecosystems, watersheds, and landscapes in moist coniferous forests is continually evolving. To date, much of the aquatic-system focus has been on relatively small spatial scales, such as stream habitat units and reaches (chap. 14). However, a variety of entities, including interested publics, interest groups, scientific review and evaluation teams, regulatory agencies, and policy and decision makers, are calling for development of policies and practices to manage aquatic and riparian resources at the watershed and landscape levels. In the disciplines of watershed and landscape ecology, significant conceptual advances have occurred since the 1980s. In this chapter, we provide a conceptual background for watershed- and landscape-scale management approaches and go on to address contemporary challenges, including the integration of landscape-ecology and disturbance-ecology frameworks in aquatic ecosystem management along with selected examples underscoring the contemporary integration of dynamic ecosystems with forest management.

Basic Concepts

Identifying the units of interest is a first step in developing management plans and actions. In the United States, a hierarchical system of hydrologic units (hydrologic unit code, HUC) is used to distinguish aquatic systems by area. For management purposes in Northwest forested landscapes, *watersheds* can be defined as 5th- to 7th-level HUCs (~5000 to 100 000 ha [~12,000 to 250,000 ac]) (e.g., RIEC and IAC 1995; Watershed Professionals Network 1999); hence these watersheds often coincide with our management designation of aquatic ecosystems. Ecosystems are vague entities with boundaries that may shift in space and time (Caraher et al. 1999). Disturbance is an integral part of any ecosystem and is required to maintain its long-term productivity (Folke et al. 2004). Therefore, ecosystems are not in a steady state but change through time and space (chap. 14; fig. 14.1) and are best managed from the perspective of resilience rather than stability (Folke et al. 2004). The range of conditions an ecosystem experiences is determined to a large extent by the frequency and magnitude of disturbances it encounters (e.g., wildfire, hurricane, timber harvest and associated activities).

A *landscape* is a heterogeneous land area composed of a cluster of interacting ecosystems with varying spatial scale, often defined with regard to human interests. Management of landscapes strives to maintain a variety of ecological states (within and among watersheds) in some desired spatial and temporal distribution by addressing the dynamics of individual ecosystems and the external factors that influence them, as well as the composite dynamics of the aggregate ecosystems (Lindenmayer et al. 2006). To achieve this complex goal, managers aim to develop a variety of conditions or states in individual ecosystems at any time (e.g., fig. 14.1), reflecting a highly variable mosaic across the larger landscape. The specific features of the ecological states and their temporal and spatial distribution will vary with the features of the landscape and the management objectives for it. These concepts are common to both the terrestrial and aquatic portions of forested landscapes.

The discipline of landscape management and ecology has matured with technological advances and mapping tools, and geographic information system (GIS) layers are now used to assess the potential implications of proposed landscape-management activities and plans (Benda et al. 2007). Several tools specific to aquatic-system conditions and processes in managed forests are now available, as discussed further below. However, Lindenmayer et al. (2006) cautioned that there are important, but seldom-recognized, limitations to this approach for forest management; one of the primary ones is the inability to assess the flow of materials and organisms and interactions among components of the landscape. They suggest that managers develop a conceptual model of the landscape of interest to guide assessment and implementation of proposed actions. Integration of the terrestrial and aquatic portions of forested landscapes is needed in such a model.

Behavior and Management in Time and Space

Understanding the broad-scale behavior of aquatic ecosystems and the landscapes in which they are embedded, especially over extended time-periods, is critical for achieving desired landscape-scale goals and objectives. Yet natural resource specialists, managers, regulators, and stakeholders may struggle with watershed (aquatic ecosystem) management at larger scales because they are not accustomed to thinking of aquatic systems at large scales or lack an adequate understanding or the tools to do so. Expectations may be formed from information developed at small spatial scales, and it may be assumed that these standards or expectations can be applied across broader areas merely by "aggregating up." This premise is not valid, however, if each spatial scale is a different level of ecological organization with its own set of behaviors and characteristics, and relevant principles governing behavior also change (Lindenmayer et al. 2006).

Failure to recognize the different levels of organization and the potential responses of each to natural disturbance or management may lead to serious problems with expectations from policies and regulations (Wiens 2002) and may incur unintended ecological consequences (Wallington et al. 2005; Mori 2011) and economic and social costs (Dale et al. 2000), such as repeated investment in the same unlikely restoration outcome. Hence a lesson emerging from the discipline of disturbance ecology is that an ecosystem demonstrates *resilience* (the ability to recover to pre-disturbance conditions) when environmental changes caused by a disturbance are within the range of conditions that the system experienced before the disturbance (Folke et al. 2004). Reduced resilience results in a decrease in the diversity of conditions of a particular ecological state, the loss of a particular ecological state, or both (Folke et al. 2004; Penaluna et al. 2016). Potential consequences of reduced resilience may include extirpation of some species, increases in species favored by available habitats, an invasion of exotic species, or a loss of ecosystem services. The less management actions resemble the disturbance regime under which an ecosystem evolved, the less resilient an ecosystem will be (Liss et al. 2006).

The obvious challenge for ecosystem management is to make management actions resemble the natural disturbance processes and regime as closely as possible. Yount and Niemi (1990) modified the definition of

Bender et al. (1984) and referred to a disturbance regime that maintains the resilience of an ecosystem as a *pulse disturbance*. A pulse disturbance occurs infrequently, and there is sufficient time between disturbances to enable the ecosystem to recover to pre-disturbance conditions. A pulse disturbance allows an ecosystem to remain within its normal bounds to exhibit the same range of states and conditions that it does naturally. A *press disturbance*, however, reduces the resilience of an ecosystem. It is a frequent or continuous impact that does not allow time for recovery to pre-disturbance conditions.

A common objective of landscape management is to maintain a variety of ecological states in some desired spatial and temporal distribution. To accomplish this, managers need to consider (1) the development of a variety of conditions or states in individual ecosystems on the landscape at any point in time; (2) the pattern resulting from the range of ecological conditions that are present; and (3) the dynamics of individual ecosystems, external factors, and the aggregate of ecosystems. Obviously, understanding the dynamics of an individual ecosystem is demanding; understanding the dynamics of the aggregate of ecosystems is much more challenging. Yet maintaining an array of ecological conditions across landscapes is critical for maintaining life-history and genetic diversity of populations (Bisson et al. 2009) so they can be resilient to natural and anthropogenic disturbances and climate change. Applying these concepts to aquatic ecosystems and watershed management is a new frontier (e.g., Rieman et al. 2015).

Management Challenges

Policy and public understanding are beginning to accept the dynamic nature of ecosystems that exhibit a range of conditions (e.g., seral stages), and as a result, many management approaches and regulations now aim to maintain this dynamism (Lindenmayer and Franklin 2002). Ideally, to maintain ecosystem resilience, the management disturbance regime should be more pulse-like and less press-like. In developing ecosystem-management plans and policies, the effect of management depends in part on how closely the management disturbance regime resembles the natural disturbance regime in frequency, magnitude, and legacy (Reeves et al. 1995).

A similar recognition of the dynamics of aquatic ecosystems and landscapes in the scientific community is not yet integrated in setting management objectives and policies. As a result, current management of aquatic ecosystems is based primarily on steady-state and spatially homogeneous concepts. These are reflected in management goals such as a single optimum set of habitat conditions (e.g., pool frequency or numbers of logs per km in streams); a narrow range of values characterizing attributes such as erosion rates or turbidity levels; or all watersheds being in "good condition." The same objectives are often applied across large geographic areas, reflecting the assumption of low spatial variability. The focus on single or narrow values comes at the expense of recognizing the ecological processes that create and maintain the diversity of freshwater habitats, such as the heterogeneous conditions occupied by Pacific salmon and the ecological contexts in which they evolved (box 15.1). In recognition of this issue, Holling and Meffe (1996) referred to the use of single threshold values for various environmental parameters as an example of a command-and-control approach to natural resource management. They contend that this approach often fails when it is applied to situations in which processes are complex, nonlinear, and poorly understood, leading to further degradation or compromising of the ecosystems and landscapes of interest (Dale et al. 2000; Rieman et al. 2006, 2015).

Recently, scientists have argued for a conceptual foundation for natural resource management based on dynamics and heterogeneity. In aquatic ecology, the disciplines of disturbance ecology and landscape ecology matured significantly from 1986 to 2008 (Stanley et al. 2010) and are now merging with respect to watershed management. Within a watershed, physical disturbances drive varying stream-channel conditions through time, leading to a range of habitats (Poff et al. 1997; Beechie et al. 2010). This variation has prompted new conceptual frameworks in river ecology that focus on patchy heterogeneity and dynamics (Fausch et al. 2002; Poole 2002; Wiens 2002). Additionally, the linear perspective of rivers has been updated by incorporating geomorphic and ecological principles that pertain to hierarchically branching river networks (Benda et al. 2004), integrating the role of heterogeneous riparian and upslope conditions with the diversity of instream conditions. Consequently, pristine aquatic systems, which are generally used to develop reference conditions for regulatory programs, may actually exhibit a wide range of conditions within and among watersheds (Rieman et al. 2006; Lisle et al. 2007).

To establish a dynamic landscape perspective, the range of natural variability must be characterized at different spatial scales. Smaller spatial scales (site, habitat unit, stream reach) generally have a wider range of variation than do large scales (watershed, landscape, Wimberly et al. 2000).

A remaining challenge is to then determine how the location of each ecological condition moves across the landscape (or watershed) through

BOX 15.1 DYNAMICS—THE PORTFOLIO EFFECT

In a recent study of populations of sockeye salmon across Bristol Bay, Alaska, Schindler et al. (2010) found that production was relatively stable through time (box fig. 15.1.1A). There was, however, tremendous asynchronous variation in the size of individual populations: some were large at one time while others were small (box fig. 15.1.1B), and those that were small increased at another time while larger ones declined. A primary reason for this pattern was the variation in environmental conditions in watersheds across the landscape. The productivity of individual watersheds varied through time, and seldom were they all in the same phase at the same time. Preserving and managing for variable landscapes and the physical processes that maintain habitat and environmental variation will help create conditions that support adaptation and a diversity of phenotypic traits, which in turn are responsible for the long-term persistence of populations of fish and other organisms.

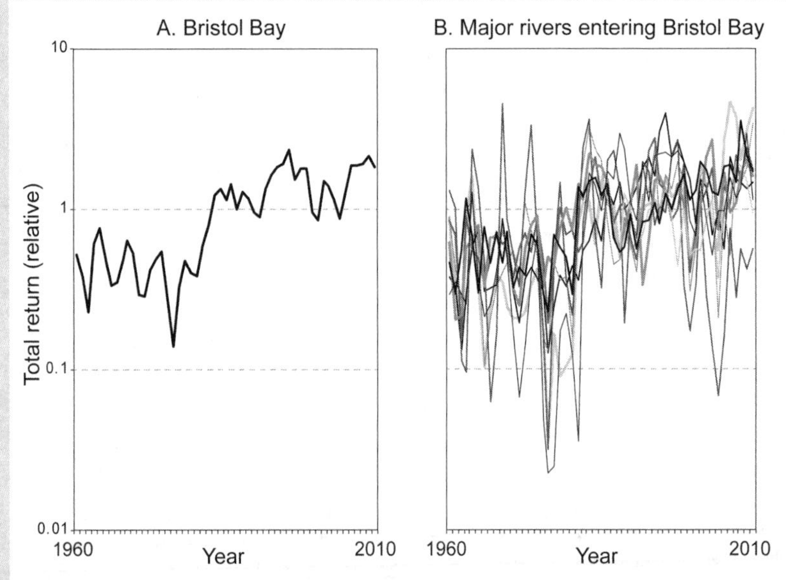

FIGURE 15.1.1. Variation in numbers of sockeye salmon returning to Bristol Bay, Alaska, and to individual streams in Bristol Bay (modified from Schindler et al. 2010).

time as disturbances occur and aquatic systems mature. The transition to ecosystem and landscape management for aquatic systems requires the articulation of principles and a conceptual basis to guide the development of policies and practices. The scientific literature provides few insights, especially in increasingly humanized systems, as scientists have generally failed to integrate time as an essential component of aquatic ecosystems. Major paradigms such as the River Continuum Concept (Vannote et al. 1980) did not consider time or its influence. Similarly, classification schemes such as Rosgen's (1994) identified a single set of conditions for a given stream or reach type; no consideration was given to how these conditions may vary over time. As a result, the dynamics of aquatic ecosystems and landscapes over long time-periods (several decades to centuries) have not been well recognized, and aquatic ecosystems have been expected to remain relatively consistent through time. As a corollary, all watersheds in the landscape are expected to be in "good condition"; any variation from "good" has been considered unacceptable.

To frame effective ecosystem- and landscape-management policies for aquatic systems in Northwest moist forests, it is essential to understand the natural disturbance regime, how it has affected aquatic habitats across scales within and among watersheds, how those scales are organized, and how they have been modified by human activities. To develop effective guidelines for management of aquatic ecosystems, it is critical to acknowledge that periodic disturbance is an integral part of these systems. Natural disturbances episodically deliver materials, sediment, and wood that form habitat over time. As a result, suitability of habitat for endemic species such as salmonids and other aquatic organisms likely varied widely through time (Reeves et al. 1995).

Is Timber Harvest a Good Mimic for the Natural Fire Regime?

Given the above conceptual frameworks of disturbance and landscape ecology, the recovery of degraded aquatic ecosystems in the Pacific Northwest depends on developing a new management system that emulates the natural disturbance regime as closely as possible across spatial scales. One of the necessary steps is to compare the elements of the natural disturbance regime to those in the human disturbance regime, with respect to legacy, frequency, successional states, and size and spatial patterns of disturbance.

How does timber harvest compare to natural disturbance? Timber-harvest strategies vary with ownership and the owner's objectives, and although generalizations will not always be relevant, they can help frame the discussion. The timber-harvest disturbance regime differs from the natural disturbance of stand-replacing wildfires that affected aquatic ecosystems in several respects, particularly with regard to legacy. Wildfires leave large amounts of standing and down wood that is delivered to streams, along with sediments, in landslides and debris torrents, accumulating in fish-bearing reaches in the valley bottoms downstream (plate 13A). The wood delivered via these hillslope processes can form a substantial portion of the total volume of wood found in streams. Once the amount of sediment declines to intermediate levels, high-quality habitats develop for Pacific salmon and other species. Landslides associated with timber harvest, on the other hand, contain primarily sediments because trees along the landslide track have been removed. As a consequence, channels are simpler following timber harvest than after wildfire.

Timber harvest and wildfire also differ in frequency. The interval between disturbances affects the range of conditions and ecological states that can develop in an ecosystem (Hobbs and Huenneke 1992). The extended time interval between natural disturbance events (e.g., fire, ~300 years) allowed a wide range of conditions to develop in aquatic ecosystems in the Oregon Coast Range. Timber harvest occurs at intervals shorter than this, generally 40 to 50 years on private lands and 80 to 100+ years on federal and state lands. The physical habitat conditions necessary to support the variety of aquatic species naturally found in coastal watersheds may not develop in such relatively short periods, especially on private lands.

Extended intervals between disturbances by timber harvest could be part of any ecosystem and landscape-management plan. Based on the limited observations of Reeves et al. (1995), it appears that favorable habitat conditions for native salmonid fishes (a regional ecosystem service of keen interest to society) in central Oregon Coast Range streams begin to appear 80 years or so following a disturbance; this is a rough approximation that requires more supporting research. The longer interval between disturbances may not have to apply to an entire watershed if riparian zones are sufficiently large and include appropriate landslide-prone non-fish-bearing and intermittent streams.

Another difference between natural and current timber-harvest disturbance regimes is the size of the disturbance events and the landscape pattern they create. Timber harvest generally occurs in small individual actions that "disturb" areas of 16 to 49 ha (40 to 120 ac) and are distributed across the landscape (plate 13B). In contrast, wildfires resulted in larger but more concentrated areas of disturbance.

Variation among watersheds in their suitability for fish and other aquatic species is reduced under the timber-harvest regime compared with the wildfire regime. Dispersal of timber-harvest activities over relatively large areas subjects more watersheds to disturbance at any point in time and has degraded streams across the landscape, while the concentration of wildfire in a relatively small proportion of the landscape has resulted in a variation in watershed conditions, ranging from poor to good, at any point in time.

The legacy of timber harvest would need to include more large wood in order to mimic natural conditions and provide for habitat diversity. This entails recognizing sources of wood: the stream-adjacent areas and non-fish-bearing streams. The former are a more chronic source of wood over time, while the latter are more episodic and potentially catastrophic. Headwater streams (plate 12), which may make up 70% to 80% or more of the stream network in many areas (Gomi et al. 2002), can contribute a large proportion of in-channel wood (Reeves et al. 2003). Leaving large trees in riparian zones along selected landslide-prone channels (Miller and Burnett 2008) will result in the delivery of more wood to fish-bearing streams, increasing the potential to develop conditions favorable for salmonids.

Implications for Forest Landscape Management

Policies and practices for landscape management could consider allowing management activities to be concentrated at the watershed (ecosystem) level rather than distributing harvest activities over wider areas (Reeves et al. 1995) (fig. 15.1). The amount of activity allowed in a watershed is often limited by rules and regulations governing cumulative effects, but ecosystem and landscape effects may be fewer if timber removals are concentrated in a smaller area rather than being dispersed.

Watershed reserves could also be considered in the development of ecosystem- and landscape-management policies, but only as short-term components. Reserves, such as the key watersheds in the Northwest Forest Plan (USDA and USDI 1994), are essential to protect watersheds that are currently in good condition. However, in dynamic environments, such reserves simply act as holding islands that persist for relatively brief ecological periods. Therefore, given the dynamic nature of aquatic and forested ecosystems in the Pacific Northwest and elsewhere, any single

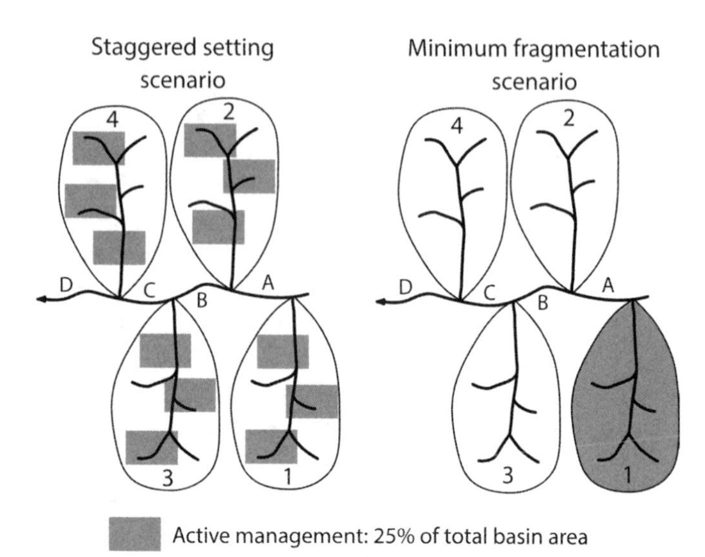

FIGURE 15.1. Examples of patterns resulting from dispersing (staggered setting) and concentrating (minimum fragmentation) land-management activities in a watershed and landscape over time (modified from Grant 1990).

watershed reserve will not and should not be expected to provide high-quality habitat for extended periods. The challenge of ecosystem and landscape management is to manage for the future generations of "reserves" so that as good habitats become less suitable, either through human or natural disturbances or through development of new ecological states (e.g., succession), others become available. This would require adaptive management of reserves.

Many hurdles must be overcome to make viable ecosystem and landscape policies for aquatic systems work. With a disturbance- and landscape-ecology framework for management of forests (box 15.2), biologists, managers, and planners need to consider much longer time frames than they are generally accustomed to, and to acknowledge and understand the dynamics of ecosystems and landscapes in space and time. Legal and regulatory constraints may have to be reconsidered. For example, current cumulative-effects regulations would prevent concentration of activities in a watershed within a certain period. Water-quality standards for temperature or suspended sediment may be temporarily violated in some watersheds following disturbances, but sediment influx is an important component of the natural disturbance process for streams, helping to maintain a range of conditions across ecosystem and landscape scales. Of course, management

BOX 15.2. MANAGEMENT TAILORED TO LOCATIONS AND ECOLOGICAL DYNAMICS

Watershed-management options (Naiman et al. 2012), including many restoration efforts (Kondolf et al. 2003), are constrained by reliance on off-the-shelf and one-size-fits-all concepts and designs, rather than on an understanding of specific features and capabilities of the location of interest. A good example of a watershed-management option that takes a tailored and dynamic approach rather than relying on a one-size-fits-all design is the Cissel et al. (1998, 1999) proposal for the Central Cascades Adaptive Management Area. Developed by many of the scientists involved in creation of the Northwest Forest Plan, the proposal was based on variation in disturbance patterns (in this case, wildfire) in the target watershed and called for harvest of some older trees and revision of the interim riparian reserves. Only limited parts of the proposal were ever implemented, however, and adjustments to riparian reserves were not made, in large part to avoid controversy and protest over the harvest of mature stands.

A primary challenge for instituting a dynamic approach is the ability to identify where critical ecological processes (wood delivery, erosion, debris flows, thermal loading) occur in a watershed or across a landscape: they do not occur everywhere, but at specific sites. New analysis tools, such as NetMap (Benda et al. 2007; Terrain Works 2016), can aid managers in identifying the specific location of a particular process, and in some cases the magnitude of the process. Examples of this are shown in box fig. 15.2.1 for wood recruitment (streamside and upslope derived) and potential for landslides and debris flows to deliver to valley bottoms. Particular attention can be given to areas where these processes are most pronounced or strongest. See Reeves et al. (2016) for more detail.

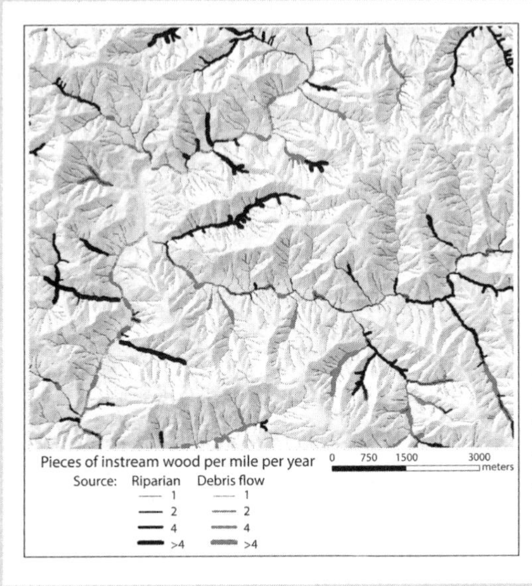

FIGURE 15.2.1. Location of sources of large wood from the streamside-adjacent riparian zone (black) and debris flows (gray).

of sediment loads has been a focus of timber harvest and road building, so maintaining the natural range of variation for processes like sedimentation is key.

There will also be social and economic considerations. Across much of the Pacific Northwest, there are multiple landowners with differing objectives. Coordinating objectives and timing of activities in such a mosaic to achieve ecosystem, landscape, and societal goals will not be easy. Recognizing and acknowledging the ecological criteria for operating at the ecosystem and landscape levels is critical if the future management of aquatic ecosystems and the resources associated with them are to meet changing societal demands and challenges amplified by a changing climate.

Focusing Ecosystem Management and Restoration on All Lands

Management options (Naiman et al. 2012) and many restoration efforts (Kondolf et al. 2003) for aquatic ecosystems are constrained by a reliance on off-the-shelf and one-size-fits-all concepts and designs, rather than on an understanding of specific features and capabilities of the location of interest. The Achilles heel of management and restoration efforts is the focus on *habitat state* rather than on *habitat-forming processes*. Physical habitat features are highly diverse within and among watersheds because, in part, of the variability in where ecological processes occur (see Reeves et al. 2016, for example). Therefore management and restoration efforts that focus on these locations of particular importance to habitat-forming processes are likely to be most effective. At the same time, spatial variability in habitat-forming processes should be anticipated, rather than being seen as conflicting with unrealistic expectations of single, optimum habitat states.

There have been few attempts to design and implement a site-specific approach because available guidelines are vague (Richardson et al. 2012), but recently developed analytical tools can assist in identifying where habitat-forming processes and other important ecological factors occur within a watershed. For instance, the intrinsic potential for coho salmon and steelhead (sea-run *O. mykiss*), based on channel gradient, valley confinement, and flow, can be used to delineate habitat variability (Burnett et al. 2007). Computer-based tools also allow for spatially explicit mapping of other factors known to be important to fish habitat, including large wood, tributary confluences, and valley width (Pess et al. 2002; Benda et al. 2007), as well as risks to habitats posed by upslope erosion and road networks (box fig. 15.2.1). Such tools allow for a much greater recognition of spatial variability in habitat-forming processes and risks posed by features such as erosion-prone landforms and road transportation systems. Additionally, they provide a watershed-level perspective that includes an ecological view of all lands within the area of interest, with potential to develop more comprehensive and effective restoration efforts in conjunction with nonfederal partners.

Literature Cited

Beechie, T. J., D. A. Sear, J. D. Olden, G. R. Pess, J. M. Buffington, H. Moir, P. Roni, and M. M. Pollock. 2010. Process-based principles for restoring river ecosystems. *BioScience* 60:209–222.

Benda, L., D. Miller, K. Andras, P. Bigelow, G. Reeves, and D. Michael. 2007. NetMap: A new tool in support of watershed science and resource management. *Forest Science* 53:206–219.

Benda, L., N. L. Poff, D. Miller, T. Dunne, G. Reeves, G. Pess, and M. Pollock. 2004. The network dynamics hypothesis: How channel networks structure riverine habitats. *BioScience* 54:413–427.

Bender, E. A., T. J. Case, and M. E. Gilpin. 1984. Perturbation experiments in community ecology: Theory and practice. *Ecology* 65:1–13.

Bisson, P. A., J. B. Dunham, and G. H. Reeves. 2009. Freshwater ecosystems and resilience of Pacific salmon: Habitat management based on natural variability. *Ecology and Society* 14:45.

Burnett, K. M., and D. J. Miller. 2007. Streamside policies for headwater channels: An example considering debris flows in the Oregon Coastal Province. *Forest Science* 53:239–253.

Caraher, D. L., A. C. Zack, and A. R. Stage. 1999. Scales and ecosystem analysis. Pp. 343–352 in *Ecological stewardship: A common reference for ecosystem management*. Vol. II. Edited by R. C. Szaro, N. C. Johnson, W. T. Sexton, and A. J. Malk. Oxford, UK: Elsevier Science.

Cissel, J. H., F. J. Swanson, G. E. Grant, and 8 coauthors. 1998. *A landscape plan based on historical fire regimes for a managed forest ecosystem: The Augusta Creek study*. General Technical Report PNW-GTR-422. Portland, OR: USDA Forest Service, Pacific Northwest Research Station.

Cissel, J. H., F. J. Swanson, and P. J. Weisberg. 1999. Landscape management using historical fire regime: Blue River, Oregon. *Ecological Applications* 9:1217–1231.

Dale, V. H., S. Brown, R. A. Haeuber, N. T. Hobbs, N. Huntly, R. J. Naiman, W. E. Riebsame, M. G. Turner, and T. J. Valone. 2000. Ecological principles and guidelines for managing the use of land. *Ecological Applications* 10:639–670.

Fausch, K. D., C. E. Torgersen, C. V. Baxter, and H. W. Li. 2002. Landscapes to riverscapes: Bridging the gap between research and conservation of stream fishes. *BioScience* 52:483–498.

Folke, C., S. Carpenter, B. Walker, M. Scheffer, T. Elmqvist, L. Gunderson, and C. S. Holling. 2004. Regime shifts, resilience, and biodiversity in ecosystem management. *Annual Review of Ecology, Evolution, and Systematics* 35:557–581.

Gomi, T., R. C. Sidle, and J. S. Richardson. 2002. Understanding processes and downstream linkages of headwater systems. *BioScience* 52:905–916.

Grant, G. E. 1990. Hydrologic, geomorphic, and aquatic habitat implications of old and new forestry. Pp. 35–53 in *Forests—managed and wild: Differences and consequences*. Edited by A. F. Pearson and D. A. Challenger. Vancouver, BC: University of British Columbia, Vancouver.

Hobbs, R. J., and L. F. Huenneke. 1992. Disturbance, diversity, and invasion: Implications for conservation. *Conservation Biology* 6:324–337.

Holling, C. S., and G. K. Meffe. 1996. Command and control and the pathology of natural resource management. *Conservation Biology* 10:328–337.

Kondolf, G. M., D. R. Montgomery, H. Piégay, and L. Schmitt. 2003. Geomorphic classification of rivers and streams. Pp. 171–204 in *Tools in fluvial geomorphology*. Edited by G. M. Kondolf and H. Piégay. New York: John Wiley & Sons.

Lindenmayer, D. B., and J. F. Franklin. 2002. *Conserving forest biodiversity: A comprehensive multiscaled approach*. Washington, DC: Island Press.

Lindenmayer, D. B., J. F. Franklin, and J. Fischer. 2006. General management principles and a checklist of strategies to guide forest biodiversity conservation. *Biological Conservation* 131:433–445.

Lisle, T. E., K. Cummins, and M. Madej. 2007. An examination of references for ecosystems in a watershed context: Results of a scientific pulse in Redwood National and State Parks, California. Pp. 118–130 in *Advancing the fundamental sciences: Proceedings of the Forest Service National Earth Sciences Conference*. Edited by M. J. Furniss, C. F. Clifton, and K. L. Ronnenberg. General Technical Report PNW-GTR-689. Portland, OR: USDA Forest Service, Pacific Northwest Research Station.

Liss, W. J., J. A. Stanford, J. A. Lichatowich, R. N. Williams, C. C. Coutant, P. R. Mundy, and R. R. Whitney. 2006. Developing a new conceptual foundation for salmon conservation. Pp. 51–98 in *Return to the river: Restoring salmon to the Columbia River*. Edited by R. N. Williams. Amsterdam: Elsevier Academic Press.

Miller, D. J., and K. M. Burnett. 2008. A probabilistic model of debris-flow delivery to stream channels, demonstrated for the Coast Range of Oregon, USA. *Geomorphology* 94:184–205.

Mori, A. S. 2011. Ecosystem management based on natural disturbances: Hierarchical context and non-equilibrium paradigm. *Journal of Applied Ecology* 48:280–292.

Naiman, R. J., J. R. Alldredge, D. A. Beauchamp, and 8 coauthors. 2012. Developing a broader scientific foundation for river restoration: Columbia River food webs. *Proceedings of the National Academy of Sciences (USA)* 109:21201–21207.

Penaluna, B. E., D. H. Olson, R. L. Flitcroft, M. Weber, J. R. Bellmore, S. M. Wondzell, J. B. Dunham, S. L. Johnson, and G. H. Reeves. 2016. Aquatic biodiversity in forests: A weak link in ecological and ecosystem service resilience. *Biodiversity and Conservation*. doi:10.1007/s10531-016-1148-0.

Pess, G. R., D. R. Montgomery, E. A. Steel, R. E. Bilby, B. E. Feist, and H. M. Greenberg. 2002. Landscape characteristics, land use, and coho salmon (*Oncorhynchus kisutch*) abundance, Snohomish River, Washington, USA. *Canadian Journal of Fisheries and Aquatic Sciences* 59:613–623.

Poff, N. L., J. D. Allan, M. B. Bain, J. R. Karr, K. L. Prestegaard, B. D. Richter, R. E. Sparks, and J. C. Stromberg. 1997. The natural flow regime. *BioScience* 47:769–784.

Poole, G. C. 2002. Fluvial landscape ecology: Addressing uniqueness within the river discontinuum. *Freshwater Biology* 47:641–660.

Reeves, G. H., L. E. Benda, K. M. Burnett, P. A. Bisson, and J. R. Sedell. 1995. A disturbance-based ecosystem approach to maintaining and restoring freshwater habitats of evolutionarily significant units of anadromous salmonids in the Pacific Northwest. *American Fisheries Society Symposium* 17:334–349.

Reeves, G. H., K. M. Burnett, and E. V. McGarry. 2003. Sources of large wood in the main stem of a fourth-order watershed in coastal Oregon. *Canadian Journal of Forest Research* 33:1363–1370.

Reeves, G. H., B. R. Pickard, and K. N. Johnson. 2016. *An initial evaluation of potential options for managing riparian reserves of the Aquatic Conservation Strategy of the Northwest Forest Plan*. General Technical Report PNW GTR-937. Portland, OR: USDA Forest Service, Pacific Northwest Research Station.

RIEC and IAC (Regional Interagency Executive Committee; Intergovernmental Advisory Committee). 1996. *Ecosystem analysis at the watershed scale. Federal guide for watershed analysis. Section II. Analysis methods and techniques*. Version 2.3. Portland, OR: Regional Ecosystem Office. http://reo.gov/documents/reports/watershed-IIr.pdf.

Richardson, J. S., R. J. Naiman, and P. A. Bisson. 2012. How did fixed-width buffers become standard practice for protecting freshwaters and their riparian areas from forest practices? *Freshwater Sciences* 31:232–238. doi:10.1899/11-031.1

Rieman, B., J. Dunham, and J. Clayton. 2006. Emerging concepts for management of river ecosystems and challenges to applied integration of physical and biological sciences in the Pacific Northwest, USA. *International Journal of River Basin Management* 4:85–97.

Rieman, B. E., C. L. Smith, R. J. Naiman, and 12 coauthors. 2015. A comprehensive approach for habitat restoration in the Columbia Basin. *Fisheries* 40: 124–135.

Rosgen, D. L. 1994. A classification of natural rivers. *Catena* 22:169–199.

Schindler, D. E., R. Hilborn, B. Chasco, C. P. Boatright, T. P. Quinn, L. A. Rogers, and M. S. Webster. 2010. Population diversity and the portfolio effect in an exploited species. *Nature* 465:609–612. doi:10.1038/nature09060.

Stanley, E. H., S. M. Powers, and N. R. Lottig. 2010. The evolving legacy of disturbance in stream ecology: Concepts, contributions, and coming challenges. *Journal of the North American Benthological Society* 29:67–83.

Terrain Works. 2016. Terrain Works: Discover and map the hidden essentials of the Earth for improved management, restoration & conservation. http://www.terrainworks.com/.

USDA and USDI (US Department of Agriculture and US Department of the Interior). 1994. *Record of decision for amendments to Forest Service and Bureau of Land Management planning documents within the range of the northern spotted owl* [plus Attachment A: Standards and Guides]. [Place of publication unknown]: US Department of Agriculture and US Department of the Interior. http://www.reo.gov/library/reports/newroda.pdf.

Vannote, R. L., G. W. Minshall, K. W. Cummins, J. R. Sedell, and C. E. Cushing. 1980. The river continuum concept. *Canadian Journal of Fisheries and Aquatic Sciences* 37:130–137.

Wallington, T. J., R. J. Hobbs, and S. A. Moore. 2005. Implications of current ecological thinking for biodiversity conservation: A review of the salient issues. *Ecology and Society* 10(1):15. http://www.ecologyandsociety.org/vol10/iss1/art15/.

Watershed Professionals Network. 1999. Oregon watershed assessment manual. Salem, OR: Governor's Watershed Enhancement Board. http://www.oweb.state.or.us/publications/wa_manual99.shtml.

Wiens, J. A. 2002. Riverine landscapes: Taking landscape ecology into the water. *Freshwater Biology* 47:501–515.

Wimberly, M. C., T. A. Spies, C. J. Long, and C. Whitlock. 2000. Simulating historical variability in the amount of old forests in the Oregon Coast Range. *Conservation Biology* 14:167–180.

Yount, J. D., and G. J. Niemi. 1990. Recovery of lotic communities and ecosystems from disturbance—a narrative review of case studies. *Environmental Management* 14:547–569.

SECTION 4

Alternative Futures for Coniferous Forests

Sections I through III of this book describe a shift in how we view moist coniferous forests, with examples from northwestern North America. We see these forests as a complex interweaving of ecological, social, and economic effects. The ecological and socioeconomic values of individual sites changes through time, and the capacity to provide different ecosystem services shifts across the landscape. How this forest ecosystem functions at any point in time depends on the interactions between controlled disturbances, management or land-use change, and uncontrolled disturbances. In a time of constant change, considering the future has never been more important.

Whereas the first three sections of this book cover how our view of moist forests incorporates a better understanding of their social and ecological dynamics, this final section provides a glimpse of the possible futures of these forests. We look forward in this section, considering changing social and economic conditions, as well as ecological uncertainties including our changing climate. Chapter 16 addresses management in a future of uncertain climate, with altered forest, water, and carbon dynamics. Chapter 17 explores how new products and economies are shifting the social and ecological values of moist forests. In chapter 18, the four types of trust emerge as a new platform for collaborative stakeholder management of natural resources. The complex sociological and philosophical processes by which a relevant public deems forest-management approaches and priorities to be appropriate or inappropriate are characterized, with an example of how to work through this era of uncertainty using an adaptive-management framework. Chapters 19 and 20 summarize collective lessons we have learned from the first decades of forest ecosystem management in the US Northwest as synthesized in this volume. We provide some ideas for how we can incorporate what we now know about ecological dynamics and social and economic systems into collaborative and learning-based planning to build a more sustainable long-term future for moist forests and the ecosystem services they provide.

Chapter 16

Climate-Smart Approaches to Managing Forests

John B. Kim, Bruce G. Marcot, Deanna H. Olson, Beatrice Van Horne, Julie A. Vano, Michael S. Hand, Leo A. Salas, Michael J. Case, Paul E. Hennon, and David V. D'Amore

The climate of Pacific Northwest moist forests is characterized by abundant rainfall (1644 mm [65 in] annually) and mild temperatures throughout the year, averaging 9°C (48°F). Most precipitation falls in winter on the two major mountains ranges, the Coast and Cascade Ranges. Global warming is bringing distinct changes to the climate of these forests. Emissions of heat-trapping gases by human activities since industrialization have sharply increased global average temperatures and altered global precipitation patterns in the last 50 years. In the Northwest, the annual average temperature has risen 0.7°C (1.3°F) in the last century. The long-term trend is unclear for precipitation, but spring precipitation has increased in the last century (Abatzoglou et al. 2014), and year-to-year variability has increased since 1970 (Dalton et al. 2013). Under a business-as-usual scenario, the average annual temperature of the Northwest is projected to increase by 6°C (~11°F) by the end of this century compared with the 1950–2005 mean. For precipitation, enhanced seasonal cycles are projected, with a small change (−5% to +14%) in the annual mean precipitation by midcentury (2041–2070) (Dalton et al. 2013). The pace of climate change is projected to be rapid: the average annual temperatures of Washington and Oregon are predicted to depart from historical ranges within the next 3 to 5 decades (plate 14).

Forests occupy a critical role in the global carbon cycle. Globally, 70% of the carbon exchanged between the land and atmosphere cycles through forests. Carbon storage per unit area in the moist forests of Oregon and Washington exceeds that of any other regional forest in the United States (Turner et al. 1995; plate 1B). What is the future of these forests under climate change? While their exact fate is unknown, a general pattern of response is emerging (Vano et al. 2015). Below, we outline broad domains in which climate change can affect Northwest moist forests, then address climate-smart approaches to forest management.

Effects of Climate Change

Climate change will fundamentally affect Northwest moist-forest ecology, species composition, and many benefits people receive from these forests.

Forest Dynamics and Biogeography

Studies of fossilized plants suggest that modern Northwest plant communities, including old-growth forests, formed only within the last ~5,000 years (Brubaker 1991). The vegetation we see today developed under the relatively cool, wet conditions of the last 700 years. In the early Holocene (10,000 to 5,000 years ago), when the climate was warmer and significantly dryer, fire probably occurred more frequently, and early-seral species were more prevalent (Whitlock 1992), with Douglas-fir, alder, oak, and giant chinquapin dominating warm, dry sites.

Moist-forest response to climate change will vary by site, by forest age, and by fire history. Dense, closed-canopy sites, where growth is limited by available light and nutrients, have increased in productivity under the warming temperatures of the last 100 years (McKenzie et al. 2001). However, for forest stands that currently experience summer water deficit, climate change is likely to intensify drought conditions, reducing productivity and increasing risk for insect outbreaks and fire damage (Littell et al. 2008). Indeed, the mortality rate of trees in the Northwest has been increasing rapidly in recent decades, altering forest composition and reducing capacity for carbon sequestration (van Mantgem et al. 2009).

Forest response to climate change is mediated by the genetic characteristics of the plant population, plant dispersal mechanisms, and the connectivity of suitable habitat. Local populations can be genetically adapted to different climates, as has been documented for wide-ranging species such as Douglas-fir (Littell et al. 2008). Strategically selecting genetically favorable seed sources for replanting can improve likelihood of success for forest restoration and regeneration efforts. Some parts of the region served as climate refugia in the past, sheltering species during periods of unsuitable climate; it may therefore be possible to identify parts of the moist forest resilient to climate change.

Vegetation modeling provides another window into possible climate effects on forests. Species-distribution models, which calculate statistical relationships between species and their environment, project less suitable habitat for many current tree species, especially in the southwest portion of the region (Crookston et al. 2010; McKenney et al. 2011). Some models that simulate mechanisms of plant growth and plant-environment interactions also project less suitable future habitat for moist forests (Coops and Waring 2011). In these simulations, hotter and somewhat dryer summers drive the loss of moist forests, which are then replaced by the warm, mixed forest of coastal Northern California to the south. These models suggest that hardwood tree species such as madrone and oak may become more commonplace farther north than their current range, and maple and alder may expand in the moister riparian zones. However, other mechanistic models show little change to the current moist forests under climate change scenarios (Rogers et al. 2011; Shafer et al. 2015). These divergent predictions result from differences in the way models simulate, or sometimes omit, mechanisms of plant-environment interactions. Plant responses to nutrient availability, fire, insects, disease, and CO_2 concentrations in the air will be critical, along with their differing phenological responses to seasonal temperatures and atmospheric moisture, yet not all models simulate these details. (Phenology is the timing of biological phenomena, such as bud burst, annual flowering, hatching of eggs, or breeding migration, as they relate to seasonal and climatic conditions.)

Disturbance Ecology

Disturbance driven by climate change will be the primary mechanism of change in Northwest moist forests (Littell et al. 2010). These forests have high fuel loads that become flammable after prolonged drought, often associated with a quasi-stationary high-pressure ridge that diverts rainstorms away from the region (Gedalof et al. 2005). Climate models unanimously project increases in extreme heat and extreme precipitation events in the Northwest (Dalton et al. 2013). Although lightning strikes are projected to increase in the United States (Romps et al. 2014), it is unclear whether that lightning will be accompanied by precipitation. If not, it is likely to ignite more fires. Simulations for Oregon and Washington forests depict warmer temperatures creating drier fuel conditions, leading to larger and more severe fires (Rogers et al. 2011). This phenomenon has already been observed (chap. 7). By removing biomass and causing soil loss, fires reduce the amount of atmospheric carbon stored by forests. In addition, steep topographies like those the Coast and Cascade Ranges are vulnerable to landslides following fire, a subsequent disturbance with effects on both terrestrial and aquatic ecosystems.

In Southeast Alaska, yellow-cedar is one species affected by the constant flux of conditions in moist coniferous forests from disturbances that are being magnified as climate patterns shift (Krapek and Buma 2015). In part of its range in the coastal forests, yellow-cedar is directly injured and killed by an altered snow regime (Hennon et al. 2012). This intensive yellow-cedar die-off in mixed-species forests creates a partial disturbance that favors other tree species, including western hemlock (Oakes et al. 2014).

Climate change will directly and indirectly affect insect outbreaks in moist forests. Warming temperatures directly affect life-cycle timing of bark beetles and their annual acquisition of cold tolerance (Bentz et al. 2010). Warmer temperatures can indirectly influence outbreaks by affecting fungi, bacteria, nematodes, and mites that have symbiotic relationships with the insects, or by altering host-tree physiology and distribution. For mountain pine beetles, a warmer climate is expected to greatly reduce areas for their outbreaks in the moist forest, shifting them upward in elevation (Littell et al. 2010). Other insects may fill in, however, as hotter summers are likely to induce greater water stress in host trees, making them more susceptible to insect attacks.

Prevalence of tree diseases is likely to increase (Sturrock et al. 2011; Weed et al. 2013). Warmer winters and wetter springs are likely to increase the severity and distribution of Swiss needle cast, a common foliar disease in the moist forest (Stone et al. 2008).

Water Resources

The effects of a warming climate on hydrology have already been observed in the Pacific Northwest (Luce and Holden 2009; Isaak et al. 2010), and there is strong evidence that climate change will alter streamflow magnitude, timing, and water quality. Currently, about half of the watersheds in the moist forests of Oregon and Washington are strongly influenced by snowmelt, but by 2080, most of those watersheds are projected to become rain-dominant systems in which monthly runoff is coincident with precipitation (Tohver et al. 2014). April 1 snowpack is projected to decrease by 38% to 46% in Washington state by the 2040s and to almost completely disappear by the 2080s (Elsner et al. 2010). More precipitation falling as rain, especially if winter precipitation were to increase, will increase flooding in the winter. For example, the magnitude of the 100-year flood is projected to nearly double in some high-elevation watersheds on the Olympic Peninsula (Tohver et al. 2014). In summer, the loss of snowpack, increased evapotranspiration, and reduced precipitation are projected to result in widespread reductions in streamflows, especially at higher elevations. Lower streamflows and higher air temperatures also lead to higher water temperatures and decreases in dissolved oxygen, both of which affect aquatic biodiversity. The projected changes in the magnitude and timing of streamflow will pose a challenge for water managers, who have to balance minimizing winter flood risk with maximizing summer water storage, when water demand is the highest.

Biodiversity

Changes in terrestrial ecosystems are expected to occur up to 10 times faster in the coming decades than at any time in the past 65 million years (Diffenbaugh and Field 2013). Species distributions in Northwest moist forests are shifting already and will shift more rapidly as climate change intensifies (Hixon et al. 2010). Species with restricted distributions and narrow habitat requirements, especially those that benefit from species-specific or site-specific (fine-filter)management approaches, are likely among the most vulnerable to climate change; these include many of the >400 rare or little-known late-successional and old-growth-dependent species in Northwest moist forests (Molina et al. 2006). Also, aquatic-dependent species tied to ephemeral waters are projected to be significantly affected (Shoo et al. 2011); these habitats include headwater streams, which may make up 70% to 80% of stream-network lengths in Northwest moist forests (Gomi et al. 2002). Changing community composition can alter predator-prey and competitive interactions, as well as key ecological functions mediated by biota. Pest, disease, and invasive species incidence also may change.

Socioeconomic Vulnerability

Climate change in moist forests will affect people and communities through two primary pathways: changes in the provision of ecosystem goods and services, and changes in the timing, frequency, and severity of hazards associated with forests. The ability of people to adapt to ecological changes is determined in part by socioeconomic conditions. Communities in and near the moist forests are highly dependent on the forest. Forestry and forest products are big business in the Northwest, with about 32% of forested land in private ownership. In Washington, lumber production has been increasing, with sawmill products valued at US$1.7 billion and pulp mill products valued at US$2.7 billion in 2006. The number of mills, however, has been in a steep decline since 1992 (Campbell et al. 2010). With climate change, regional pulp mills are vulnerable to falling prices and increasing competition from the tropics on the global market. In coastal Oregon forests, landownership is a complex mosaic of private, state, and federal agencies, and coordinating adaptations to climate change while pursuing various landowner priorities will be a challenge. Land value itself may shift. With urbanized land valued at more than 100 times that of forestland in the Pacific Northwest (Alig and Plantinga 2004), much of the Puget Sound region and the Willamette Valley is likely to urbanize further (Kerns et al. 2016), especially as perceived pleasantness of the climate begins to worsen in other parts of the country by midcentury (Egan and Mullin 2016). Although warmer temperatures will reduce opportunities for winter recreation, warmer spring and fall seasons combined with increasing population in nearby urban areas may increase summer recreation in the moist forests.

Wildfires are likely to pose increasing hazards to communities in and near the moist forests, as wildfire frequency, size, and severity are projected to increase under climate change (Rogers et al. 2011). Wildfires can damage residential structures, municipal watersheds, energy infrastructure, and recreational infrastructure and can temporarily create hazardous air quality (Thompson et al. 2011).

Climate-Smart Approaches

Climate-smart conservation is "the intentional and deliberate consideration of climate change in natural resource management, realized through adopting forward-looking goals and explicitly linking strategies to key climate impacts and vulnerabilities" (Stein et al. 2014). Key characteristics of climate-smart conservation include management that is well informed through partnerships with science across a broad landscape context; conservation goals that prepare the landscape and its wildlife for the future; and strategies that embrace and confront uncertainties.

Science-Management Partnership

Climate change adaptation in moist forests faces a multitude of challenges. Climate change is likely to overlap with multiple environmental stressors, including urban development pressures, wildfire risk, and insect and disease outbreaks (plate 14; Kerns et al. 2016). Climate niches (climate factors defining a species' habitat) are shifting at a rapid pace (Loarie et al. 2009), but landownership is diverse and spatially interspersed, and environmental regulations, policies, and laws are static relative to climate.

Multiagency science-management partnerships and adaptive management (chap. 8) can foster agile, science-based planning and coordinate long-term monitoring and adaptation campaigns across landownership boundaries. An early implementation of a science-management partnership for climate change adaptation in the United States has begun in the Olympic National Forest, Washington, and has expanded to various forested lands of the western United States. These partnerships build upon an existing network of managers and scientists and recruit participants to create a climate change adaptation collaboration for a specific region, spanning ecosystem types, landownership boundaries, agency and organizational boundaries, and scientific and management disciplines. Keys to successful adaptation partnerships include building trusted professional relationships, communicating recent advances in climate change science, and collaboratively identifying vulnerabilities and management options.

Managing Forests for Resistance, Resilience, and Response

Recent management practices in Northwest moist forests were predicated upon a stable climate. They are therefore not well designed to address climate change and may have made some forests more vulnerable to its adverse effects (Spies et al. 2010). Although conserving biodiversity and ecosystem services is commonly accepted as a core motivating principle, climate change adaptation is ultimately rooted in socially determined values of harm versus benefit (Noble et al. 2014). No single approach will suit all moist forest ecosystems, owing to varying levels of vulnerability to climate change and differing management goals.

To date, the dominant paradigm for climate change adaptation has been to use the historical conditions of the forest as a reference (Stein et al. 2013) and to aim to keep rates, scales, and intensities of change within the historical range of variability (Noss 2001). This approach reflects some optimism for the future, in which adaptation practices help buy time while effective, global-scale mitigation efforts stabilize the climate sometime this century (Hansen et al. 2003). Two main strategies consistent with this paradigm are to manage forests to be more resistant and more resilient to climate change effects. Managing for increased resistance of a moist forest may include fuel treatments to reduce likelihood of extreme fires or insect outbreaks; it may have only short-term effects and is best applied to forests with low sensitivity to climate change. Managing for increased resilience takes a longer and broader view: those practices aim to facilitate the forest's return to a prior condition after disturbance (chap. 14, fig. 14.1) and may include surplus seed-banking and other intensive management of the regeneration phase (Millar et al. 2007). Some forests may exhibit tipping-point and threshold responses to increasing disturbances and changing climate stressors (chap. 14, fig. 14.1; plate 14). For example, increases in the frequency and extent of canopy fires could shift the forest community into a new, unanticipated stable state (Fletcher et al. 2014).

Adaptation to long-term, large-scale changes can include assisted migration of species, or spreading risk by diversifying age, structure, and genetic stock both within stands and across landscapes. Maintaining seed collections is particularly important, so that varieties planted can withstand challenges from warmer and possibly drier summers, larger burned areas, or northward expansion of insect pests and diseases. Research to understand natural variation in tree growth pattern, phenology, and genetics can be used to project the value of using seed from trees locally adapted to climate conditions to the south to produce seedlings for climate-resilient trees at higher latitudes or altitudes. On the Olympic National Forest, climate change adaptation management includes diversifying the species and genotypes planted by selecting for stock from a wide variety of locations and implementing thinning and harvest treatments and fuel-reduction treatments to increase landscape diversity (Littell et al. 2012). Forest-management activities may have cascading effects, such as enhancing snow retention (Lundquist et al. 2013) and affecting water supply downstream. Some effects may be undesirable, such as reduction of habitat, carbon emissions, reduction of air quality (from prescribed fire), and increases in invasive species (Spies et al. 2010). On the policy side, land-management goals may need to be revised to be consistent with the dynamic, uncertain realities of climate change.

Managing Forests for Wildlife

Habitat-management tools include extending the persistence of ephemeral waters (ponds, streams); restoring aquatic organism passage across stream-crossing culverts; and retaining cool and moist microhabitats via shading or augmenting refugia (e.g., Shoo et al. 2011). Because the majority of water in the region filters through forests, forests provide an opportunity to mitigate the negative consequences of climate change on water resources. Maintaining or replanting well-shaded riparian areas can help retain cooler temperatures, lessening effects on salmonids and other cool-adapted aquatic species (chap. 14).

Habitat connectivity may be needed for wildlife dispersal into new areas. Connecting fragmented areas of mature or old-growth forest cover may help conserve some forest carnivores and species associated with late-successional conditions. Similar management is needed for the spatial and temporal distribution of early-successional habitat. For animals, relocation, reintroduction, translocation, and assisted migration may be useful but are controversial (Hagerman and Satterfield 2014) and largely untested for wildlife in the Northwest, with a few exceptions such as with the reintroduction of Pacific fishers to the Olympic Peninsula (Lewis et al. 2016).

Climate-smart approaches include prioritization of species or locations with maximal biological value or co-occurrence of multiple ecosystem stressors (plate 14). However, climate change effects on wildlife will vary greatly from taxon to taxon. For example, 26 bird species are expected to vary in their response to five possible future scenarios in the Northwest (fig. 16.1). Modeling software tools, such as MARXAN (Ball et al. 2009) and Zonation (Moilanen 2007) help find optimized solutions for prioritizing habitat corridors and connections across the landscape. Future projections of stream temperatures such as NorWeST models (Isaak et al. 2010) can aid identification of refugia for cold-water-adapted species.

Other views of modeling results also help understand and manage for uncertainty. For example, individual taxon differences aside, figure 16.1

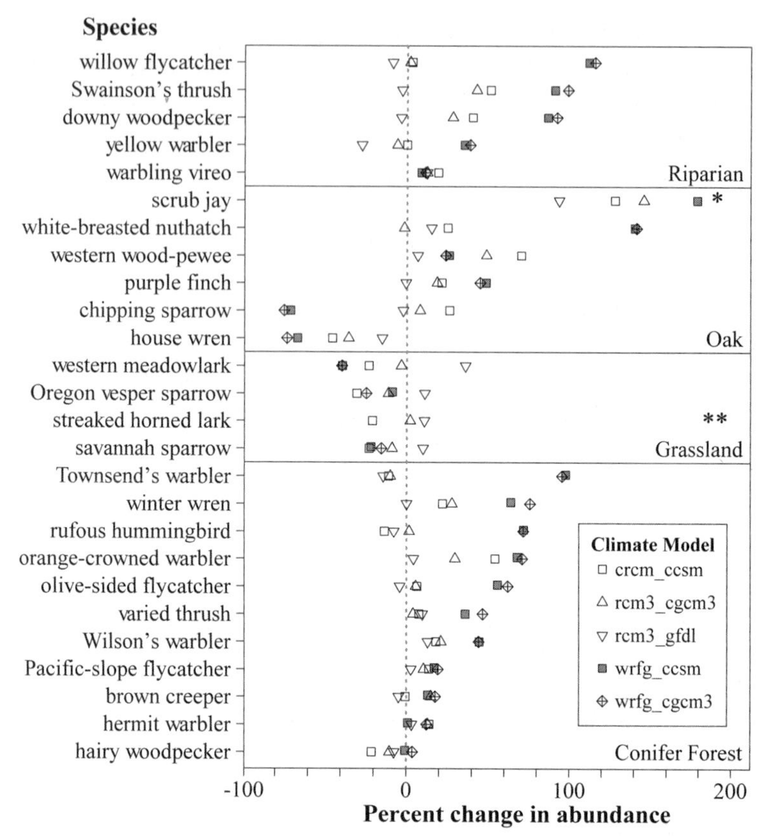

FIGURE 16.1. Change in abundance (%) with respect to historical values across the landscape, under 5 different future climate scenarios (different symbols), for 26 bird species of the Pacific Northwest. A larger spread of symbols, as for willow flycatcher and white-breasted nuthatch, reflects greater uncertainty regarding the outcome. * or ** = value not shown for >300% change. [Data set described in Veloz et al. 2015]

shows that uncertainties about climate change effects on bird abundances vary among individual species and across habitat types. Conservation planning may need to focus on species or habitat types with greater uncertainties, especially those with declines in abundance.

Species-specific climate-niche models project expanded ranges for some species. Species with phenotypic plasticity (able to adapt to a variety of conditions), including generalist species, may fare well. Altered climate can be a strong selective force, and some taxa may be able to adapt to the new conditions. However, many species will not be able to adapt rapidly enough to meet the challenge.

Confronting Uncertainty

Effective long-term management requires projecting the effects of climate change on future forest conditions while explicitly accounting for uncertainties. The degree to which these forest systems can continue to provide natural resource products, biodiversity conservation, opportunities for recreational experiences, cultural values, and other ecosystem services depends in large part on their resilience in the face of changing disturbance regimes. Our ability to gauge resilience over time and across geographic scales using the tools of risk analysis and risk management, such as trade-off analysis (Sample et al. 2014; Wagner et al. 2014), depends in part on the degree of uncertainty about future conditions.

Two types of uncertainty create challenges for the management of moist forests in the face of climate change. The first challenge comes from unpredictable changes inherent in system variability, or the degree to which the forest system itself varies over space or time, either randomly or in response to poorly known or unknown influences. This case seems to fit the moist coniferous forests in transition under changing climates and associated disturbance regimes. We are only beginning to unravel the complex relationships in soil ecology; the key ecological roles of invertebrates, fungi, and nonvascular plants; and the long-term effects of major shifts in fire, drought, and other stressors in determining the response of forest biota and the emergence of novel ecological communities. The second challenge is in how we measure, model, and predict those changes and comes in part from uncertainty in how well the structure and the ecological processes of the forest ecosystem itself are understood. There is also uncertainty over the influence and values of parameters for modeling forest growth, structure, and productivity and how those parameters are described and linked. Measurement error, which can be random or systematic, can add to uncertainty.

The implications of both classes of uncertainty extend to projecting the amounts and quality of natural resources and ecosystem services we derive from forests. In short, greater uncertainty means greater inability to provide for those resources and services at any given expected rate or level. This could lower their perceived value, leading to overexploitation in the near term.

Three key questions can help frame how we approach major issues of the uncertain near-term future: What are the specific uncertainties that pertain to managing moist coniferous forests? What are the implications of those uncertainties for meeting management goals and societal expectations? And how can we craft a sustainable future for moist forests, accounting for those uncertainties?

One effective tool may lie in explicitly accounting for the various dimensions of uncertainty in a framework based on decision science. Decision science is the set of concepts, tools, and models that integrate uncertainty in risk analysis and risk management. It entails four main stages: problem structuring, problem analysis, choice of action, and implementation and monitoring (Marcot et al. 2015). Tools such as the Ecosystem Management Decision Support modeling framework, which integrates GIS with fuzzy-logic modeling, have proven value in supporting forest environmental analysis and planning in the Northwest and beyond (Reynolds et al. 2014).

Decision science has been used in climate change adaptation planning (Cross et al. 2012; Stein et al. 2013), including planning for forest ecosystem resilience and sustainability (Wolfslehner et al. 2005; Blennow and Salinas 2006; Ogden and Innes 2009). Sample et al. (2014) used a decision-advisory approach to developing federal forest-management adaptation to climate change in a risk analysis of effects on forest, water, wildlife, biodiversity, ecosystem services, and other resources. Their approach invited a wide array of potential management responses to the types and degrees of uncertainty, addressed costs, and helped to set management priorities.

Decision-science tools, such as causal modeling with sensitivity and influence analyses, can be used to hypothesize which aspects of the forest system can be controlled by management and the degree of that control. Such analysis is vital for helping to set realistic expectations for forest planning and management and determining the degree to which specific levels of forest resources and services can be promised under a changing climate. Decision-science methods, such as analyzing the expected value of additional information, can help prioritize management activities under finite budgets. Methods such as multicriteria analysis have been used to engage stakeholder groups and public participation for developing sustainable forest plans (Mendoza and Prabhu 2005; Sheppard and Meitner 2005). However, benchmarks such as historical range of natural variation will become less useful for setting management objectives and future expectations when forest ecosystems and landscapes manifest new states in response to multiple stressors (Ravenscroft et al. 2010; plate 14).

Conclusions

Northwest moist forests face an uncertain future under climate change. With a business-as-usual scenario, the projected temperature increase has the potential to significantly alter forest composition and productivity, intensify wildfire regimes and insect and disease outbreaks, and transform landscapes and wildlife habitats. In turn, these changes may adversely affect human communities that rely on the moist forests for various ecosystem services, such as timber, water, recreation, and biodiversity conservation. Management of these forests is a substantial investment of public and private funds in the future welfare of the region and deserves evaluation for its long-term, aggregated effect on the forested landscape under a changing climate. The ability of forests to supply desirable ecosystem services will depend on managers' ability to incorporate what is known about increased risks to each of these services into planning and management activities. Climate-smart approaches include both a historical and a forward-looking perspective. Adaptation strategies may be selected to build resistance to change in the short term and resilience to change in the longer term. Accepting inevitable change may be appropriate in some circumstances, and prioritizing management of projected climate refuges is a prudent course of action. Understanding and confronting the uncertain future is essential, and employing decision science can help us to wade through complex quandaries. As with any economic investment, ongoing monitoring of key attributes and their response to both climate change and management practices is essential. The complexity of the moist forest ecosystem, the longevity of dominant tree species, and future uncertainty require that we act in anticipation of changes in ecosystem services rather than waiting until effects resonate throughout these forests to cause irreparable damage.

Literature Cited

Abatzoglou, J. T., D. E. Rupp, and P. W. Mote. 2014. Seasonal climate variability and change in the Pacific Northwest of the United States. *Journal of Climate* 27:2125–2142.

Alig, R. J., and A. Plantinga, A. 2004. Future forestland area: Impacts from population growth and other factors that affect land values. *Journal of Forestry* 102:19–24.

Ball, I. R., H. P. Possingham, and M. Watts. 2009. Marxan and relatives: Software for spatial conservation prioritisation. Chap. 14. Pp. 185–195 in *Spatial conservation prioritisation: Quantitative methods and computational tools.* Edited by A. Moilanen, K. A. Wilson, and H. P. Possingham. Oxford, UK: Oxford University Press.

Bentz, B. J., J. Régnière, C. J. Fettig, E. M. Hansen, J. L. Hayes, J. A. Hicke, R. G. Kelsey, J. F. Negrón, and S. J. Seybold. 2010. Climate change and bark beetles of the western United States and Canada: Direct and indirect effects. *BioScience* 60:602–613.

Blennow, K., and O. Salinas. 2006. Decision support for active risk management in sustainable forestry. *Journal of Sustainable Forestry* 21:201–212.

Brubaker, L. B. 1991. Climate change and the origin of old-growth Douglas-fir forests in the Puget Sound lowland. Pp. 17–24 in *Wildlife and vegetation of unmanaged Douglas-fir forests*. Edited by L. F. Ruggiero, K. B. Aubry, A. B. Carey, and M. H. Huff. General Technical Report PNW-GTR-285. Portland, OR: USDA Forest Service, Pacific Northwest Forest and Range Experiment Station.

Campbell, S., K. Waddell, and A. Gray. 2010. *Washington's forest resources, 2002–2006: Five-year forest inventory and analysis report*. General Technical Report PNW-GTR-800. Portland, OR: USDA Forest Service, Pacific Northwest Research Station.

Coops, N. C., and R. H. Waring. 2011. Estimating the vulnerability of fifteen tree species under changing climate in northwest North America. *Ecological Modelling* 222:2119–2129. doi:10.1016/j.ecolmodel.2011.03.033.

Crookston, N. L., G. E. Rehfeldt, G. E. Dixon, and A. R. Weiskittel. 2010. Addressing climate change in the forest vegetation simulator to assess impacts on landscape forest dynamics. *Forest Ecology and Management* 260:1198–1211.

Cross, M. S., E. S. Zavaleta, D. Bachelet, and 18 coauthors. 2012. The adaptation for conservation targets (ACT) framework: A tool for incorporating climate change into natural resource management. *Environmental Management* doi:10.1007/s00267-012-9893-7.

Dalton, M. M., P. Mote, and A. K. Snover. 2013. *Climate change in the Northwest: Implications for our landscapes, waters, and communities*. Washington, DC: Island Press.

Diffenbaugh, N. S., and C. B. Field. 2013. Changes in ecologically critical terrestrial climate conditions. *Science* 341:486–492.

Egan, P. J., and M. Mullin. 2016. Recent improvement and projected worsening of weather in the United States. *Nature* 532:357–360.

Elsner, M. M., L. Cuo, N. Voisin, J. S. Deems, A. F. Hamlet, J. A. Vano, K. E. Mickelson, S. Y. Lee, and D. P. Lettenmaier. 2010. Implications of 21st century climate change for the hydrology of Washington state. *Climatic Change* 102:225–260.

Fletcher, M.-S., S. W. Wood, and S. G. Haberle. 2014. A fire-driven shift from forest to non-forest: Evidence for alternative stable states? *Ecology* 95:2504–2513.

Gedalof, Z. E., D. L. Peterson, and N. J. Mantua. 2005. Atmospheric, climatic, and ecological controls on extreme wildfire years in the northwestern United States. *Ecological Applications* 15:154–174.

Gomi, T., R. C. Sidle, and J. S. Richardson. 2002. Understanding processes and downstream linkages of headwater systems: Headwaters differ from downstream reaches by their close coupling to hillslope processes, more temporal and spatial variation, and their need for different means of protection from land use. *BioScience* 52:905–916.

Hagerman, S. M., and T. Satterfield. 2014. Agreed but not preferred: Expert views on taboo options for biodiversity conservation, given climate change. *Ecological Applications* 24:548–559.

Hansen, L. J., J. L. Biringer, and J. R. Hoffman. 2003. *Buying time: A user's manual for building resistance and resilience to climate change in natural systems*. Berlin: World Wide Fund for Nature Climate Change Program.

Hennon, P. E., D. V. D'Amore, P. G. Schaberg, D. T. Wittwer, and C. S. Shanley. 2012. Shifting climate, altered niche, and a dynamic conservation strategy for yellow-cedar in the North Pacific coastal rainforest. *BioScience* 62: 147–158.

Hixon, M. A., S. V. Gregory, W. D. Robinson, and 17 contributing authors. 2010. Oregon's fish and wildlife in a changing climate. Chap. 7. Pp. 266–358 in *Oregon Climate Assessment Report*. Edited by K. D. Dello and P. W. Mote. Corvallis: Oregon Climate Change Research Institute, College of Oceanic and Atmospheric Sciences, Oregon State University.

Isaak, D. J., C. H. Luce, B. E. Rieman, D. E. Nagel, E. E. Peterson, D. L. Horan, S. Parkes, and G. L. Chandler. 2010. Effects of climate change and wildfire on stream temperatures and salmonid thermal habitat in a mountain river network. *Ecological Applications* 20:1350–1371.

Kerns, B. K., J. B. Kim, J. D. Kline, and M. A. Day. 2016. U.S. exposure to multiple landscape stressors and climate change. *Regional Environmental Change* 16:2129–2140. http://link.springer.com/article/10.1007%2Fs10113-016-0934-2.

Krapek, J., and B. Buma. 2015. Yellow-cedar: Climate change and natural history at odds. *Frontiers in Ecology and Evolution* 13:280–281.

Lewis, J. C., K. J. Jenkins, P. J. Happe, D. J. Manson, and M. McCalmon. 2016. Landscape-scale habitat selection by fishers translocated to the Olympic Peninsula of Washington. *Forest Ecology and Management* 369:170–183.

Littell, J. S., E. E. Oneil, D. McKenzie, J. A. Hicke, J.A. Lutz, R. A. Norheim, and M. M. Elsner. 2010. Forest ecosystems, disturbance, and climatic change in Washington State, USA. *Climatic Change* 102:129–158.

Littell, J. S., D. L. Peterson, C. I. Millar, and K. A. O'Halloran. 2012. US National Forests adapt to climate change through science–management partnerships. *Climatic Change* 110:269–296.

Littell, J. S., D. L. Peterson, and M. Tjoelker. 2008. Water limits tree growth from stand to region: Douglas-fir growth-climate relationships in northwestern ecosystems. *Ecological Monographs* 78:349–368.

Loarie, S. R., P. B. Duffy, H. Hamilton, G. P. Asner, C. B. Field, and D. D. Ackerly. 2009. The velocity of climate change. *Nature* 462:1052–1055.

Luce, C. H., and Z. A. Holden. 2009. Declining annual streamflow distributions in the Pacific Northwest United States, 1948–2006. *Geophysical Research Letters* 36(16):L16401. doi:10.1029/2009GL039407.

Lundquist, J. D., S. E. Dickerson-Lange, J. A. Lutz, and N. C. Cristea. 2013. Lower forest density enhances snow retention in regions with warmer winters: A global framework developed from plot-scale observations and modeling. *Water Resources Research* 49:6356–6370.

Marcot, B. G., L. A. Fisher, M. P. Thompson, and M. Tomosy. 2015. *Applying the science of decision making: A survey of use and needs in the National Forest System. A USDA Forest Service joint internal report from the National Forest System Ecosystem Management Coordination Program and the Research and Development Mission Area*. Washington, DC: USDA Forest Service. http://www.fs.fed.us/emc/nepa/includes/FS_SurveyofDecisionScience2015.pdf.

McKenney, D. W., J. H. Pedlar, R. B. Rood, and D. Price. 2011. Revisiting projected shifts in the climate envelopes of North American trees using updated general circulation models. *Global Change Biology* 17:2720–2730.

McKenzie, D., A. E. Hessl, and D. L. Peterson. 2001. Recent growth of conifer species of western North America: Assessing spatial patterns of radial growth trends. *Canadian Journal of Forest Research* 31:526–538.

Mendoza, G. A., and R. Prabhu. 2005. Combining participatory modeling and multi-criteria analysis for community-based forest management. *Forest Ecology and Management* 207:145–156.

Millar, C. I., N. L. Stephenson, and S. L. Stephens. 2007. Climate change and forests of the future: Managing in the face of uncertainty. *Ecological Applications* 17:2145–2151.

Moilanen, A. 2007. Landscape zonation, benefit functions and target-based planning: Unifying reserve selection strategies. *Biological Conservation* 134:571–579.

Molina, R., B. G. Marcot, and R. Lesher. 2006. Protecting rare, old-growth, forest-associated species under the Survey and Manage program guidelines of the Northwest Forest Plan. *Conservation Biology* 20:306–318.

Noble, I. R., S. Huq, Y. A. Anokhin, J. Carmin, D. Goudou, F. P. Lansigan, B. Osman-Elasha, and A. Villamizar. 2014. Adaptation needs and options. Pp. 833–868 in *Climate Change 2014: Impacts, adaptation, and vulnerability. Part A: Global and sectoral aspects. Contribution of Working Group II to the Fifth Assessment Report of the Intergovernmental Panel on Climate Change*. Edited by C. B. Field, V. R. Barros, D. J. Dokken, and 13 coeditors. Cambridge, UK: Cambridge University Press.

Noss, R. F. 2001. Beyond Kyoto: Forest management in a time of rapid climate change. *Conservation Biology* 15:578–590.

Oakes, L. E., P. E. Hennon, K. L. O'Hara, and R. Dirzo. 2014. Long-term vegetation changes in a temperate forest impacted by climate change. *Ecosphere* 5(10): 1–28.

Ogden, A. E., and J. L. Innes. 2009. Application of structured decision making to an assessment of climate change vulnerabilities and adaptation options for sustainable forest management. *Ecology and Society* 14(1):11. http://www.ecology andsociety.org/vol14/iss1/art11/.

Ravenscroft, C., R. M. Scheller, D. J. Mladenoff, and M. A. White. 2010. Forest restoration in a mixed-ownership landscape under climate change. *Ecological Applications* 20:327–346.

Reynolds, K. M., P. F. Hessburg, and P. S. Bourgeron, eds. 2014. *Making transparent environmental management decisions*. Berlin: Springer.

Rogers, B. M., R. P. Neilson, R. Drapek, J. M. Lenihan, J. R. Wells, D. Bachelet, and B. E. Law. 2011. Impacts of climate change on fire regimes and carbon stocks of the U.S. Pacific Northwest. *Journal of Geophysical Research* 116:G03037. doi:10.1029/2011JG001695.

Romps, D. M., J. T. Seeley, D. Vollaro, and J. Molinari. 2014. Projected increase in lightning strikes in the United States due to global warming. *Science* 346: 851–854.

Sample, V. A., J. E. Halofsky, and D. L. Peterson. 2014. U.S. strategy for forest management adaptation to climate change: Building a framework for decision making. *Annals of Forest Science* 71:125–130.

Shafer, S. L., P. J. Bartlein, E. M. Gray, and R. T. Pelltier. 2015. Projected future vegetation changes for the northwest United States and southwest Canada at a fine spatial resolution using a dynamic global vegetation model. *PLoS ONE* 10(10):e0138759.

Sheppard, S. R., and M. Meitner. 2005. Using multi-criteria analysis and visualisation for sustainable forest management planning with stakeholder groups. *Forest Ecology and Management* 207:171–187.

Shoo, L. P., D. H. Olson, S. K. McMenamin, and 19 coauthors. 2011. Engineering a future for amphibians under climate change. *Journal of Applied Ecology* 48:487–492.

Spies, T. A., T. W. Giesen, F. J. Swanson, J. F. Franklin, D. Lach, and K. N. Johnson. 2010. Climate change adaptation strategies for federal forests of the Pacific Northwest, USA: Ecological, policy, and socio-economic perspectives. *Landscape Ecology* 25:1185–1199.

Stein, B. A., P. Glick, N. Edelson, and A. Staudt, eds. 2014. *Climate-smart conservation: Putting adaptation principles into practice*. Washington, DC: National Wildlife Federation.

Stein, B. A., A. Staudt, M. S. Cross, and 8 coauthors. 2013. Preparing for and managing change: Climate adaptation for biodiversity and ecosystems. *Frontiers in Ecology and the Environment* 11:502–510.

Stone, J. K., L. B. Coop, and D. K. Manter 2008. Predicting effects of climate change on Swiss needle cast disease severity in Pacific Northwest forests. *Canadian Journal of Plant Pathology* 30:169–176.

Sturrock, R. N., S. J. Frankel, A. V. Brown, P. E. Hennon, J. T. Kliejunas, K. J. Lewis, J. J. Worrall, and A. J. Woods. 2011. *Climate change and forest diseases.* US Department of Agriculture, Forest Service / University of Nebraska Lincoln Faculty Publications. Paper 143. http://digitalcommons.unl.edu/usdafsfac pub/143.

Thompson, M. P., D. E. Calkin, M. A. Finney, A. A. Ager, and J. W. Gilbertson-Day. 2011. Integrated national-scale assessment of wildfire risk to human and ecological values. *Stochastic Environmental Research and Risk Assessment* 25:761–780.

Tohver, I. M., A. F. Hamlet, and S.-Y. Lee. 2014. Impacts of 21st-century climate change on hydrologic extremes in the Pacific Northwest region of North America. *Journal of the American Water Resources Association* 50:1461–1476. doi:10.1111/jawr.12199.

Turner, D. P., G. J. Koerper, M. E. Harmon, and J. J. Lee. 1995. A carbon budget for forests of the conterminous United States. *Ecological Applications* 5:421–436.

van Mantgem, P. J., N. L. Stephenson, J. C. Byrne, and 8 coauthors. 2009. Widespread increase of tree mortality rates in the western United States. *Science* 323:521–524.

Vano, J. A., J. B. Kim, D. E. Rupp, and P. W. Mote. 2015. Selecting climate change scenarios using impact-relevant sensitivities. *Geophysical Research Letters* 42:5516–5525. doi:10.1002/2015GL063208.

Veloz, S., L. Salas, B. Altman, J. Alexander, D. Jongsomjit, N. Elliot, and G. Ballard. 2015. Improving effectiveness of systematic conservation planning with density data. *Conservation Biology* 29:1217–1227.

Wagner, S., S. Nocentini, F. Huth, and M. Hoogstra-Klein. 2014. Forest management approaches for coping with the uncertainty of climate change: Trade-offs in service provisioning and adaptability. *Ecology and Society* 19(1):32. http://dx.doi.org/10.5751/ES-06213-190132.

Weed, A. S., M. P. Ayres, and J. A. Hicke. 2013. Consequences of climate change for biotic disturbances in North American forests. *Ecological Monographs* 83: 441–470

Whitlock, C. 1992. Vegetational and climatic history of the Pacific Northwest during the last 20,000 years: Implications for understanding present-day biodiversity. *Northwest Environmental Journal* 8:5–28.

Wolfslehner, B., H. Vacik, and M. J. Lexer. 2005. Application of the analytic network process in multi-criteria analysis of sustainable forest management. *Forest Ecology and Management* 207:157–170.

Chapter 17

Next-Generation Products and Greenhouse Gas Implications

Eini C. Lowell, Vikram Yadama, Laurence R. Schimleck, and Kenneth E. Skog

The social and economic benefits of wood harvest from forests depend on the demand for different ecosystem services, including products desired from trees. Although uses of species such as Douglas-fir and western hemlock for finished lumber, plywood, and paper have dominated past demand in moist coniferous forests of the Pacific Northwest, the emergence of new uses and technologies has expanded opportunities for wood utilization. This changes what, when, and how wood is harvested from the forest and how different management activities may intersect with restoration targets and ecosystem services—some of which are only just emerging.

The products most typically associated with trees are things we see or use every day, such as furniture, paper, flooring, and building materials. Commonly, logs cut from trees are processed in a number of ways to produce an array of products, including lumber, veneer, and chips. Lumber and veneer then can be reassembled to make engineered wood products such as cross-laminated timber or laminated-veneer lumber that have improved mechanical properties and can be used in specialized applications. Chips and wood flour may be combined with other materials, such as plastic or concrete, to create composites. Alternately, wood can be broken down into molecular and elemental components to produce chemicals, specialty products, and fuels.

The coniferous trees of moist forests have long been prized for their structural and appearance properties in (solid) wood products. A diversity of land-management objectives in response to changing societal demands for our forests has led to a shift in raw material available to the forest-products industry. The properties and quality of these raw materials differ from those previously supplying the manufacturing infrastructure. The economics of manufacturing traditional wood products on a global scale have changed the playing field, and companies constantly look for ways to stay competitive. More complete utilization efficiency that captures high-value wood products and co-products is essential.

At the same time that the base resource available for wood-products manufacturing is changing, national attention is being directed toward energy independence, reducing greenhouse gas emissions, and ensuring economically resilient communities. Trees are becoming a key resource for producing nontraditional forest products such as aviation jet fuel and carbon nanocrystals, playing a role in carbon calculations, and entering into the discussion on national energy security. More is being demanded from our forests. In general, focusing on a single use or issue is not as sound economically as integrating a variety of products throughout the value chain. For example, the chemical market, with its international reach, can be part of the overall economic picture. Scale and product diversity are key. This chapter provides an overview of wood-based products that can be produced by integrating technologies and growth opportunities for the wood-products industry and addresses the influence of increasing our use of wood in carbon-management strategies.

Mass Timber Products

Although technological developments continue to identify new products for previously undervalued woody materials from moist coniferous forests, there is still a place for products manufactured from the solid-wood-processing stream. Housing an increasingly urban society presents humanity with one of its greatest challenges. High population densities dictate that tall buildings be used—structures that traditionally have been built using concrete and steel, both of which have large carbon footprints. With growing concerns over carbon dioxide (CO_2) emissions and with society increasingly asking questions about the sustainability of its building materials, mass timber construction (sometimes referred to as tall or solid timber construction) is experiencing a renaissance as an alternative for tall buildings. Wood, when used as a building material in place of steel and concrete, can reduce fossil-fuel emissions and also stores carbon for the life of the building. Mass timber construction provides opportunities for greatly increased carbon sequestration and low embodied energy in multistory buildings. Previously, wooden structures were limited to four stories, but buildings ten stories tall have been built with mass timber construction, and plans exist for buildings up to thirty stories.

Mass timber construction addresses the issue of changing resource quality, facilitating the use of low-quality construction lumber by upgrading it to a higher-value product with improved mechanical properties (stiffness and strength). This engineered wood product offers better insulation properties and intrinsic fire safety because the massive timber elements burn slowly.

Solid timber construction systems include glued and nonglued wood products (Smith et al. 2015). Glued-wood products include (1) glue-laminated timber (glulam), lumber glued together to form a beam (plate 15); (2) structural composite lumber, including laminated-veneer lumber built of multiple layers of veneer glued together to form a panel that is then cut into beams; and (3) cross-laminated timber, a composite-panel product consisting of three to nine layers of dimension lumber arranged perpendicular to each other and glued. Nonglued products include (1) dowel-laminated timber, similar to glulam, with dowels fastening the lumber; (2) nail-laminated timber, as above, with nails holding the lumber; (3) cross-nail laminated timber, a panel with nails rather than glue holding layers together; and (4) interlocking cross-laminated timber, cross-laminated timber with interlocking layers.

Glulam and laminated-veneer lumber have been used in construction for many years (plate 15), but the emergence of cross-laminated timber-type systems has been relatively recent. The first commercial cross-laminated timber was produced in Switzerland in 1995 and since has given rise to a new approach to building, using prefabricated wall, floor, and ceiling elements. Panel prefabrication simplifies building construction on-site and provides several advantages compared with steel and concrete. Smith et al. (2015) summarized lessons learned about solid timber construction (table 17.1).

Construction time and cost savings for solid timber construction were estimated for four hypothetical designs of a four-story commercial building: mass timber, structural steel-frame system using open-web steel joists, structural steel incorporating precast hollow-core panels, and reinforced concrete (Canadian Wood Council 2015). Speed of construction was fastest for mass timber (4.5 months) and slowest for steel (7.25 months). Overall, the reinforced-concrete structure provided the lowest cost per

TABLE 17.1. Advantages and disadvantages of solid timber construction (adapted from Smith et al. 2015)

Advantages	*Disadvantages*
Speed: Because building elements are prefabricated, construction time is reduced. Also, a subcontractor can begin work once the first floor is completed, unlike concrete and steel buildings where this cannot be done efficiently.	**Logistics:** Panels are delivered by truck or in a container in order from foundation to final component, with the first piece to be used on top of the shipment. Storing and rearranging panels is costly and time consuming.
Labor costs: Owing to off-site prefabrication and shorter site-preparation time, labor needs on-site are reduced.	**Job displacement:** Fewer people (and labor hours) are required on-site owing to ease and speed of construction.
Raw material: Most solid timber products are delivered fully finished and can be exposed as an interior surface.	**Planning:** Designing solid timber construction buildings is completely different, and all design work must be completed (front-loaded) before panels are prefabricated, for example, location of electrical and mechanical systems must be decided before fabrication of panels.
Remote sites: Use of prefabricated panels that are delivered to the site allows buildings to be assembled quickly, which is important in remote locations, locations with a narrow window of time for construction, and locations with minimal labor force.	**Research:** Owing to lack of experience with solid timber construction, information related to construction methods, connection systems, and delivery methods is unfamiliar or not readily available.
Carbon reduction: Wood sequesters carbon, providing a reduction in the carbon footprint of the building compared with concrete and steel.	**Acoustics and vibration:** Owing to the rigid nature of panels, buildings are susceptible to sound and vibration that can be transferred through walls and floors. To mitigate these effects, extra soundproofing is required.
Precision: Tolerances are small, resulting in tight connections and envelope. Energy efficiency of the building is also increased.	**Component flexibility:** Panels require heavy machinery to install, and the size of the panels limits on-site adjustment.

TABLE 17.1. Continued

Advantages	*Disadvantages*
Weather: Utilizing a "dry process" (building takes place under a large, temporary protective structure), solid timber construction buildings can be erected in any weather.	**Knowledge and labor:** Solid timber construction is very different from traditional forms of construction and is unfamiliar to the majority of designers, contractors, and engineers in North America.
Safety: Because finished panels are delivered and there are fewer parts to assemble and transport, there is less potential for injury.	**Wind:** Can be a concern when moving panels with a crane from truck to site, owing to their large surface area.
Weight: Buildings weigh less, hence foundations can be smaller and buildings can be taller for a similar cost as compared with traditional methods of construction. Reduced weight can also be an advantage on sites unfavorable to concrete and steel structures.	**Code and permits:** Solid timber construction is relatively new in North America, so permitting officials are generally not familiar with important aspects of panels related to structural, fire, and acoustic performance. As a consequence, permitting can be delayed while documentation and engineering data are obtained.

square meter, while the costs of the other structures were higher by 1% (mass timber), 1.5% (structural steel with open-web steel joists), and 2.5% (structural steel with precast concrete hollow-core panels).

Mass timber construction offers an alternative to traditional methods of erecting tall buildings. In Europe, cross-laminated timber is well established as an option for multistory buildings, and strong demand supports approximately 30 fabrication facilities that are concentrated in Austria, Germany, and Italy. In North America, Canada has led development, while in the United States, use of the product is starting to increase, with a number of cross-laminated timber building projects recently announced across the nation. One limitation has been access to cross-laminated timber panels, but three facilities in North America now produce commercial-grade cross-laminated timber, and a fourth also produces a variety of cross-laminated timber products, including crane and oil field mats and furniture. Many questions about solid timber construction (table 17.1) are being addressed by research.

Raw materials unsuitable for solid timber products are being used in novel applications. Other integrated technologies provide alternative pathways for products that can be manufactured from by-products of the mechanical breakdown process, as well as wood harvested in land-management activities that does not have the necessary quality for particular wood products.

Biorefinery Concept

Biorefining can produce high-value products like chemicals, fuels (solid, liquid, or gas), and high-value specialty products such as automotive parts and biomedical supplies—not what traditionally comes to mind when you think of wood products.

A biorefinery is analogous to a petroleum refinery, where feedstock (crude oil) is converted or refined into fuels (gasoline, jet fuel, or diesel) and a variety of products, including lubricants, waxes, asphalt, plastics, solvents, and other chemicals. Many of these chemicals are used in everyday products such as ink, tires, eyeglasses, medicines, shampoo, and toothpaste. In a wood-based biorefinery complex, several conversion options need to be integrated, as appropriate, to progressively derive value from the woody biomass at various stages. Feedstock (logs, wood chips, bark, forest slash, sawdust, and planer shavings) can be converted into a variety of products, such as lumber, veneer, composite-panel products, energy, wood pellets, cellulose, lignin, sugars and their derivatives, polymers, and other chemical products (fig. 17.1).

The chemical structure of wood enables production of these varying commodities. Chemical components of wood include the macromolecular cell-wall components cellulose (30%–50%), hemicellulose (17%–35%), and lignin (15%–30%), low-molecular-weight components categorized as extractives (2%–8%), and ash or inorganic mineral substances (0.5%–3%). Wood is therefore classified as a lignocellulosic material, which is one of the most abundantly available materials on earth. Cellulose is a linear high-molecular-weight polymer composed of glucose molecules and is the main structural component providing strength and stiffness to wood. The main constituents of hemicellulose include 6- and 5-carbon sugars: glucose, mannose, galactose, xylose, and arabinose. Polymer chains composed of

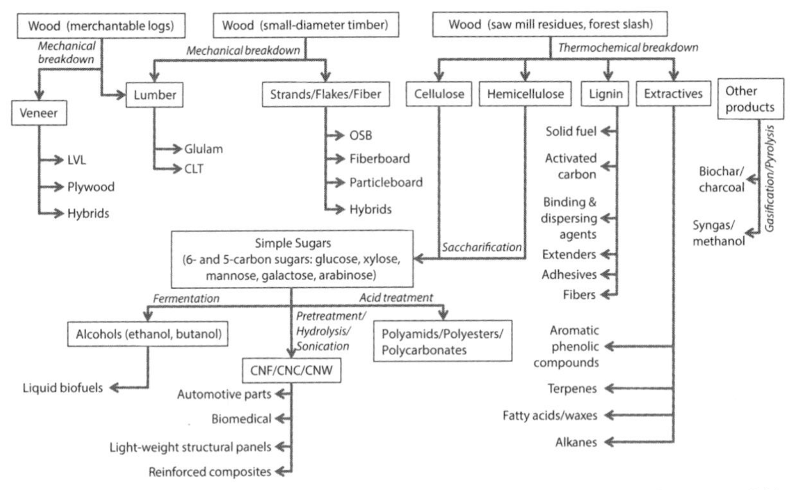

FIGURE 17.1. Examples of processes and products that constitute an integrated bio-based processing facility. Processes are in italics, and not all processes are shown.

these sugars are shorter than those of cellulose and can be branched. Cellulose and hemicellulose, together, can yield sugars for production of biofuels and biopolymers in a biorefinery.

Lignin is an amorphous macromolecule composed of an aromatic system of phenylpropane units. Lignin is incorporated into the cell-wall structure for strengthening, as well as between wood fibers to serve as a natural binder holding the fibers together. The structure of lignin varies between softwoods and hardwoods: generally, softwoods contain more lignin. Lignin can be used to formulate dispersants, binders, emulsifiers, stabilizers, and extenders. Chemical products such as phenols, vanillin, methane, and benzene can be derived from lignin isolated by different processes. Recent developments have led to lignin-based co-products such as carbon fibers for composite-fiber applications (Kubo and Kadla 2005; Ruiz-Rosas 2010).

Extractives, although a minor component of wood's chemical structure, can influence wood properties (for example, durability) and processing of wood (for example, serving as lubricants during extrusion of wood-plastic composites). Low-molecular-weight extractives in wood include aromatic phenolic compounds, terpenes, fatty acids, waxes, and alkanes. Potassium, calcium, and magnesium are some of the inorganics found in wood at very low concentrations.

To get from tree to products, wood must undergo initial processing (mechanical breakdown) and then conversion. To produce the aforementioned products from wood in a biorefinery complex, several conversion options need to be integrated to progressively derive value from the woody biomass at various stages. Wood saccharification involves hydrolysis of the carbohydrate components of wood, cellulose and hemicellulose, to 5- and 6-carbon sugars, such as xylose and glucose. Several pretreatment options, primarily involving the presence of acids, organic solvents, or enzymes, have been developed to isolate lignin from the carbohydrate components for pulp and paper production, and hydrolyze wood into fermentable sugars (Jørgensen et al. 2007; Zhao et al. 2009; Zhu et al. 2009; Pelaez-Samaniego et al. 2013). Sugars isolated from cellulose and hemicellulose by hydrolysis can be fermented into alcohols for biofuels (e.g., ethanol and iso-paraffinic kerosene) and bioplastics (e.g., polyhydroxy alkanoates), commercial acids (acetic acid, lactic acid, gluconic acid), acetone, and yeast. These sugars can also be converted through acid treatments into commercial products such as polyamides, polyesters, and polycarbonates. Sugars derived from hydrolysis of hemicellulose can similarly be converted into value-added commercial products. Isolated lignin in different forms (kraft lignin, lignin sulfonates, organosolv lignin) is mainly used for production of energy via direct combustion. However, potential for deriving high-value low-molecular-weight chemicals and other products from this amorphous polymer (lignin) has provided impetus for developing innovative technologies. With advances in conversion technology, we are able to further fractionate the cellulosic component of wood fiber to the microfibril level, thus enabling production of cellulosic nanofibrils and cellulosic nanocrystals or nanowhiskers. Research on cellulosic nanofibrils began more than 20 years ago (Eichhorn et al. 2010). Cellulose nanofibrils found naturally in wood have unique properties and sizes that differ from synthetically produced ones, such as carbon nanotubes (Khalil et al. 2012). Wood-based nanoscale feedstock can be used, for example, to produce biomedical products, fiber-reinforced composites, lightweight structural panels, and automotive parts.

Wood also can be fractionated and converted into a variety of other products, including energy, by means of mechanical, thermal, and chemical techniques. Heat can be produced by burning wood. A kilogram of wood can yield on an average 19 megajoules (MJ) (lower heating value, LHV) of energy, with low ash and sulfur contents. Higher heating value (HHV) of wood, which also includes the latent heat contained in the water vapor, ranges between 18 and 21 MJ. However, burning wood for heat releases stored carbon and precludes implementing other processes to derive chemical products that can extend the benefits of this renewable resource.

Besides being directly burned for energy, wood can also be thermally degraded at temperatures up to 500°C (932°F) under an inert atmosphere to produce solid components (charcoal and biochar); condensable volatile compounds (tar); and noncondensable volatile gases (CO_2, CO, methane, and other hydrocarbons). This process is referred to as pyrolysis (Fengel and Wegener 1983; Lange 2007). Gasification of wood (Fengel and Wegener 1983; Lange 2007) at temperatures between 850°C and 1000°C (1,562°F and 1,832°F) can yield different products depending on process conditions and water content. Gasification of wood in the presence of air yields CO_2, CO, methane, hydrogen, and nitrogen. Gasification in the presence of oxygen and additional steam can yield gas that can be subsequently converted by cleaning and adding hydrogen to produce synthesis gas.

Conversion of woody biomass to wood products, chemicals, energy, and higher-value composites can, in some cases, reduce fossil-fuel emissions if those products substitute for nonwood products that generate higher fossil-fuel emissions in their production. In cases in which the production of wood products reduces fossil-fuel emissions but increases wood-fuel emissions, there will be lower overall carbon emissions over time as the wood-fuel carbon emissions are reabsorbed by forest growth. When there is a reduction in fossil-fuel emissions, use of wood products reduces wastage, energy consumption, emissions during processing, and dependency on fossil fuels. Integrating these manufacturing technologies (e.g., a facility that produces pulp and paper, concentrated sugar solution for biofuels, co-products from extracted lignin, and combined heat and power) could further reduce environmental impacts. These impacts should be objectively determined by conducting cradle-to-grave life-cycle assessment of production systems from raw material extraction through materials processing, manufacturing, distribution, and end-of-life disposal.

Greenhouse Gas Implications of Wood-Products Production and Use

Sequestering carbon (C) in forests and offsetting carbon emissions with use of wood for energy and products are two of a range of objectives in managing forests. Increasing carbon storage and carbon offsets across a range of carbon pools and emission sources contributes to stabilizing atmospheric concentrations of CO_2.

Chapter 12 discusses how forest management, in combination with forest-products and wood-energy production, can influence carbon storage over time. Here we review how forestry and forest-products production and use form part of a larger system of processes—including production of fossil-fuel energy and nonwood products. To determine the full effects of forest-products production or wood-energy use we need to evaluate carbon storage and greenhouse gas emissions from this larger system of processes over time.

Evaluating the Effect of Forest-Products Production and Use on Carbon Storage and Greenhouse Gas Emissions

Numerous processes are associated with forest management and wood-products production and use that can affect carbon storage and emissions (fig. 17.2). The effect of these processes on the atmosphere can be evaluated by tracking carbon flows across system boundaries over time.

The *forest sector* includes forest, wood products, and wood-energy processes (processes inside the dashed-line box, fig. 17.2). The forest sector

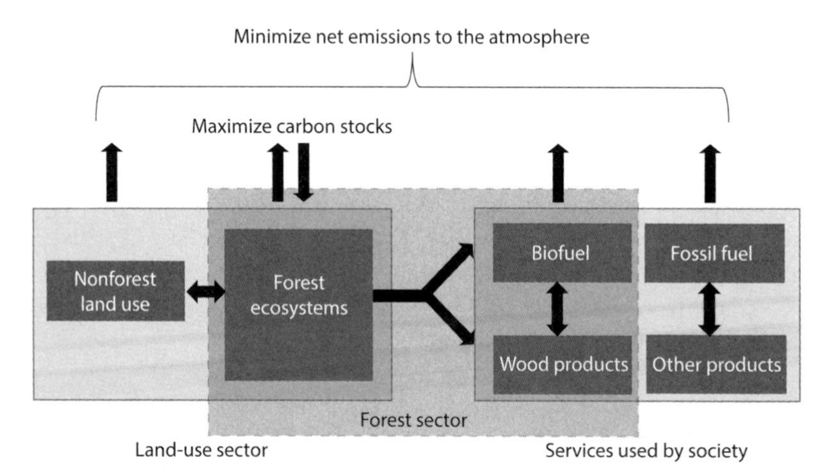

FIGURE 17.2. Forest-sector and non-forest-sector greenhouse gas emissions and stock changes that are influenced by forest management and forest-products production and use.

can be defined to include only carbon fluxes to and from forests and/or wood products, or it may also include carbon fluxes from equipment used to manage forests, to harvest and transport logs, and to manufacture wood products.

Forest management and production of wood products can also affect greenhouse gas emissions beyond the forest sector. If energy is produced from wood, this would displace fossil-fuel emissions. Similarly, emissions from the manufacture of wood products could replace emissions from producing substitute nonwood products. Forest-products production can also indirectly cause land-use change by changing market prices for wood products. Higher prices may induce landowners to keep land in forest that they may have otherwise converted, or to convert land to forest, resulting in increased carbon storage.

Evaluation of carbon-management strategies associated with forests and forest products requires projecting carbon-stock changes and emissions for a scenario of altered forest management and/or product production, compared with a baseline (or reference case) projection without such changes (box 17.1). The results of a carbon assessment for forest management and wood-products production will be determined in large part by how system boundaries are defined, including the time period over which carbon storage and emissions are evaluated. Currently, no standard approach exists for selecting system boundaries. Including more processes would give a more complete understanding of impacts. However, as more processes are included, the uncertainty in the estimates of effects on carbon storage and emissions would likely increase.

The effectiveness of an economic policy strategy, such as incentives to increase wood use for energy or solid-wood products in place of nonwood products, is determined by changes in forest and nonforest landowner behavior. Decisions relative to harvesting or land-management (e.g., planting) activities could provide more wood for those uses. For example, increased timber demand for products and building construction can alter some landowners' management practices. Although some may change rotation age, others may convert nonforest land to forest plantations. In either case, the additional wood demand changes the time profile for carbon stored on the land. With increased wood demand for energy, landowners may thin more frequently or sell logging residue from existing timber sales, which will also change the time profile of carbon on the land relative to a reference case without such demand.

BOX 17.1. CARBON-STOCK CHANGES WITH WOOD AND NONWOOD PRODUCTS

To evaluate the effect on net carbon emissions (or net carbon storage) of an increase in timber harvest to make wood products, we need to compare a reference-case projection of carbon stored on the land, in products, or in fossil fuels with an alternate case that involves an increase in wood harvesting. Carbon storage changes with either production of a wood product or a substitute nonwood product *n* years following production (all variables in tons carbon) can be calculated by comparing these two cases.

Carbon storage will be greater as a result of wood products production after *n* years if

$$-EWP - H + W + GNWP + GWPextra > -EWP - ENWPextra + GNWP$$

(for parameter definitions, see table 17.2); or, after simplification,

$$GWPextra - H > -ENWPextra - W \quad (1)$$

GWPextra, the extra forest growth after harvest in the case of wood-product production, likely will approach the harvest amount, H, over time. GWPextra may fall short of H by an amount that is the sum of extra emissions needed to make nonwood products and the amount of carbon stored in wood products after *n* years.

Thus the carbon storage advantage of wood products after *n* years can be improved by increasing the rate of forest regrowth, decreasing the energy needed to make wood products relative to nonwood products (increasing ENWPextra) and increasing the amount of carbon stored in wood products over a long time period. Another way to increase carbon storage advantage from producing wood products is more complete utilization (less waste) of harvested wood. For a given amount of harvest, a decrease in the right side of equation (1), leaving the left side unchanged, could result in more storage of carbon in wood products (W).

Implementing Strategies to Increase Carbon Storage

Carbon-storage strategies involving forest management and forest-products production and use may be ineffective because of flaws in incentive structures or policies, not owing to the biophysical attributes of the strategy itself. For example, an incentive program might favor harvesting large trees that produce lumber, based on the assumption that the increase in lumber produced would replace building materials that emit more carbon in manufacturing. However, if this incentive strategy were imple-

TABLE 17.2. Parameters used to calculate carbon-storage differences with either production of a wood product or a substitute nonwood product *n* years following production

EWP = fossil emissions from the manufacturing of wood products
ENWPextra = extra fossil emissions needed to make nonwood products versus wood products
H = amount of wood carbon removed from a forest to make wood products
W = amount of carbon stored in wood product in use or landfills after *n* years
GNWP = growth in forests, over *n* years, when nonwood product is produced (no harvest for forest products)
GWPextra = extra forest growth, over *n* years, in the case of forest harvest and wood-products production rather than no harvest

Parameter	*Alternate case: One unit of wood product is produced*	*Reference case: One unit of nonwood product is produced (functional equivalent to one unit of wood product)*
Fossil fuel emissions in manufacturing	EWP	EWP + ENWPextra
Timber harvest required (year zero)	H	0
Carbon storage in wood products after *n* years	W	0
Forest growth over *n* years after product production	GNWP + GWPextra	GNWP
Change in carbon stored over *n* years in forests, wood products in use, fossil fuels in ground	–EWP – H + W + GNWP + GWPextra	–EWP – ENWPextra + GNWP

mented, the lumber could go to nonbuilding uses, or an increase in harvest by one landowner could be offset with a decrease by another. This is a flaw of the incentive system, not in the underlying wood-substitution strategy. If there were incentives for builders to use wood in structures instead of other materials, the strategy could be effective in reducing overall fossil emissions from manufacturing; however, the effectiveness depends on the assumed changes in forest management.

Evaluation of carbon-management strategies often focuses on understanding how they change carbon storage per unit area of forest under management over time. It is also possible to evaluate strategies by focusing on the change in carbon storage associated with producing one additional unit of wood energy or one additional unit of wood product. Evaluation of these effects could include the effect of changing demand for several products, including nonwood products, on emissions.

In order to compare the carbon-mitigation effect of different strategies to alter forest management or alter forest-products production and use, it will be critical to use a common set of system boundaries (processes) and a common metric for evaluating the effects on the atmosphere.

Economic and Ecological Implications

Novel technologies that efficiently and sustainably use natural resources and reduce demand for fossil fuels are needed to mitigate environmental impacts ranging from net greenhouse gas emissions to unsustainable forest practices. In communities with abundant supplies of lignocellulosic biomass (crop based to forest based), there is also a need for value-added manufacturing opportunities that are economically viable and suitable for small- to medium-scale operations. Furthermore, there is a national desire to move from reliance on fossil fuels to renewable fuels to reduce fossil emissions. These conditions provide an opportunity to develop a fully integrated biomass-based technology system that meets growing demand for building materials, chemicals, and energy.

Producing cellulosic biofuels from woody biomass can reduce emissions and energy dependence significantly more than crop-based biomass (Lippke et al. 2012). Current sources of biofuels, such as ethanol, from crop-based feedstock (corn or sugarcane) have severe limitations in net displacement of fossil fuel. These limitations include reallocation of land from food to biofuel production and net energy efficiency. Efforts toward sustainable management of our forests and an abundant supply of underutilized forest-based biomass (woody biomass) have elevated cellulosic-based biofuels to one of the primary means to achieve our national goal of energy independence. However, because established infrastructure is lacking, economic viability of cellulosic-based biorefineries depends on low-cost feedstock, improvements in transportation systems and conversion technologies, subsidized costs, and value-added products. High-value-added co-products enable the greatest possible economic return from biorefinery feedstock in an overall global climate change initiative.

Vertically integrated manufacturing and value-chain considerations that make use of raw materials harvested from both natural stands and intensively managed plantations, along with the by-products of harvesting and manufacturing processes, can maximize wood's value, considering its life cycle. Biorefining can occur at different scales integrating two or more aforementioned conversion technologies to produce two or more products at one location. Scale and product depend on the available form of raw materials, accessibility, and market demands. The biorefinery concept provides a means to convert low-value lignocellulosic biomass into an array of higher-value products. Woody biomass can be fractionated into its main components by sequential treatments to give separate streams that may be used for different product applications, allowing a maximization of the benefits from a renewable and complex resource that will become increasingly important as fossil sources become more constrained or expensive. Economic benefits accrue predominantly to local communities close to the resource base, while the environmental benefits accrue region-wide through the substitution of renewable-based products for fossil-based products. These products also allow for more complete use of harvested trees of many types and sizes. This, in turn, makes the array of forest thinning strategies that can be economically feasible much more flexible as we adapt to a continually changing environment.

Literature Cited

Canadian Wood Council. 2015. *Mountain Equipment Co-op Head Office: A case study*. http://cwc.ca/wp-content/uploads/2015/05/Mountain-Equipment-Co-Op-Case-Study_.pdf.

Eichhorn, S. J., A. Dufresne, M. Aranguren, and 20 coauthors. 2010. Review: Current international research into cellulose nanofibres and nanocomposites. *Journal of Materials Science* 45:1–33.

Fengel, D., and G. Wegener, eds. 1983. *Wood: Chemistry, ultrastructure, reactions*. Berlin: Walter de Gruyter.

Jørgensen, H., J. B. Kristensen, and C. Felby. 2007. Enzymatic conversion of lignocellulose into fermentable sugars: Challenges and opportunities. *Biofuels, Bioproducts and Biorefining* 1:119–134.

Khalil, H. A., A. H. Bhat, and A. I. Yusra. 2012. Green composites from sustainable cellulose nanofibrils: A review. *Carbohydrate Polymers* 87:963–979.

Kubo, S., and J. F. Kadla. 2005. Lignin-based carbon fibers: Effect of synthetic polymer blending on fiber properties. *Journal of Polymers and the Environment* 13:97–105.

Lange, J. P. 2007. Lignocellulose conversion: An introduction to chemistry, process and economics. *Biofuels, Bioproducts and Biorefining* 1:39–48.

Lippke, B., R. Gustafson, R. Venditti, P. Steele, T. A. Volk, E. Oneil, L. Johnson, M. E. Puettmann, and K. Skog. 2012. Comparing life-cycle carbon and energy impacts for biofuels, wood product, and forest management alternatives. *Forest Products Journal* 62:247–257.

Pelaez-Samaniego, M. R., V. Yadama, E. Lowell, and R. Espinoza-Herrera. 2013. A review of wood thermal pretreatments to improve wood composite properties. *Wood Science and Technology* 47:1285–1319.

Ruiz-Rosas, R., J. Bedia, M. Lallave, I. G. Loscertales, A. Barrero, J. Rodríguez-Mirasol, and T. Cordero. 2010. The production of submicron diameter carbon fibers by the electrospinning of lignin. *Carbon* 48:696–705.

Smith, R. E., G. Griffin, and T. Rice. 2015. *Solid timber construction: Process, practice, performance*. University of Utah, Integrated Technology in Architecture Center, College of Architecture and Planning. http://itac.utah.edu/ST_Perform_files/STC%20PPP%20V1.1.pdf.

Zhao, X., K. Cheng, and D. Liu. 2009. Organosolv pretreatment of lignocellulosic biomass for enzymatic hydrolysis. *Applied Microbiology and Biotechnology* 82:815–827.

Zhu, J. Y., X. J. Pan, G. S. Wang, and R. Gleisner. 2009. Sulfite pretreatment (SPORL) for robust enzymatic saccharification of spruce and red pine. *Bioresource Technology* 100:2411–2418.

Chapter 18

Enhancing Public Trust in Federal Forest Management

Michael Paul Nelson, Hannah Gosnell, Dana R. Warren, Chelsea Batavia, Matthew G. Betts, Julia I. Burton, Emily Jane Davis, Mark Schulze, Catalina Segura, Cheryl Ann Friesen, and Steven S. Perakis

The connections between social and biophysical sciences are being forged in new ways as researchers and practitioners of natural resources seek to understand how lands can be managed for the benefit of human societies and the broader biotic community. Increasingly, we recognize that social and physical systems are tightly integrated, with human actions and decisions both shaping and shaped by the ecological systems in which they are embedded (e.g., Carpenter et al. 2009). In this context, a variety of social actors, including scientists, managers, policy makers, and the public, are collectively playing a larger role in decisions about environmental governance (e.g., collaboratives, chap. 9), drawing upon an accumulating body of knowledge describing the dynamics of complex socioecological systems. Learning-based approaches using adaptive-management experiments (chap. 8) represent one particular type of formal tool that can be appropriated to this process of adaptive environmental governance.

Consideration of ethics is another important if underappreciated part of environmental decision making (Doak et al. 2008). Analysis of the implicit values and ethical frameworks underlying natural resource management can help us understand, for example, how the influence of science on environmental policy changes over time, or how the public response to management decisions may shift. In this chapter we consider the integration and feedbacks between social and biophysical data, providing ideas about how to more fully understand and design planning and implementation processes on public lands.

In the Pacific Northwest, current social and environmental changes (e.g., growing human population, changing land use, climate change) appear to be intensifying pressure on natural resources such as forests (Hays 2006; Spies and Duncan 2009; Spies et al. 2010). At the same time, public knowledge, values, and perceptions intersect in social goals and expectations for federal forests, at times even propelling movements to effect profound institutional change in the laws, policies, and science that govern federal forest management (Franklin and Johnson 2014; Winkel 2014). In these dynamic times, policy prescriptions can become confused, confusing, and even aberrant when they fail to thoughtfully engage with public attitudes, perceptions, and values. Potential outcomes include decreased stakeholder involvement, failure to address critical environmental or sociopolitical issues, and loss of trust in the process of natural resource management.

In particular, trust is an increasingly important factor underlying successful management of public forestlands. When the public trusts natural resource agencies, public approval of management decisions increases, resistance to planning efforts is minimized, and managers have more latitude to experiment and engage in adaptive management (Lachapelle and McCool 2012). By reducing social resistance to public forestland management, trust can also accelerate management actions and reduce total project cost. Conversely, without the trust of the public, social acceptability of active forest management on public lands tends to falter, creating delays and additional costs that generally impede management activities. Therefore active and effective management requires that the public, and key stakeholders in particular, trust land managers, granting them the latitude to pursue the management interventions most likely to maintain ecosystem health and resilience in light of changing social and biophysical conditions.

Trust can be defined as "a psychological state in which one actor (the trustor) accepts some form of vulnerability based upon positive expectations of the intentions or behavior of another (the trustee), despite inherent uncertainties in that expectation" (Stern and Coleman 2015). It is widely held that, beginning with controversies surrounding the logging of old-growth forests on federal lands, the cycle of trust (fig. 18.1)—with the public as trustor and forest managers as trustees—has broken down across the moist coniferous forest landscape of the US Northwest, resulting in a loss of social acceptability of many active forest-management practices on public lands (Spies and Duncan 2009). Thus this region offers a unique opportunity to explore trust in timber-harvesting practices, and more broadly, the cycle of trust around management of public lands.

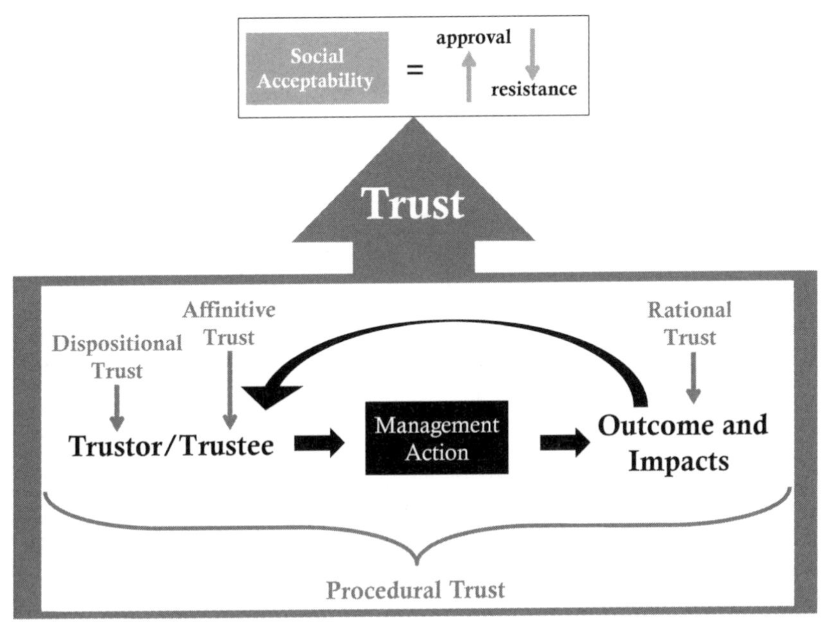

FIGURE 18.1. The trust cycle. Increased approval and decreased resistance generally follow from social acceptability. Trust is a necessary condition for social acceptability but can break down in a variety of ways—each in turn implying unique conditions for remediation. There are four components of the trust cycle: *Procedural trust* is based on the institutional structures regulating a management action. *Affinitive trust* is based on the character or qualities of the trustee. *Rational trust* is based on the calculated utility—outcomes and impacts—of a management action, as well as the trustee's ability to deliver those outcomes and impacts. *Dispositional trust* is the predisposition of a trustor to be trusting. In a natural resource context, these four dimensions of trust interact to influence management actions, the outcomes of which subsequently inform future trust dynamics. Cycles that foster trust generally reduce social resistance and increase social acceptability, while cycles that degrade trust or cultivate distrust do the opposite. Inspired by Stern and Coleman (2015).

Although many people believe public trust in federal agencies and land managers has diminished, the actual process by which the public has lost trust in federal land-management agencies remains unclear. That is, we currently lack a sophisticated scientific understanding of *why* the public does not trust federal stewardship and management. It stands to reason that, unless we know *why* public trust has diminished, we will fail to understand *how* public trust might be regained. Therefore we suggest the research community needs to continue building upon a small but robust body of work on the social acceptability of different management approaches in the Northwest (e.g., Shindler et al. 2002). Researchers also need to develop and empirically verify theoretical frameworks explaining how different kinds of trust interact with one another to influence overall trust in federal agencies, in order to contribute to practical, policy-relevant solutions.

In the following sections, we review four types of trust and incorporate these into a conceptual framework—a *trust cycle*—allowing us to partition and explore where and how trust may be lost or regained in the management of public lands. We focus in particular on the extensive moist coniferous forests of the Pacific Northwest, providing an example of how the social dynamics of trust around federal forest management in this region could be integrated into an interdisciplinary adaptive-management experiment. The goal of this hypothetical experiment would be not only to describe the multidimensional construct of trust, but also to understand how trust might most effectively be restored, in an effort to improve and facilitate management of federal forests in the Northwest.

Dynamics of Trust

Trust is not a monolithic phenomenon. Instead, it can be conceived as an interplay among at least four components (fig. 18.1), each of which has dynamic elements (box 18.1). Based on research dating at least to the mid-1970s, managers and scientists have developed an understanding of the relationship between public involvement in the management process on the one hand, and trust, social license (social acceptance for different management actions and objectives), and management success on the other. Wengert (1976:23) framed this well: "Participation and involvement . . . may induce modifications of values and opinions and increase confidence and trust. . . . Group discussions and exchanges of ideas are said to minimize hostility and may permit constructive collaboration." Although in this chapter we focus exclusively on the positive role of trust, we also point out that *distrust* has been identified as a critical factor motivating people to engage with participatory management processes (e.g., Parkins and MacFarlane 2015). From this angle, trust may be seen not as a key to social acceptability and cooperation, but instead as a precursor to public complacency, counteracting broader initiatives to more fully democratize public land management (e.g., Parkins and Mitchell 2005). This distinction empha-

BOX 18.1. TYPES OF TRUST IN A MANAGEMENT RELATIONSHIP

Four types of trust can be discerned in management relationships (Stern and Coleman 2015).

1. *Dispositional trust.* Some people demonstrate a "predisposition to trust others" and accept higher levels of vulnerability. Although largely an innate psychological trait, dispositional trust also can be rooted in a number of contextual factors, including past or current interactions with agency managers; agency reputation; and other social or cultural norms.

2. *Affinitive trust.* Affinitive trust deals with a trustor's willingness to accept vulnerability based on an assessment of the trustee's character or qualities, such as benevolence or integrity. Basically, if members of the public do not trust managers as people or do not believe managers have their best interests at heart, trust is likely to falter. Like dispositional trust, affinitive trust can be based on a number of variables, including past experiences with managers and beliefs about the manager's values.

3. *Procedural trust.* The structures surrounding and supporting a management action or direction form the basis for procedural trust. Largely a function of perceived legitimacy and credibility, procedural trust can be influenced by factors at a number of scales, from specific outreach and engagement efforts in local agency districts to trends in national politics. For example, recent scholarship reveals a perception by a large majority of agency scientists that political pressures are inappropriately influential in federal agencies (Goldman et al. 2015). This view likely contributes directly to the participating scientists' procedural trust (or lack thereof) in agency management.

4. *Rational trust.* This cognitive dimension of trust is based upon a trustor's calculation of the utility of a particular action, as carried out by the trustee. Rational trust is based not only on an expectation of beneficial outcomes but also on beliefs about the competence of the trustee to deliver those benefits. Thus members of the public may believe variable-retention harvest can be used to maintain late-successional habitat for endangered species, but if they do not believe agency personnel are capable of designing and executing relatively complex variable-retention harvest prescriptions, they may lack rational trust. Rational trust inherently requires sound information about likely outcomes, so the trustor has sufficient information to compare actions.

sizes the complexity of trust and the need to better understand it in shades and subtle nuances rather than broad strokes (Stern and Coleman 2015).

Genuine forms of citizen participation in planning, modifying, and even executing management actions appear to be critically linked to trust and social license (e.g., Gray 1989). The qualification that participation be "genuine" is worth highlighting: people are more satisfied with public involvement processes when they believe their input actually informs management decisions (Daniels and Walker 2001). On the other hand, the public is likely to become disillusioned (and distrustful) if they feel the outcomes of participatory processes are not meaningfully incorporated into management decisions (Irvin and Stansbury 2004).

It is important to understand the multivariant nature of trust (box 18.1), because each component of trust is promoted by different efforts and different agents. Unless we understand where in the cycle (fig. 18.1) trust breaks down, we will not understand how to maintain or restore it. For example, if distrust is motivated predominantly by a lack of faith in the system, remediation likely requires a modification of that system to enhance its perceived trustworthiness. If distrust arises primarily out of the trustor's assessment of a trustee's character, that trustee might have some outreach or character-building work to do. If distrust arises because of the public's uncertainty about ecological impacts associated with a proposed action, certain research questions may warrant investigation, with results communicated through concerted education and public engagement efforts. In short, anyone interested in building social license to pursue active forest management by restoring public trust in federal land management must first identify the specific sources and implications of both trust and distrust.

Building Trust in Management: A Social-Ecological-Ethical Study

Having established trust as a powerful dynamic affecting management of federal forests in the Northwest, we provide an example of how a hypothetical social-ecological-ethical study could be tied to an adaptive-management experiment or pilot demonstration and be used to examine and potentially improve trust relationships.

General Experimental Design

Large-scale silvicultural adaptive-management experiments or demonstrations in Northwest forests carry high credibility among scientists, managers, and other stakeholders, including collaborative groups (chap. 9). These studies, conducted at operational scales, are used to test the approaches that might be used across a spectrum of timber-harvest operations. While such studies often examine a forest's vegetation and structural components, such as down wood and snags, they vary in the degree to which other variables are measured and monitored. Social perception, in particular, is included in only about one-third of these experiments (Poage and Anderson 2007).

Large-scale silvicultural experiments have other limitations as well. For example, it may take a long time (on the order of decades) to amass interpretable data. Disturbances such as fire or disease can unevenly affect replicates or otherwise confound results. Treatment plots can be damaged or studies discontinued owing to fluctuating budgets, lack of agency commitment, or loss of personnel. As with any field study, results may not be generalizable, especially if site selection is biased or if there are not enough sites included in the experimental design for robust statistical analysis. Still, the credibility of such studies is heightened by common agreement that they address important questions; conform to many real-world constraints; are conducted at practically relevant scales; and have high degrees of inclusivity—for example, by involving both scientists and managers in study designs. They also serve as demonstration sites where people can openly view treatments and the resultant stands. Overall, large-scale silvicultural experiments have high potential to contribute to adaptive-management processes (Poage and Anderson 2007).

By effectively gauging public responses to new and existing demonstrations of active management, managers and scientists may be able to develop a better understanding of the trust dynamics underlying social acceptance on public lands. Silvicultural treatment plots can demonstrate a range of existing or proposed management prescriptions for more extensive implementation on federal forests in the Northwest. For example, restoration thinnings designed to accelerate development of late-successional forest attributes have already been widely employed on federal lands throughout the region (e.g., Manning and Friesen 2013). More recently, managers and scientists have become interested in implementing variable-retention harvests on federal lands, often to create early-seral, pre-forest conditions (e.g., Johnson and Franklin 2012). (Variable-retention harvest retains dispersed or aggregated live trees, and dead trees, to create environmental values associated with structurally complex forests [Franklin et al. 1997].) Demonstrations of these various approaches could create a spectrum of forest types at the broader landscape scale, which could then serve as a valuable backdrop for an interdisciplinary social-ecological-ethical research project.

Measurement of Biophysical Response Variables

Variable-retention harvests have been proposed as a method to create complex early-seral habitat in landscapes where decades of fire suppression and plantation establishment have reduced its occurrence (Takaoka and Swanson 2008). Such treatments are postulated to increase diversity, as compared with no treatment or a relatively uniform thinning, by encouraging the development of complex early-seral vegetation and habitat structure while retaining shade-tolerant, late-seral species. A number of plant and animal species depend upon early-seral habitat, including several migratory songbird species currently experiencing population declines (Betts et al. 2010). However, while the potential benefits to some biota and ecosystem processes are increasingly well understood (e.g., Mori and Kitagawa 2014; Seidl et al. 2014), variable-retention harvest is still controversial in the Northwest, at least in certain social circles where it is perceived as just a "sloppy clearcut" (e.g., Kerr 2013).

Creation of early-seral habitat could be a good candidate for adaptive-management experiments. However, even establishing scientific experiments to evaluate the effects of creating early-seral habitat requires public trust—as evidenced by the recent controversy surrounding pilot projects designed to demonstrate variable-retention forestry in western Oregon (Johnson and Franklin 2012). Trust in the conservation objectives of this type of management treatment is likely to be influenced by people's beliefs about whether there is a real need to create early-seral habitat in addition to that produced by wildfire and industrial forest management, as well as their acceptance of claims that variable-retention treatments do in fact produce "high-quality" early-seral habitat. Including a biophysical component is essential, with data collection on a suite of variables that will allow for integrated assessment of feedbacks and linkages. Key variables for assessment include *vegetation* (e.g., overstory and understory composition and densities); *legacy forest structure* (such as standing and down dead wood); *forest soils* (standing stocks at a minimum, and potentially including nutrient dynamics and processing rates of important elements, such as carbon); associated *aquatic environments* (discharge, water chemistry, water temperature, and nutrient processing rates; see box 18.2 on research needs for water resource management); and the *biota* that occur in both terrestrial and aquatic areas across the forested landscape.

Quantifying a response in biota may be particularly important, because the maintenance and restoration of biodiversity in the context of ecosystem management may lie at the heart of many public trust concerns. Research-

BOX 18.2. WATER RESOURCES AND TRUST

There are substantial concerns among stakeholders regarding the effects of forest management on water resources (Barten et al. 2008), but although many stream-reach studies have been undertaken, the effects of forest-management treatments at larger spatial scales are still poorly understood (chap. 15). Hence there is a pressing need for watershed-scale studies, perhaps best achieved through replicated, entire-watershed comparisons. Although small watersheds would be most feasible to study, knowledge gaps around water resources in larger streams are important to address as well. At minimum, paired watershed studies should be used to test one treatment against a control riparian management area, with additional watersheds incorporated to test the effects of additional treatments. As with biodiversity, there are many aspects of the aquatic ecosystem that can be measured. However, three key measurements are relatively easy and inexpensive to monitor: (1) water temperature; (2) turbidity (associated with erosion and sedimentation); and (3) nitrate concentrations. These three metrics begin to address the question of how water quality will change in response to treatments such as variable-retention harvest and other riparian management alternatives, which might in turn be critical in influencing public perceptions of and trust in active forest management.

ers could use a number of information sources (including a Q-method study of the sort discussed below) to elucidate trust dynamics surrounding questions about management effects on species of concern and their habitat. Unfortunately, complete assessment of treatment impacts on biodiversity is often cost prohibitive. A common practice is to focus on a single taxonomic group as a biotic indicator and to evaluate changes in abundance and composition within that group. For example, birds could be an effective group for this method, as most species are easily surveyed and bird communities span the range of ecological associations, from early-seral to old-growth obligates; however, when birds are used as a response variable, larger treatments are generally required to detect potential treatment signals.

Because short-term responses of biotic, hydrological, and other ecological variables are likely to shift over time (in theory generating parallel shifts in social trust as well), it is crucial to host such an experiment in a location that can draw upon long-term measurements. This will require long-term commitment of funding, agency resources, and personnel. A citizen-science approach, involving public groups in rapid assessments of select key biophysical factors through multiparty or collaborative monitoring not only cuts costs, but allowing the public to participate in data collection and analysis has been shown to build social capital (and cultivate trust) among participants (e.g., Wagner and Fernandez-Gimenez 2008). Along with providing data and engaging site visitors, these rapid assessments would provide an opportunity to evaluate the effectiveness of quick assessment methods and to gauge the extent to which they correlate in a broad sense with more rigorous research outcomes.

Measurement of Social Response Variables

Along with information about the forest's biophysical response to management actions, people may also draw upon more fundamental perceptions of reality and the values they associate with the natural world (collectively known as their *worldview*) in forming opinions about forest management. It is reasonable to assume that these various worldviews are correlated in important ways with other variables influencing trust in forest management. If this is the case, understanding the worldview of relevant stakeholders becomes critical for restoring trust and earning social license. Although traditionally the purview of philosophy, theology, and ethics, worldviews can also be studied empirically by employing the tools of social science (e.g., Vaske and Donnelly 1999; Gore et al. 2011). We might theorize that differences in worldview among certain individuals or groups will be linked to other variables, such as their willingness to accept certain forms of active management or to trust the agent of a management action. For example, it seems reasonable to hypothesize that more morally inclusive groups will be less willing to accept risk or trust management actions (i.e., groups that attribute intrinsic value to more things may perceive more to be at risk, and so be more reluctant to trust). With careful research design, this is a hypothesis that can be tested.

A Q-method design could be used to investigate the interplay between public trust and active management, and more broadly, to describe the current discourses around environmental governance with greater precision (box 18.3). Empirical data from the biophysical assessments discussed above could also be incorporated into focus-group study design in a controlled way, allowing researchers to begin disentangling the influence and interplay of the various dynamics of trust (box 18.1). Information derived from these empirical data would enable richer discussions of trade-offs and

BOX 18.3. USING THE Q-METHOD FOR UNDERSTANDING THE TRUST RELATIONSHIP

The Q-method is used to collect qualitative and quantitative data about subjective viewpoints and is a powerful tool in environmental social science. It highlights major differences and agreements among social perspectives, and the underlying philosophical arguments used to justify claims, without imposing biases of structured survey questionnaires (Robbins and Krueger 2000). The Q-method could be used to develop a nuanced, multifaceted understanding of the "landscape of beliefs" surrounding active management in the Pacific Northwest. This rich understanding could then be used to inform and facilitate meaningful dialogue with stakeholders by those seeking to cultivate or restore a cycle of trust.

The Q-method has four main steps. First, researchers identify the "concourse" of possible feelings or reactions related to a particular domain (e.g., active forest-management project). Usually this includes archival analysis and informal interviews. Archival analysis could focus on newspaper editorials, published debates and reports, and websites of regional organizations that address forest management. Researchers can conduct informal interviews with organization representatives, relevant scientists, agency personnel, and local residents. These interviews may explore cognitive and affective antecedents that might influence trust in active management, including disposition; past experiences with agency personnel, policies, and management actions; social memory; and vulnerability to risks related to management actions. Interviews could focus on questions addressing three related topics that may bear importantly on trust or distrust of active management (table 18.1). Answers could shed light on the types of information that might be salient to rational trust, affinitive trust, and procedural trust.

Second, participants are recruited for a sorting exercise. Participants would sort quotes drawn from the "concourse" discussed above, from "most agree" (+4) to "most disagree" (−4). This process would be paired with open-ended interviews eliciting explanations for sorting responses. Third, sorted data are analyzed. Finally, researchers identify distinguishing statement sets per significant factor from the results (Robbins and Krueger 2000). Follow-up semistructured interviews would be conducted with a subset of the original sorters to validate the researchers' preliminary interpretations of factors. The Q-method analysis could be followed by field trips and focus groups with the original interview participants, building toward a rich theoretical understanding of trust while simultaneously working to rekindle trust among key players in the Pacific Northwest.

TABLE 18.1. Topics with bearing on trust or distrust of active forest management, and relevant questions to be asked during stakeholder interviews during implementation of the Q-method to examine social viewpoints

Topic	*Description*
Environmental knowledge regarding forest dynamics	How do interviewees (both trustors and trustees) understand forest dynamics related to landscape pattern and process in general and seral stages in particular? Which forest-health indicators are meaningful to them? What kind of management (e.g., active vs. passive) do they believe is necessary to ensure a healthy forest? If active management is desirable, which specific interventions or treatments are viewed as most effective?
Perceptions of stakeholders	How do members of the public characterize federal land managers? What key characteristics do they believe render (or would render) them trustworthy as decision makers? On the flip side, how do land managers perceive the public? To what extent and on what basis do they value and trust public opinion as a guide for management decisions? Are there key historic interactions that have significantly influenced or altered public perceptions of managers, or vice versa?
Institutional preferences	What sorts of governance arrangements do people believe can most effectively ensure management of healthy federal forests? Who are the necessary actors (e.g., who should authorize, oversee, and carry out treatments), and which rules and behaviors should guide the process (e.g., what kind of environmental review is necessary, and is the National Environmental Policy Act process adequate)?

synergies in managing public forest landscapes for a diversity of seral stages and forest products and could provide a nexus around which trust might begin to be restored.

Integration and Trust

As we hope is clear by now, an exploration of trust in forest management is highly interdisciplinary, requiring inputs from both ecological and social sciences and drawing insights from the humanities disciplines of philosophy and ethics. Typical multidisciplinary projects operate by creating a disciplinary division of labor, wherein each discipline operates autonomously alongside some other set of autonomously operating disciplines on a question of common interest (sometimes jokingly called interdisciplinarity by stapler). In contrast, a broadly interdisciplinary experiment (including aquatic ecology, philosophy and ethics, soil science, hydrology, outreach, plant ecology, social science, biodiversity, landscape ecology, and forest management) comes together as a matter of necessity when the phenomenon to be studied (e.g., public trust) not only allows but indeed *requires* the participation and collective wisdom of many disciplines. For example, it is reasonable to believe—and consistent with the concept of trust portrayed in Stern and Coleman (2015)—that the ecological response of a system undergoing active management is highly relevant to public perception and rational trust (fig. 18.1). Hence, in natural resources arenas, one cannot adequately understand the social phenomenon of public trust without understanding the ecological response of a system. Likewise, while dispositional trust, affinitive trust, and procedural trust likely vary significantly among individual members and larger groups of the public (fig.18.1), we anticipate that this variation could be categorized according to discrete sets of variables studied by (among others) social psychologists, sociologists, political scientists, and ethicists. In short, we suggest that the complex and multifaceted concept of trust can best be understood through an analysis and synthesis of its various components, a process that is inherently, comprehensively interdisciplinary in nature. An adaptive-management experiment offers an excellent opportunity for such an analysis.

Conclusion

Much effort is expended in conceptualizing, planning for, and justifying various management actions. This is perhaps especially true for moist temperate forests in the Northwest, given the iconic status of old-growth forests, strong social connections with forests, and the relatively recent history of highly contentious forest-management approaches. Recognizing that any proposed approach will be enacted only if met with approval by a relevant public, researchers can incorporate social dynamics (such as trust) into their theoretical frameworks and conceptual models as they continue to think about future management in these forests. The process by which that relevant public deems a particular forest-management approach appropriate or inappropriate is complex—certainly more complex than we have recognized in the past—and likely responds to a number of ecological, sociological, and philosophical variables and their interactions.

Although researchers have attended to some of the ecological variables in the past, they have spent far less effort understanding and accounting for the sociological and philosophical variables that influence trust and therefore social license. At present, they have only a generalized understanding of public trust, and it seems apparent that it may be easily lost among key social demographics, contributing to an unraveling of management partnerships. As efforts to remediate this situation proceed, it may befit managers and researchers to partner in designing multidisciplinary adaptive-management projects, which might allow them to begin making efforts to restore public trust not on the basis of guesswork and intuition, but rather on the basis of the best available social and ecological science.

Acknowledgments: Support for this work was provided by the H.J. Andrews Experimental Research Program, funded by the National Science Foundation's Long-Term Ecological Research Program (DEB 1440409), US Forest Service Pacific Northwest Research Station, and Oregon State University.

Literature Cited

Barten, P. K., J. A. Jones, G. I. Achterman, and 11 coauthors. 2008. *Hydrologic effects of a changing forest landscape*. Washington, DC: National Academies Press.

Betts, M., J. Hagar, J. Rivers, J. Alexander, K. McGarigal, and B. McComb. 2010. Thresholds in forest bird occurrence as a function of the amount of early-seral broadleaf forest at landscape scales. *Ecological Applications* 20:2116–2130.

Carpenter, S. R., H. A. Mooney, J. Agard, and 13 coauthors. 2009. Science for managing ecosystem services: Beyond the Millennium Ecosystem Assessment. *Proceedings of the National Academy of Sciences of the United States of America* 106: 1305–1312.

Daniels, S. E., and G. B. Walker. 2001. *Working through environmental conflict*. Westport, CT: Praeger.

Doak, D. F., J. A. Estes, B. S. Halpern, and 11 coauthors. 2008. Understanding and predicting ecological dynamics: Are major surprises inevitable? *Ecology* 89:952–961.

Franklin, J. F., D. R. Berg, D. A. Thornburgh, and J. C. Tappeiner III. 1997. Alternative silvicultural approaches to timber harvesting: Variable retention harvest systems. Pp. 111–137 in *Creating a forestry for the 21st century: The science of ecosystem management*. Edited by K. A. Kohn and J. F. Franklin. Washington, DC: Island Press.

Franklin, J. F., and K. N. Johnson. 2014. Lessons in policy implementation from experiences with the Northwest Forest Plan, USA. *Biodiversity and Conservation* 23:3607–3613.

Goldman, G., M. Halpern, D. Bailin, A. Olali, C. Johnson, and T. Donaghy. 2015. *Progress and problems: Government scientists report on scientific integrity at four agencies*. Report for the Union of Concerned Scientists. Cambridge, MA: Center for Science and Democracy at the Union of Concerned Scientists. http://www.ucsusa.org/sites/default/files/attach/2015/09/ucs-progress-and-problems-2015.pdf.

Gore, M. L., M. P. Nelson, J. A. Vucetich, A. Smith, and M. Clark. 2011. Exploring the ethical basis for conservation policy: The case of inbred wolves on Isle Royale, USA. *Conservation Letters* 4:394–401.

Gray, B. 1989. *Collaborating: Finding common ground for multiparty problems*. San Francisco: Jossey-Bass.

Hays, S. P. 2006. *Wars in the woods: The rise of ecological forestry in America*. Pittsburgh, PA: University of Pittsburgh Press.

Irvin, R. A., and J. Stansbury. 2004. Citizen participation in decision making: Is it worth the effort? *Public Administration Review* 64:55–65.

Johnson, K. N., and J. F. Franklin. 2012. *Southwest Oregon Secretarial Pilot Projects on BLM lands: Our experience so far and broader considerations for long-term plans*. http://www.blm.gov/or/news/files/pilot-report-feb2012.pdf.

Kerr, A. 2013. *Senator Ron Wyden's Oregon and California Land Grant Act of 2013: The good, the mediocre, the bad, and the ugly*. http://www.andykerr.net/wydenoandc/.

Lachapelle, P., and S. McCool. 2012. The role of trust in community wildland fire protection planning. *Society & Natural Resources* 25:321–335.

Manning, T., and C. Friesen, eds. 2013. *The young stand thinning & diversity study (YTSDS): Establishment report, study plan, and key findings*. USDA Forest Service and Oregon State University. http://ecoshare.info/projects/central-cascade-adaptive-management-partnership/forest-studies/young-stand-thinning-and-diversity-study/.

Mori, A. S., and R. Kitagawa. 2014. Retention forestry as a major paradigm for safeguarding forest biodiversity in productive landscapes: A global meta-analysis. *Biological Conservation* 175:65–73.

Parkins, J. R., and B. L. McFarlane. 2015. Trust and skepticism in dynamic tension: Concepts and empirical refinements from research on the mountain pine beetle outbreak in Alberta, Canada. *Human Ecology Review* 21:133–153.

Parkins, J. R., and R. E. Mitchell. 2005. Public participation as public debate: A deliberative turn in natural resource management. *Society & Natural Resources* 18:529–540.

Poage, N. J., and P. D. Anderson. 2007. *Large-scale silviculture experiments of western Oregon and Washington*. General Technical Report PNW-GTR-713. Portland, OR: USDA Forest Service, Pacific Northwest Research Station.

Robbins, P., and R. Krueger. 2000. Beyond bias? The promise and limits of Q method in human geography. *Professional Geographer* 52:636–648.

Seidl, R., W. Rammer, and T. A. Spies. 2014. Disturbance legacies increase the resilience of forest ecosystem structure, composition, and functioning. *Ecological Applications* 24:2063–2077.

Shindler, B., M. Brunson, and G. Stankey. 2002. *Social acceptability of forest conditions and management practices*. General Technical Report PNW-GTR-537. Portland, OR: USDA Forest Service, Pacific Northwest Research Station.

Spies, T., and S. Duncan, eds. 2009. *Old growth in a new world*. Washington, DC: Island Press.

Spies, T. A., T. W. Giesen, F. J. Swanson, J. F. Franklin, D. Lach, and K. N. Johnson. 2010. Climate change adaptation strategies for federal forests of the Pacific Northwest, USA: Ecological, policy, and socio-economic perspectives. *Landscape Ecology* 25:1185–1199.

Stern, M., and K. Coleman. 2015. The multidimensionality of trust: Applications in collaborative natural resource management. *Society & Natural Resources* 28: 117–132.

Takaoka, S., and F. Swanson. 2008. Change in extent of meadows and shrub fields in the central western Cascade Range, OR. *Professional Geographer* 60:1–14.

Vaske, J. J., and M. P. Donnelly. 2010. A value-attitude-behavior model predicting wildland preservation voting intentions. *Society & Natural Resources* 12: 523–537.

Wagner, C. L., and M. E. Fernandez-Gimenez. 2008. Does community-based collaborative resource management increase social capital? *Society & Natural Resources* 21:324–344.

Wengert, N. 1976. Citizen participation: Practice in search of a theory. *Natural Resources Journal* 16:23.

Winkel, G. 2014. When the pendulum doesn't find its center: Environmental narratives, strategies, and forest policy change in the US Pacific Northwest. *Global Environmental Change* 27:84–95.

Chapter 19

Human-Forest Ecosystem Sustainability

Deanna H. Olson, Beatrice Van Horne, Bernard T. Bormann, Robert L. Deal, and Thomas H. DeLuca

In this book we have woven a socioecological synthesis to describe how forests and communities have changed over the last two to three decades, especially in the moist coniferous forest zone of the US Pacific Northwest. Lessons have emerged from the social, physical, and biological studies of these forests, from contemporary forest resource management, and from traditional, indigenous resource and environmental management (table 19.1). In this chapter, we take a broad perspective on what we have learned and highlight the emerging principles from our cross-disciplinary review, with an eye toward improving management of moist coniferous human-forest ecosystems. In the next and final chapter, we continue this theme and focus on some tangible steps to sustain the best these landscapes have to offer into the future.

As forest sciences have developed and managers have applied the new information over the last three decades, there have been considerable advances, as well as struggles and pitfalls. Two currents that might seem to be in opposition but that have been gaining strength in both scientific and social sectors are (1) an increased appreciation for the complexity resulting from the spatial and temporal dynamics of forest ecosystems and (2) the reality that we can no longer talk about how best to sustain forests without considering people. These currents underpin the growing acceptance of what we describe as a human-forest ecosystem-sustainability goal. In this goal, we see the promise of a middle ground that will damp the pendulum swing of priorities from the historical "people-first" wood-production

TABLE 19.1. Key points from chapters 1–18

Chapter	*Key Points*
1	Creation of today's human-forest ecosystems has altered forest sustainability objectives from a focus on timber to providing a broader array of ecological services. We are seeing an increase in the amplitude and frequency of change to forests because of human actions, climate change, fire, and other disturbances. This book provides baseline historical context, new knowledge since the early 1990s, and insights into possible future directions for the next phases of forest management.
2	Complex vegetation patterns of today's Northwest moist coniferous forests set the stage for a flexible approach to focused commodity production and policies supporting multiple silvicultural methods.
3	Tending forest plants for human uses yields intergenerational knowledge that can help in managing ecosystems for structural, compositional, and functional diversity. Opportunities for collaboration can integrate cultures and perspectives.
4	The social and economic conditions of forest-based human communities have not been stable over the past century. Dealing with relentless change is a challenge for community residents and those who value working forested landscapes.
5	Valuation systems are being developed to describe forest ecosystem services. But societal agreement is needed on priority services so that these can be used to evaluate human benefit trade-offs and to design management for key forest resources and their supporting components.
6	Historical forest management in the Pacific Northwest has led to a bimodal landscape with variable priorities that depend upon landownership. An ecosystem-services framework can be used to describe forest benefits for people and to evaluate potential trade-offs between services following management actions.
7	Effects of past ownerships and harvest patterns, reserves, and fires are visible on today's forest landscape. Today's decisions will be reflected in future patterns.
8	While some elements of adaptive management are evident in forest-management practices over the years since 1994, several planned programs have been forestalled, limiting learning opportunities.
9	Collaborative groups addressing novel community-scale forest governance at larger spatial scales are being convened regionally. Whether they will succeed has yet to be determined.
10	A pulse of alternative silvicultural approaches has been tested in the last few decades, with goals to accelerate development of late-successional forest conditions, achieve ecological objectives during active forest management, and manage fire and other risks.

TABLE 19.1. Continued.

Chapter	*Key Points*
11	Productivity of moist coniferous forests is associated with soil diversity; atmospheric conditions (growing seasons, CO_2); nonconiferous vegetation; and harvest practices, including soil compaction and nutrient retention. The relationship of productivity to fire, down wood, and active management techniques is more variable.
12	Northwest forests sequester large amounts of carbon, preventing the release of gases that would contribute to the greenhouse effect. Management to retain and restore forest lands, maintain and improve ecosystem and soil productivity, and increase the life span of carbon in live and dead wood as well as forest products would maintain or increase the size of this carbon store.
13	Biodiversity elements as ecosystem services of moist coniferous forests have become a dominant management priority since the early 1990s. Hundreds of species, their life-history variation, and their key ecological functions are being actively managed for long-term persistence. Important lessons include the need to consider multiple disturbances to species, not just effects of timber harvest, and connectivity of their habitat across all lands.
14	Aquatic-riparian systems are multistate, with considerable spatial and temporal variation across the region. Use of stream buffers to forestall adverse effects of timber harvest on critical habitat attributes (temperature, down wood, landslides) and sensitive species is being tested. Mixed-width riparian buffers are being developed to protect and restore riparian zones for multiple upland and aquatic objectives.
15	With multiple ecosystem components being aggregated at the larger spatial scales of watersheds and landscapes, management of key structures, functions, and processes for ecological resilience is a priority. Technological and modeling advances in landscape ecology have stimulated active development of tools to bridge science advances and management needs.
16	Climate change patterns in Northwest forests are being documented, spurring model development, projections for key ecosystem services, and new management considerations. Climate-smart forest management includes managing for shade, fire risk, and restoration of vegetation for altered climate conditions.
17	A wealth of new forest products relies on component parts of trees for use in construction, chemical engineering, and consumer goods. This is changing and diversifying ways people think about and conduct timber harvesting.
18	Trust is essential to effective collaboration among stakeholders with an interest in management of moist forests. Distinguishing four types of trust helps in understanding barriers to conflict resolution. Trust can be built with collaborative adaptive-management projects in which partners learn together how systems respond to management treatments.

focus (a philosophy that still resonates with some landowners) to the more recent "ecology-first" ecosystem-management goal.

Swinging Pendulums

In the US Pacific Northwest, including Southeast Alaska, the swing away from people-first toward adoption of an ecology-first priority on public lands was largely a reaction to the perceived unsustainability of the people-first priority, along with the recognition that forests provide a broader set of ecosystem services beyond wood products. It was an attempt to slow the fragmentation of the forest and retain species, habitats, processes, and functions that appeared to be on the verge of loss (e.g., chaps. 13, 14). Federal forests took on a protective role for a wide range of forest ecosystem elements associated with late-successional forest conditions under the 1994 US federal Northwest Forest Plan. This also may have forestalled the perceived growing burden on nonfederal lands to make more significant ecological contributions; since Plan implementation, many management decisions on nonfederal lands may have been made assuming the persistence of ecology-first priorities on federal forestlands (Suzuki and Olson 2008).

The 1994 Plan led to a need for new information. Anticipating future forest conditions and the ramifications of the new land-use allocations (e.g., 84% of federal forestlands within the range of the northern spotted owl were placed in reserved land-use allocations, versus 16% in matrix, where regeneration harvest was allowed), scientists and managers examined different silvicultural treatments (chap. 10) using adaptive-management (chap. 8) and applied-research experiments (chap. 14). These studies aimed to accelerate development of late-successional forest conditions; retain critical aquatic-riparian resources; achieve other ecological objectives; and reduce fuel loading in fire-prone areas. Also, interactions of different resource mitigations of the Plan with dynamic processes within the Plan area merited attention. For example, management recommendations for reducing fuels in areas with sensitive late-successional forest-associated species were developed for implementation in the wildland-urban interface near human communities (USDI and USDA 2002). During this post-Plan period, many people anticipated a reduction in federal wood production, and indeed the volume of large logs from federal lands greatly decreased. Largely in response to global market demand, private industrial lands shifted their harvests to more frequent rotations, often of 30 to 40 years, helping to spawn a pulse of development of new wood products (chap. 17).

A bifurcated landscape began to develop, with people-first profit maximization on private lands and ecology-first retention of old-growth forest and management for late-successional structure on federal lands. Overall, in the dramatic move toward ecology-first goals on US federal forestlands during the 1990s, community- and forest-health efforts lost ground to protective philosophies. With only 16% of the Plan area in the harvestable matrix allocation and a portion of that area in recent clearcuts, timber production from federal lands was reduced. Matrix lands included a high proportion of second-growth forests, but some areas with mature trees. Because of social pressures and new recommendations for spotted owl management under the recovery plan, many federal land managers have deferred harvest of the older-forest component of the landscape.

Some environmental organizations and individuals now advocate a philosophy of protectionism, pushing for no commercial harvest on federal lands. Harvests proposed in matrix lands containing older forests are often appealed, protested, and litigated. Both federal land managers and environmental organizations have had mixed success in court, but the entire process is costly in time and funds, with federal land managers bearing major financial costs even when they prevail in court. Continued lawsuits, perceived threats of challenges, and other constraints have prevented federal forest managers from taking full advantage of the flexible provisions of the Northwest Forest Plan that might have allowed for ecologically designed timber harvests. For example, the Plan allowed for varying the widths of riparian buffers according to results of a watershed analysis, but this flexibility was rarely realized. Uncertainties in how to achieve Aquatic Conservation Strategy objectives added to managers' hesitancy in removing trees from riparian reserves. Both uncertainties and protective philosophies likely fostered an initial risk-averse approach to riparian and upland forest management; more recently, however, adaptive management based on new knowledge is being planned.

The bimodal forest has been modeled into the future for a subset of the area and its ecosystem services (e.g., Spies et al. 2007). For the Oregon Coast Range province, landownership was the strongest predictor of biodiversity patterns. The diversity of habitat-patch types declined slightly overall but declined steeply within certain ownerships. Overall, the province moved toward but did not achieve the historical range of variation of forest age classes. Spies et al. (2007) examined alternative policies to reduce the spatial contrast among ownerships. One policy alternative—increasing retention of wildlife trees on private lands—reduced the contrast and provided benefits to both early- and late-seral-associated species.

In this vein, more managers, scientists, community members, and stakeholders have begun to think about moving from a static protective stance to more active management to foster a greater variety of dynamic ecological conditions. Forest-restoration goals that could benefit from active forest management include mitigating the loss of early-seral vegetation; improving riparian and aquatic systems; managing fire and fuels across broader lands (not just the wildland-urban interface); addressing all-lands solutions to concerns for declining species; and adapting to climate change. At the same time, many have recognized the need to slow or stop the economic decline of forest-based rural communities, reawakening concerns about socioeconomic well-being. The pendulum is swinging away from ecology-first toward a middle ground combining forests and people. The latter view is more consistent with traditional resource- and environmental-management practices of peoples indigenous to the region (chap. 3).

Furthermore, since the 1990s, we have increased our understanding of how forest resources respond to different management actions. New knowledge has resulted in additional protections on state and private lands managed primarily for commodity production. For example, research and monitoring findings since 1994 have resulted in consideration of wider riparian management zones for state and private lands, and down-wood provisions and water-temperature buffering for streams on all lands. Hence provisioning of multiple forest resources has expanded beyond federal lands to provide more ecological benefits from lands managed for profit. Together, current trends for more active management on federal lands and some specific additional protections on nonfederal lands appear to be softening the bifurcation of forest practices.

The ecology-first and people-first perspectives come to the forefront as multiple landowners convene to plan management strategies across their adjoining boundaries. Often, polarized perspectives become clear as collaborative groups seek resolutions to address multiple-resource provisioning (chap. 8). For example, across the Northwest, forest collaborative groups that include a diverse array of stakeholders (chap. 9) have been addressing how the common goal of sustainable forest management could be implemented to promote both ecological and community well-being. Many collaborative meetings in the last decade have discussed win-win solutions to achieve this goal, such as by promoting management for different ecosystem services in different portions of watersheds or the broader landscape. Yet resolving conflicts is not easy with diverse value systems, forest management priorities, and legal status of jurisdictions in play. Furthermore, trust between people who have different knowledge and hold different views about forests is a complex issue, with no simple formula for resolution (chaps. 8, 9, 18). The efficacy of collaborative group governance for forest management is not fully established and continues to develop. A full solution to conflict resolution regarding the diverse values that society places on natural resources has not yet been realized.

A New Sustainability Goal

The debate about ecological and social priorities in setting an agenda for a sustainable future of the human-forest ecosystem has been influenced by a number of global events. As discussed in chapter 1, sustainability has been adopted around the world in one form or another by many conservation groups, the United Nations, and the World Bank. Views on sustainability have evolved since the 1980s, with the concept coming under increasing scrutiny owing to our lack of progress in dealing with the growing impacts of the Anthropocene, the epoch of a human-dominated Earth. From climate change and atmospheric pollution to nitrogen saturation and marine hypoxia, to resource exploitation and ecosystem collapse, sustainability can appear to be unattainable. In this light, some have proposed an emphasis on seeking ecological resilience in an ever-changing landscape (e.g., Stoll-Kleemann and O'Riordan 2002; Benson and Craig 2014), with the hope that we might find a better means of realizing harmony in an uncertain future. This concept of resilience is appealing because it acknowledges that ecosystems naturally occur in multiple states (chap. 14, fig. 14.1; see also chap. 15), whereas some scientists and managers have focused on concepts of thresholds or constant, static target metrics for maintaining idealized attributes (e.g., Holling and Meffe 1995), rather than a range of attributes. In a dynamic forest environment, maintaining both ecological and economic sustainability points toward managing heterogeneous conditions including disturbances and change through multiple successional stages to provide resilience. Communities in particular desire a predictable and sustained income from dynamic forests and fluctuating markets; flexible harvest guidelines and a broader base of forest products offer a partial solution.

The goal of human-forest ecosystem sustainability offers a means to integrate the ecology of moist forests with the people who love and depend on them. The goal of sustainability is laudable and may be locally realistic when the definition is constrained to meeting human needs while inducing minimal environmental degradation. Hence we propose that human-forest ecosystem sustainability be considered the paramount goal to be supported by continued learning and adaptation at both local and regional scales. Such learning contributes to resilience.

Making sustainability of both ecological and social systems an overarching goal can be helpful in evaluating major changes in community and ecosystem well-being that have taken place in these moist forests since the 1980s. With current landownership priorities and an increasingly bimodal landscape focused on either production of timber (private lands) or forest ecological integrity (public lands, chaps. 6, 7), a point has been reached where the costs of working in opposition are high enough to motivate change. For example, ecological costs such as risks to some species or populations occur where key habitats are relatively unprotected on some private lands (e.g., some Pacific salmon, Reeves et al. 2016; wildlife habitat and connectivity concerns, chap. 13), whereas economic costs occur on federal timberlands where timber removal is restricted. On private forestlands with a primary focus on wood production, a secondary focus is emerging for riparian systems to retain water quality and quantity and aquatic-dependent species, and for terrestrial areas to support habitat for at-risk wildlife by retaining snags, preferred trees, and broad-scale connectivity. On federal lands, novel approaches for timber resource extraction are developing (chap. 10).

On federal lands under the Northwest Forest Plan, the scientific impetus was to restore the full distribution of successional stages that were characteristic of the region (especially old-growth forest conditions) and to support their associated plants and animals. Although the Plan recognized that development and disturbance would continue to shift stages through time, the fixed boundaries between areas with differing management objectives have made coherent management of this changing landscape difficult. A reknitting of the landscape could allow us to take a new perspective on forest management for diverse objectives that include enhancing ecological integrity and resilience, as well as promotion of societal sustainability.

Implementing the Goal

The key points of this book (table 19.1) highlight several general principles that, if incorporated into collaborative agreements and policies, could significantly contribute to a renewed goal of human-forest ecosystem sustainability for moist coniferous forests in the northwestern United States and elsewhere. We have emphasized the importance of evaluating sustainability at a landscape scale, along with the importance of considering a variety of social and ecological scales. Different scales encompass different ecological states and processes, ecosystem services, and societal interest groups. Where differences in goals at these different spatial scales require trade-offs in planning management activities, we might consider what processes or incentives could facilitate compromise, negotiation, and conflict resolution. Flexibility to allow forest management to vary in different landscape areas is essential.

Integrating across Spatial Scales

The location of forest-management boundaries on a regional map reflects an underlying understanding of the processes and functions at play, whether they are ecological or socioeconomic. Multiple stakeholders and ecosystem services are at the heart of large human-forest ecosystems, thus different boundaries can be used to define the system from multiple perspectives (fig. 19.1). Also, different species and ecosystem processes operate at different spatial scales. Box 19.1 describes boundaries used in resource assessments in relationship to scaling and other planning issues. Because disturbances and issues such as connectedness cut across ownership, sociopolitical, provincial, watershed, county, and state boundaries, however, none of these is a panacea, and an all-lands approach (Vilsack 2009) is needed to bridge time and geography. Ideally, a systems approach that encompasses both biological and ecological consequences, costs, and trade-offs of management activities could be used to highlight how and where particular activities can best be implemented.

Using Local Knowledge

Bridging ecological and socioeconomic priorities across all lands is a tall order. A first step could be to develop finer-scale management aims and define specific objectives and metrics of ecological and community well-being critical to sustainability for a particular place. Differences can be bridged by encouraging groups to prioritize local goals in terms of ecosystem

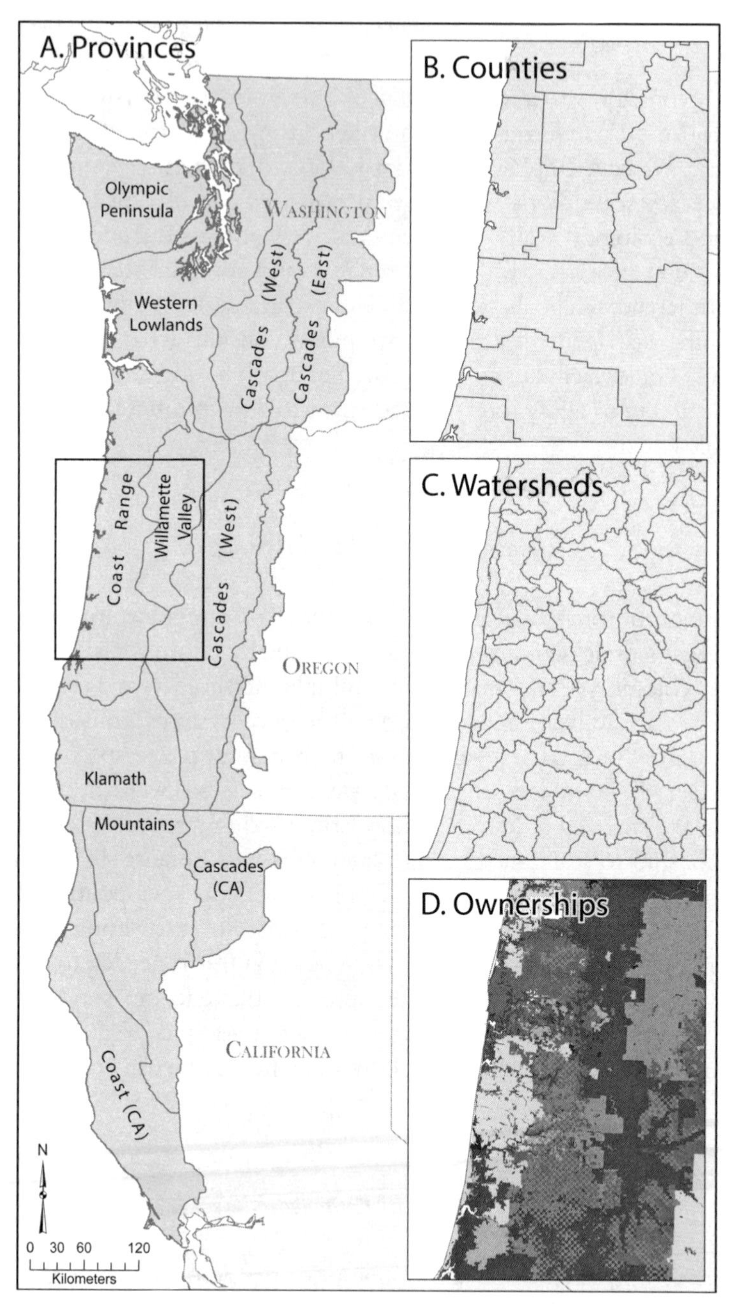

FIGURE 19.1. Examples of alternative jurisdictional boundaries used for management considerations in moist coniferous forests, including physiographic provinces within the Northwest Forest Plan area of Oregon and Washington, USA (A) and inset boxes from a smaller area showing counties (B), 5th-field watersheds (10-digit hydrologic units, C), and numerous landownerships (D).

BOX 19.1. BOUNDARIES AND SCALE

Among the boundaries used for land-management prescriptions, landownership boundaries (fig. 19.1D) may seem to have fewer ties to forest ecology. However, some of these boundaries have direct links with forest productivity or ease of access for timber harvest, facilitating profitable resource extraction in lands dominated by private ownerships. They also drive decisions about changes in governance or land-use policies, such as through land aggregation, sales, or swaps. For this reason, the map of current protected reserves reflects the legacy of prior landownership and allocation decisions (chap. 7).

In the development of the Northwest Forest Plan, *physiographic* (*vegetative*) *provinces* (fig. 19.1A) were used to define important ecological boundaries. These provinces have the advantage of capturing differing vegetative types and physiographic conditions so that forest management can focus on the differing ecosystem services these conditions support. Historical disturbance regimes vary between and within the physiographic provinces, with a general trend of increasing fire frequency from north to south and from the coast to east of the Cascade Range. As fire disturbances rose to prominence in the last three decades, differences in fire behavior between provinces became evident with recent large wildfires (chap. 7), supporting the suitability of province-based forest planning. Linkages among pests, pathogens, and invasive species are complex but also can align with provinces as well as administrative boundaries

In the Pacific Northwest, *watershed boundaries* (fig. 19.1C) are commonly used as planning units for forest management, with a focus on hydrological and other processes that support aquatic ecosystem integrity. Watersheds also integrate across vegetative types and sometimes across provinces; they are inclusive of forested uplands and lowland rural-urban settings, so they include human and forest elements. In his 1890 *Map of the Arid Region of the United States, showing Drainage Districts*, John Wesley Powell advocated using regional watershed boundaries as the essential units of government (Powell 1891).

In contrast to these ecological or physiographic approaches, the use of borders based on *county and state lines* (fig. 19.1B) fits the goal of human-forest ecosystem sustainability, in which social (political) processes can be accommodated across larger geographies. Broad-ranging taxonomic groups (e.g., owls, murrelets, Pacific salmon, forest carnivores) require habitat connectivity over large areas regardless of political boundaries and would benefit from larger-scale planning across counties and states as well as watersheds, provinces, and ownerships.

services, then use conceptual reverse-engineering approaches to establish the highest priorities for management activities to support both those final and their requisite intermediate ecosystem services (chaps. 5, 6). When whole-system consequences are defined partly in terms of ecosystem services, the processes that need management attention may become more explicit. Managing for ecosystem services involves understanding the broad suite of benefits a forest provides to society and clarifying relationships between the quantity and quality of services provided and the condition of ecosystems that provide them. Systems-based thinking can combine diverse ecological conditions and human perspectives to address the concept of sustainability. People and forests can be considered together using such a systematic framework, in which we can recognize the internal interactions of ecological and social processes as drivers of whole-system consequences.

Listening to and respecting local ecological and community knowledge is essential for informing priorities based on ecosystem services goals. Variation in soils, climate, habitats, biota, and human needs can occur at small scales, so local knowledge is important for understanding vulnerabilities and developing strategies that address both social and ecological concerns.

Defining areas where success in reaching different goals (chaps. 6, 8) is tested using adaptive-management approaches could form the backbone of novel research and monitoring programs, such as those mandated in the recent Forest Service planning rule (USDA 2012). Initial steps—to achieve and sustain ecological and community well-being in either the old- or young-forest areas—could focus debate among collaborative groups where local knowledge and innovation may be tapped.

Forest collaboratives (chap. 9) have specifically recognized and implemented a local focus—where both local people and those from outside with major stakes in the outcome have a voice and take leadership, assisted by professional input from managers and researchers. Regional or national federations of such groups may offer stability and economies of scale provided by shared approaches, ideas, and recommendations (Pretty 2002). Ultimately, defining sustainability for both rural and urban ecosystems and making the ecological and socioeconomic flows between them explicit can foster mutually respectful and inclusive solutions.

Integrating across Temporal Scales

Knowing that the landscapes, societies, and our understanding of both will change through time owing to a variety of factors (and not always in ways we expect) argues strongly for a dynamic perspective and for the corresponding concepts underpinning disturbance ecology (multistate systems and resilience, fig. 14.1) as fundamental components of the human-forest ecosystem. This also supports an adaptive-management approach that includes multidecadal studies to monitor long-term trends, with an understanding that different ecosystem services have different response times. A shifting-mosaic concept contributes to this perspective by recognizing community patches (e.g., forest stands and local human communities) within a complex of larger patches where change is the rule, not the exception.

An example of applying this shifting-mosaic concept to restoration management within the area of the Northwest Forest Plan could focus management activity on both large, even-aged, old harvest units and newly disturbed patches. Sequential patch management at a planned pace across spatially contiguous areas can help to address the need for early-seral vegetation (chaps. 14, 20). Also, local managers could take advantage of unanticipated disturbances (wind, fire, disease) for objectives such as early-seral provisioning. The shifting-mosaic concept could be applied to older-forest stands to reduce fuels or address specific habitat objectives. Harvest of specialty forest products from successional patches, including mushrooms, floral greens, firewood, and even large down logs near roads could accommodate small local industries. Tending practiced by tribal communities (chap. 3) is an example of how this patch management has been used in the past to promote human-forest ecosystem sustainability.

Changes in patches can be slow, such as by growth and succession, or rapid, following disturbance. A variety of types and intensities of disturbance interact with a large array of pre-disturbance conditions in ways that defy prediction, but these interactions lead to a changing kaleidoscope of times since disturbance, degrees of development, and patches of many sizes and shapes. Opportunities for adaptive management of stands in different locations to meet priority goals can be identified from this shifting landscape mosaic.

Integrating across Disciplines

Research retains a fundamental role in developing strategies for human-forest ecosystem sustainability. Working across scales of time and space and incorporating social and ecological outcomes into these strategies requires interdisciplinary efforts of multiple research units working with practitioners, agencies, citizens, organizations, and legislators. The complexity of watershed and landscape models and decision-support systems demands a stronger alliance between the developers of such tools and their end users, so the tools can be designed and adapted for specific area contexts. The combination of breadth, depth, integration, learning, outreach, research, and scientific freedom allows academic and research institutions to provide an additional dimension to the adaptive-management cycle. When monitoring demonstrates that existing efforts are not yielding expected or desired outcomes, these institutions provide a means of refreshing the existing approach and developing a new set of operating procedures that align with the goal of long-term sustainability.

Implementing Management Changes

Developing agreement on priorities and strategies for management changes among interest groups is difficult, but so is implementation of agreed-upon management changes. The risk-averse culture of many stakeholders and bureaucracies leaves them sluggish in the face of rapid change. Complex interacting factors, including availability of funds from appropriations or commodity receipts, legal injunctions, environmental and economic policies, road and hydrologic infrastructure, and sociopolitical demands can all limit management flexibility. When issues come to a crisis point, special agreements such as the Northwest Forest Plan may spark major changes in management. The healthy business model of periodic reflection, such as we have done here, can help coalesce diverse adaptive-management considerations (e.g., relative to an evolution of regional planning and guidance, box 19.2). Given rapidly changing biological and social conditions, however, prescriptions that fail to embrace change will not suffice over time. The next chapter explores some options for an improved future for moist coniferous forests that builds upon the principles for implementation that we have described here—integration across space, time, and disciplines, and using local knowledge—while considering resilience of whole landscapes undergoing disturbances. Our overarching goal remains human-forest ecosystem sustainability, whatever the social and ecological future may bring.

BOX 19.2. EVOLUTION OF NORTHWEST REGIONAL-SCALE PLANNING AND GUIDANCE

After over 20 years of US federal Northwest Forest Plan implementation, important questions have arisen:

1. Is a larger evolution of regional-scale forest-management planning and guidance worthwhile?

Several new perspectives merit consideration for region-scale forest adaptive management: (1) goals for human-forest ecosystem sustainability are being redefined and differ from those identified in 1994; (2) a new appreciation for dynamic socioeconomic, ecological, and scientific processes supports the need for guidance that takes these dynamics into account, especially as novel outcomes are projected for future landscapes; (3) moist forests are a multistate ecosystem, yet we may not be managing for heterogeneity or resilience—for which regional-scale assessments and planning may be useful; (4) some ecosystem services are not well provisioned (fish and wildlife habitat, early-seral communities, sustainable human communities) and may be best accommodated across larger spatial scales—reknitting across all lands (e.g., reassessing land-use allocations) could provide better ecological and socioeconomic solutions.

2. Is now the time for regional planning evolution?

In 1993, the assembly of the Forest Ecosystem Management and Assessment Team (FEMAT 1993) signaled a new era of regional interagency coordination and guidance for forest-ecosystem management planning. Regionally, since then, there have been some iterative changes for selective Plan components (chap. 8). Downscaled forest plans by smaller administrative units (e.g., Oregon BLM, national forest plans) are revised every one to three decades, and revisions will be unfolding into the 2020s. Such smaller-scale plan revisions have been the institutional mechanism for substantial change in agency management. There is an opportunity now to have a synchronized larger-scale "ratchet" of forest-ecosystem adaptive management, and another such opportunity may not arise for a decade or more. Minimally, integration of new science and coordination across diverse individual forest-plan revisions is of paramount importance because many principles and elements to be addressed have broad scope, interrelating forest-management decisions across lands. Hence the time is ripe for evolution of regional ecosystem forest-management guidance.

Literature Cited

Benson, M. H. and R. K. Craig. 2014. The end of sustainability. *Society & Natural Resources* 27:777–782.

FEMAT (Forest Ecosystem Management Assessment Team). 1993. *Forest ecosystem management: An ecological, economic, and social assessment. Report of the Forest Ecosystem Management Assessment Team*. Portland, OR: US Department of Agriculture; US Department of the Interior [and others].

Holling, C. S., and G. K. Meffe. 1996. Command and control and the pathology of natural resource management. *Conservation Biology* 10:328–337.

Powell, J. W. 1891. *Map of the arid region of the United States, showing drainage districts*. Washington, DC: US Geological Survey.

Pretty, J. 2002. People, livelihoods and collective action in biodiversity management. Chapter 4. Pp. 61–86 in *Biodiversity, sustainability, and human communities*. Edited by T. O'Riordan and S. Stoll-Kleemann. Cambridge, UK: Cambridge University Press.

Reeves, G. H., B. R. Pickard, and K. N. Johnson. 2016. *An initial evaluation of potential options for managing riparian reserves of the Aquatic Conservation Strategy of the Northwest Forest Plan*. General Technical Report PNW-GTR-937. Portland, OR: USDA Forest Service, Pacific Northwest Research Station.

Spies, T.A., B. C. McComb, R. S. H. Kennedy, M. T. McGrath, K. Olsen, and R. J. Pabst. 2007. Potential effects of forest policies on terrestrial biodiversity in a multi-ownership province. *Ecological Applications* 17:48–65.

Stoll-Kleeman, S., and T. O'Riordan. 2002. Enhancing biodiversity and humanity. Chapter 13. Pp. 295–310 in *Biodiversity, sustainability, and human communities*. Edited by T. O'Riordan and S. Stoll-Kleemann. Cambridge, UK: Cambridge University Press.

Suzuki, N., and D. H. Olson. 2008. Options for biodiversity conservation in managed forest landscapes of multiple ownerships in Oregon and Washington, USA. *Biodiversity Conservation* 17:1017–1039.

USDA (US Department of Agriculture, Forest Service). 2012. National Forest System land management planning. Final rule and record of decision. 36 CFR Part 219. *Federal Register* 77:21162–21267. http://www.fs.usda.gov/Internet/FSE_DOCUMENTS/stelprdb5362536.pdf.

USDI and USDA (US Department of the Interior and US Department of Agriculture). 2002. Survey and manage management recommendation amendments for fuel hazard reduction treatments around at-risk communities: Group 2—certain mollusks, amphibians, and red tree vole. http://www.blm.gov/or/plans/surveyandmanage/files/mr-fire_amendment-rtv-ig-2003-02-att1.pdf.

Vilsack, T. J. 2009. Agriculture secretary Vilsack announces new direction and vision for America's forests. USDA Office of Communications Release No. 0383.09. Washington, DC: US Department of Agriculture. http://www.usda.gov/wps/portal/usda/usdahome?contentid=2009/08/0383.xml.

Chapter 20

Visions: 20 Years Hence

Beatrice Van Horne, Deanna H. Olson, and Thomas Maness

In this chapter we peer into an admittedly misty future, again with a lens on the northwest portion of the moist coniferous forest landscape in the United States. Serious hazards are apparent for these forests and the people who depend on them if we continue to follow our current trajectory. Alternatively, changes in political, policy, ecological, or economic conditions could produce other trajectories. We know that management policies have changed in unexpected ways in the past and that the future holds many uncertainties. This point resonates in the sentiments we have heard from luminaries who have built their careers from these forests and their resources, who say things like "Never in a million years would I have dreamed of a riparian reserve as wide as two site-potential tree-heights"; "Who would have thought that our forests would be managed by the judicial system and court injunctions?"; and "How crazy was it that the bottom dropped out of the housing market and overseas wood markets changed exactly when the new federal land allocations were implemented?"

To illustrate the likely constraints or costs of current policies and management actions, we portray two potential futures, followed by how bold steps over the next few decades could be taken to incorporate principles from previous chapters to make these forests more resilient and sustainable. We know that restoring the forest ecosystems of the past is neither possible nor desirable—the forests of the future will likely include an increasing wildland-urban interface, with more people expanding urban boundaries and populating rural landscapes, along with an increase in permanent roads to accommodate expanding human populations. These forests will continue to be dominated by myriad anthropogenic effects. We have also learned that the forest landscape is not static; it cannot be frozen in time—disturbance is needed to maintain important ecological components, goods, and services. Future disturbance trajectories will need to be continually evaluated as the human presence changes in a dynamic ecological context. To contrast costs and benefits of different paths, we outline two alternative futures here.

Trajectory 1 presents a business-as-usual continuation of local federal forest planning with regional guidance as currently offered by the Northwest Forest Plan. Public tax revenues would continue to be used for annual management (e.g., fire control, road and recreation maintenance, law enforcement, forest-health management) and for conducting below-cost harvest operations for forest restoration on federal lands. This trajectory subsidizes local communities for lost revenue from reduced timber harvests through the rural schools initiative. There are two major constraints in this scenario: (1) the amount of funding approved by the US Congress to subsidize management is too small to accomplish the management objectives; and (2) the increasing cost of fire control reduces the budget available for annual forest management—including fire-hazard reduction. In this downward-spiral situation, several management objectives (e.g., protecting sensitive species, reducing wildfire, and preserving jobs in rural communities) are not being met.

Trajectory 2 deals with these shortcomings by optimizing wood production over the entire landscape, across all ownerships, to increase overall benefits, reduce costs, and employ more people in the enterprise. It relaxes restrictions on selective cutting and adaptive forest management to produce higher-value timber products while improving ecosystem conditions according to the best available science. This path can produce more revenue for communities, reduce the need for subsidies, and increase revenue to management agencies to help them do their jobs. Trajectory 2 emphasizes research, adaptive management, and trust building to facilitate a sustainable vision. Several legal restrictions would need to be addressed, but these concepts could be considered for an alternative pathway for all-lands forest management of all stressors, all species, and values benefiting all people.

Trajectory 1: Status Quo and Diminishing Returns

Today's regional landscape is vastly different from the landscape of the past. Ecological information from the past can help guide the creation of more resilient ecosystems, but we know that the evolving landscape must function in the climatic, demographic, and social conditions of the future. Our concern is that today's passive forest management may not lead to a sustainable track for functional ecosystems. Based on current trends, the future landscape will differ even more from the desirable features of the past, as trees are long-lived and vegetation patterns carry a long legacy of the signature of past human activities.

A human footprint that will last for generations is established by even-aged forest patches that have been logged on all ownerships. Future patches of forest that are managed primarily for wood production will continue to respond in concert with market and sociopolitical conditions, producing a mix of short-rotation and longer-rotation forests and often disconnected set-asides (areas preserved for other uses). Given the relatively similar ages of these forest stands today, a pulsed pattern of timber harvest could continue. This is especially true on federal matrix lands within the Northwest Forest Plan area that have been managed for 80-plus-year rotations, synchronized by the post-World War II harvest pulse. Mobile human communities might capitalize on such a mosaic of conditions over time, but fixed local communities may be prone to boom-and-bust production. Rising transportation costs and loss of wood-production infrastructure in some rural communities further constrain local capacity to maintain socioeconomic resilience in the face of pulsed harvest schedules.

Trajectories of state and federal lands will also depend on current land-use allocations and site contexts. Those lands in federal reserves in the Northwest Forest Plan area (largely administered by the Forest Service) could progress into old-growth status. Although there may be social challenges, some such stands might be thinned to accelerate development of old growth, encourage minority tree development, or produce uneven-aged stands for wildlife and recreational values. None of these management scenarios creates early-seral communities; early-seral species management could become an additional priority for public lands and become a challenge to integrate with existing timber and species-biodiversity goals. Forest Service, state, and Bureau of Land Management (BLM) lands would maintain a significant focus on biodiversity retention, with riparian reserves protecting habitats for a suite of aquatic and terrestrial species (Olson and Burnett 2013) and upland management for persistence of a variety of rare and little-known species. Plantation thinning and regeneration harvests on matrix land-use allocations in the Plan area could produce wood over longer time scales (harvested at 80-plus years) under the guidelines of the Northwest Forest Plan, as subject to the more recent northern spotted owl recovery plan (US Fish and Wildlife Service 2011). The BLM checkerboard lands are not designed for efficiency in multiple-resource production and protection, but several designs might be considered to balance habitat management across mobile patches on these dispersed lands with timber-harvest priorities (box 20.1; plate 16A).

Although demand is increasing for high-value building materials as taller wood buildings are constructed around the world, under this status-quo trajectory the long-term contribution of federal- and state-managed lands to timber production becomes more challenging in the region for several reasons. First, these lands are carrying an increasing burden to provide for all ecosystem services, with demands for clean water, recreation, and natural areas increasing as concern for endemic species grows. Opposition to timber harvest on public lands has increased in recent decades and will continue to be a sociopolitical challenge. Countering this is a renewed concern for rural economies and focus on sustainable production of the region's resources. Second, because of factors underlying past allocations of lands to public and private ownerships, state and federal forestlands are generally inherently less productive than private industrial lands and hence are expected to continue to show lower productivity.

Third, as the once-integrated timber industry divests from processing, owing to US tax laws, and private timberland management is separated from wood-products milling and manufacturing, private forest owners may not support timber production on federal lands, as federal timber often competes with private timber. Also, with the loss of mills since 1994 and more frequent rotations on private lands, there is a concern that existing mills have been retooled to accommodate smaller logs. The relatively infrequent 80-plus-year rotation harvests from federal lands could produce larger logs, but this is likely to require unrealistic transportation costs to the fewer, scattered, labor-intensive large-log mills. These mills are now found mostly in southern Oregon, where their access to larger logs from BLM lands has now diminished; they often supplement their supply with large logs from British Columbia. This international source will likely dry up in the next decade or two, as old-growth and large beetle-killed trees become rare. Another potential source of larger logs is private family forests managed for long-term timber production and other values, where rotations of ≥60 years are a common practice.

On private industrial forestlands, rotations as short as 32 years stand out today in an aerial view (plate 16B). Many management decisions on these lands are brokered through timber investment management organizations (TIMOs) beholden to investors looking for short-term gains.

BOX 20.1. THE BUREAU OF LAND MANAGEMENT (BLM) CHECKERBOARD—A MANAGEMENT OPPORTUNITY

The checkerboard landscape, a legacy of nineteenth-century railroad land allocations, provides an excellent example of management inefficiencies introduced by artificial boundaries. Reconfiguring the checkerboard area to preserve habitat connectivity along existing riparian systems could enhance the efficiency of conservation as well as the efficiency of timber harvest and associated infrastructure. Short of that goal, on the BLM checkerboard landscape, the diagonals might become key to overland habitat connectivity for both early- and late-seral species (box fig. 20.1.1). Staggered regeneration patches across diagonals over time could be used to test early-seral sustainability; the wandering gaps could march among connected squares and sustain a mobile early-seral community. For taxa requiring forest structure or canopy for dispersal, such diagonals could be used to promote forest habitat contiguity. However, just as European hedgerows do not provide habitat for the full complement of native species, ribbons of connected forest along BLM-square diagonals would likely aid only a portion of native forest species. Yet the benefits to biodiversity of both riparian management areas and connectivity management would seem to significantly favor biodiversity on these lands relative to industrial forestlands.

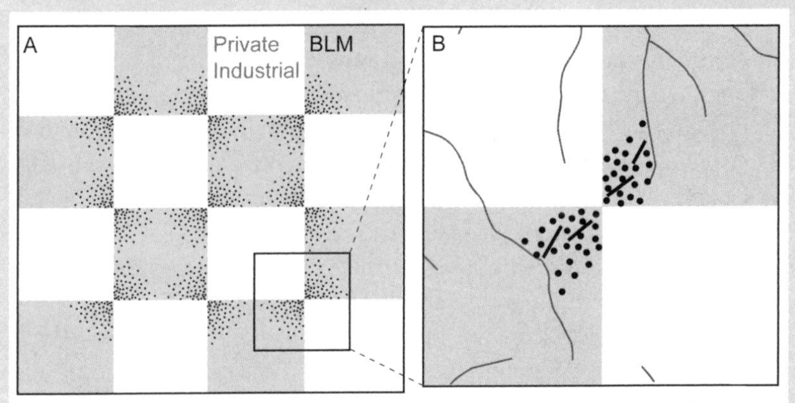

FIGURE 20.1.1. Conceptual designs for late-seral species connectivity across Bureau of Land Management checkerboard landscapes, where retention of habitats across diagonals could aid species persistence by (A) weighting tree retention toward diagonals to favor species associated with canopy structure or shaded, cooler ground microhabitats, encouraging connectivity at corners; (B) extending riparian buffers to corners, via down wood and/or overstory tree retention to promote cross-corner dispersal. Reprinted from Olson and Burnett (2013).

Under this trajectory, these lands would collectively continue to function as they do now, with relatively lower contributions to biodiversity but with more certain contributions to local economies than federal and state lands or other ownerships managed with longer rotations or no harvest. There are current pressures for private lands to increase protection of aquatic resources and wildlife habitat, but these services have economic costs, and private landowners may not have incentives to comply without legal directives. Furthermore, the trend of increasing harvest mechanization (plate 5B) will reduce the economic contribution of timber harvesting to local communities, as would a loss of productivity from repeated harvesting or increases in fire, erosion, pest or disease outbreaks, or other disturbances. Soil erosion and changing climate may cause some of these patches to stall production of merchantable timber, especially toward the southern end of the Douglas-fir range. Unless alternative markets are found (e.g., biomass), some areas could be abandoned or converted to nonforest uses. Abandoned areas would be vulnerable to fire, especially at lower elevations or near lands with more abundant fuels and greater fire risk, increasing firefighting costs.

Across all ownerships, the single-minded focus on Douglas-fir production in Oregon and Washington and lack of prescribed burns would result in a continued loss of oak woodlands, madrones, alder, and other deciduous species, and the diversity of species associated with them. Future fires would increase the representation of early-successional vegetation patches, adding to current levels of plant and animal diversity, yet large-scale fires may create an abundance of such habitat all at once in a restricted area. Early-seral vegetation is likely to be concentrated in certain areas, such as the drier, fire-prone areas of the southern and eastern portions of the moist forests. This may add further management challenges to long-term sustainability of well-distributed early-seral communities across the region.

Overall, we project losses of biodiversity on both public and private lands associated with the status-quo trajectory. Prime habitat for some species occurs on unprotected lands: some of the most productive areas for marbled murrelets and coho salmon in Oregon, Washington, and California do not lie on federally reserved lands (murrelets, Raphael et al. 2016; salmon, Reeves et al. 2016). Moreover, the short rotations on private lands are expected to have cumulative adverse effects on late-seral-associated species, especially those with low mobility. Fragmentation of old-growth habitat conditions across landscapes due to different ownerships and disturbance patterns will reduce connectivity and gene flow, isolating populations that are at a higher risk of extinction from stochastic processes and a host of disturbances.

Current social and policy trajectories would increase the contribution of federally managed forests to meeting the needs of urban dwellers and people interested in wildlife and recreation, especially as these populations continue to grow, but would come with a compensatory decrease in economic production from these forests. Reduced income from federal forests and a concomitant decline in federal budgets could spell problems for maintenance of infrastructure, sustainability of federal timber programs, and environmental monitoring. Any increase in future wildfires could further cut into federal and state budgets for other priorities.

Embracing the future calls for active forest management, but the capacity to manage on public lands has been significantly reduced by a number of factors, including increases in the costs of fighting fires, loss of timber-harvest income (receipts), and overall reductions in nondefense spending. Clearly, declines in operating budgets have exerted a heavy toll on all areas of land-management capacity in the moist forest regions. For example, most forest plan environmental documents at the scale of individual projects include resource-monitoring plans that are never actually implemented (DeLuca et al. 2010). With these limitations, enhanced capacities such as learning, which is central to managing a sustainable human-forest ecosystem, are minimized. In terms of ecosystem services, this trajectory would produce declining timber revenues and biodiversity.

We recognize that we are spinning a yarn of "could-be" scenarios with status-quo management, acknowledging that some outcomes are uncertain and that some things will change the trajectory. Nevertheless, the balance of costs and benefits shows a variety of diminishing returns for biodiversity, wood production, and some rural communities, raising concerns about this current path.

Trajectory 2: Adaptive Collaborative Vision

Trajectory 2 focuses on removing or modifying one-size-fits-all regulations and attempts to optimize the overall landscape to achieve more benefits. Active management is increased in some areas to produce revenue to help achieve the objectives. Restrictions are tightened in some other areas where connectivity or other benefits will make a major difference to habitat goals. The system is built around a theme of adaptive management, in which monitoring feeds back into management plans. Additional revenue input into the system from increased logging in lower-risk areas helps provide funding to achieve other objectives. The overall goal is to have a feasible management plan given available resources and to constantly improve the system through continuous adaptation.

Imagine a forested landscape managed collaboratively to support sustainable ecosystem services, less constrained by ownership boundaries: How might the resulting forest function socially and ecologically? In this scenario, our new knowledge base would help us to identify areas across all lands that are most valuable for producing large-diameter trees; salmon habitat; a kaleidoscope of bird, mammal, and amphibian species and their associated communities; valued hardwoods; tribal cultural and subsistence resources; or water for drinking, irrigation, power generation, recreation, and other uses. Landscape areas with focal management priorities for key services would then be developed, with an eye to accommodating each service in its priority areas. For instance, instead of mandating a one-size-fits-all buffer along streams by stream size or fish presence, a tiered approach could be developed to match watershed conditions, whereby the most valuable portions of these streams would be managed to produce the best possible salmon habitat, and protections could be eased in less productive areas in favor of other priorities. Although tenets of this approach are integral to best management practices for streams, adaptive management to refine approaches with emerging new knowledge has lagged. In this landscape, we ask: Where and on what processes should we focus management to get the most efficient and effective results? Which places support the ecosystem services that are most valued and most at risk? Where resources are at risk, areas can be managed either to help reduce risk or to enhance their resilience when subject to disturbances.

The resulting landscape would differ from our status-quo projection in a number of attributes that contribute directly to multiple goals for ecosystem services and human-forest ecosystem sustainability, but monitoring and adaptive management are part of the design. Monitoring can be the foundation of trust among stakeholders, as it will show the results of management and form the basis of adaptive management. Monitoring can be expensive, so it would benefit from a scientific and economic focus on what information is needed to assess the status of the most-valued ecosystem services and on how this information can be most efficiently obtained while maintaining sufficient reliability. Landscape-scale monitoring for threats to forested ecosystems, such as from insects, disease, changed fire regimes, or loss of important fish and wildlife habitat, would be a priority to support policy and management decisions at larger scales.

What Tools Do We Have to Help Realize This New Landscape?

It is difficult to see how we can make much progress on this trajectory without several important tools. First, a group would need to be established to coordinate assessment and management across ownerships by periodically reviewing the status of key ecosystem services and processes of change. Although some elements are currently monitored, a broader suite of elements would be considered across ownerships, including disturbances (fire, landslides, windthrow); aquatic-riparian and sensitive salmon status; status of other species of concern; early-successional and late-successional vegetation patterns; invasions of diseases, pests, or weeds; tree mortality from drought; socioeconomic trends; adaptive-management results; collaborative management approaches; and other landscape-scale ecosystem services. Monitoring does not need to be at the same frequency for all elements. Regular communication with local collaboratives could help to resolve issues and set priorities among ecosystem services and local monitoring efforts in support of adaptive management. A close partnership with researchers could aid development of monitoring designs; project and manage early- and late-seral vegetation patterns through time; identify priority areas for ecosystem services; and provide decision support.

Second, support for creative research in forestry would advance the socioeconomic and biophysical aspects of this trajectory. For example, a renewed focus on large-log wood products can be envisioned. High-quality wood from Douglas-fir and a variety of other species could form the backbone of a new approach to product manufacturing in the Pacific Northwest, reducing competition with southeastern US pine forests and industrial forests in other countries for "low-end" commodity wood products. This innovation runs counter to the current federal focus on increasingly low-quality timber resources, but longer rotations and selective cutting could help to protect ecosystem services. In this way, managers could concentrate on receiving more value from what they are already cutting, instead of relying on the strategy of cutting more timber. High-value products are more labor intensive and generally support higher-paying jobs. Improved coordination and collaboration could foster demonstrations in model "working forests" by supporting selective cutting that is consistent with ecosystem priorities. Markets for high-value large-log timber products and different wood types could open up, especially if production costs can be kept in check and more mills can be modified to meet this need.

Along with a demonstration of product-focused selective cutting, there is a need to further demonstrate and test novel, ecologically based approaches to timber harvest in uplands and riparian areas using adaptive-management approaches. With 20-plus years of alternative silvicultural approaches and riparian buffer studies to inform next-generation designs, science has identified some promising approaches to active forest management. There is now ample baseline information, and the modeling tools to use it, to create on-the-ground designs to test the efficacy of novel approaches. Stand-scale clearcut harvests are not the only means of wood production, and one-size-fits-all riparian buffers are not the only way to retain aquatic-riparian conditions. Blended watershed- and site-based management activities and monitoring can inform priorities for both meeting key upland and aquatic conservation goals and providing forest products.

Validation of models and designs in field trials using an adaptive-management approach is important and helps to build trust. Some of the questions that could be addressed by site- to landscape-scale testing are How can riparian areas in existing even-aged stands best be treated to improve stream-riparian conditions? Are multiple-entry thinned stands an effective long-term approach for upland management for wood and ecological priorities? Can landscape-scale ecological resilience be retained with a mix of approaches across a broad landscape (plate 16B)? Can a mix of approaches in proximity to rural communities sustain them over the long term?

The intensive watershed- and site-scale demonstration and testing can best inform management and collaboratives by being placed in the context of landscape-scale assessments based on overlays of local priorities for forest planning. This will help stakeholders and managers to focus on conflicting values or priority topics (e.g., landslide potential, wood recruitment, thermal loading, fish habitat, Reeves et al. 2016; ecological forestry design, Franklin and Johnson 2012) across the larger landscape. Where climate change, fire risk, roads, and sensitive species are particular concerns for forest management of upland vegetation and instream conditions, overlaying those geographic information system (GIS) layers or models can provide additional baseline knowledge to identify priority areas for restoration activities.

Along with landscape-scale assessment and multiscale research and collaboration, a toolbox of policies or incentives to encourage private landowner participation is needed. This toolbox could include benefits to private landowners from ecosystem services on federal lands and land swaps to consolidate lands with certain long-term priorities. These would be particularly important in enhancing connectivity of prime fish and wildlife habitat, or to enhance sustainability of local rural communities. Incentives could be designed to support existing efforts to connect habitat on US federal lands: dispersal habitat is being considered as an element for management of red tree voles, and old-growth habitat links between the Coast and Cascade Ranges are being considered during designation of set-asides (USDI 2016). Land-swap benefits could be used on the BLM checkerboard forest landscape (box 20.1) to consolidate lands for both timber and nontimber resource management (plate 16B). The current requirement that land swaps involve lands of "equal value" is a significant barrier that could be revised by congressional action.

Overall, this potential future trajectory, supported by organizational, scientific, and regulatory changes, would focus efforts and financial resources in those forestlands that would provide the best outcome for priority ecosystem services, including those that produce revenue. It would focus on producing the desired landscape of the future instead of restoring some previous landscape that is no longer attainable. It would allow for different contributions by different land-management entities to the resulting mix of services provided by the landscape and would support management of a dynamic landscape in which management would take advantage of planned and unplanned disturbances. Steps toward this future have already been taken, with some innovative approaches being negotiated by watershed stewardship councils and other collaborative groups. Increased flexibility could result in solid gains for declining native species and commodity production and a recognition that not all ecosystem services are delivered by every stand. At minimum, trial landscapes could be used to test this approach. Over the next 20 years, a renewed commitment to collaboration and openness to new ideas could assure the continued delivery of highly valued ecosystem services.

Epilogue

Is our second trajectory hopelessly idealistic? Certainly the hurdles to achieving such a path are numerous and complex. But we have reached a critical social and ecological juncture where we clearly see different potential futures, and it is timely to consider changing directions. Developing a truly sustainable human-forest ecosystem in these forests will require leaders who are willing to offer a vision that allows for dynamic interaction with the landscape and interest groups. These leaders would need political and legislative support for their vision, with the recognition that any single change may fix key issues of today, but that continued adaptive management will be needed in the future. A practical start on breaking roadblocks would be to develop performance outcomes tied to social and political priority ecosystem services—both monetized (e.g., forest products) and tied to well-being (e.g., biodiversity, recreation, clean water). Actions supporting these outcomes would be designed based on long-term ecosystem science predictions of measurable goals contributing to landscape-scale outcomes. A healthy dose of humility warns us to include adaptive-management designs, with the knowledge that some such trials will succeed and others may not. Encompassing a range of alternatives and replication of treatments in adaptive-management designs can accelerate the learning process where outcomes are uncertain. Economic, social, and biological outcomes will need to be monitored and managers held accountable for them, including for their contribution to learning and adaptive management.

Managing moist coniferous forests intelligently in the future will require adjusting to continually changing socioeconomic and ecological contexts, necessitating a considerable expenditure of time, energy, and funds. Sustainability of these forest ecosystem services, including biodiversity and economic benefits, depends on wise and efficient management actions supported by knowledgeable and scientifically informed interest groups. In the Pacific Northwest, these groups include tribal, federal, state, and private forest managers; regulators providing legal oversight; and people using the forests to provide wood, recreation, hunting, fishing, mushroom harvesting, hiking and backpacking, and wildlife viewing. It is likely that they can all agree that the forests and associated rural communities are magnificent assets and that sustaining these assets is a worthy goal.

Decision makers generally understand local socioeconomic and ecological conditions, while at the same time recognizing the role of larger regional conditions and priorities. They may consider this local knowledge across a multigenerational time span in committing to learning to assure sustainability. Coming up with solutions is a very tall order. We have outlined elements of a possible approach. Whichever path we choose to walk, integration of people into the land, resource, and wildlife equation is of paramount importance—socially, economically, and ecologically—to achieving a sustainable future.

Sustaining people and nature in moist conifer-dominated human-forest ecosystems is a goal to which we can all aspire. We see this goal, and the hard work needed to achieve it, as an enduring vision of management for sustainability in the twenty-first century. We can contribute ingenuity, humility, appreciation of science and beauty, and dedication to ecosystem sustainability—and persevere to find a new balance of approaches for a resilient, persistent ecosystem, including the people who depend on it.

Literature Cited

DeLuca, T. H., G. H. Aplet, H. B. Wilmer, and J. Buchfield. 2010. The unknown trajectory of forest restoration: A call for ecosystem monitoring. *Journal of Forestry* 108:288–295.

Franklin, J. F., and K. N. Johnson. 2012. A restoration framework for federal forests in the Pacific Northwest. *Journal of Forestry* 110:429–439.

Olson, D. H., and K. M. Burnett. 2013. Geometry of forest landscape connectivity: Pathways for persistence. Pp. 220–238 in *Density management for the 21st century: West side story.* Edited by P. D. Anderson and K. L. Ronnenberg. General Technical Report PNW-GTR-880. Portland, OR: USDA Forest Service, Pacific Northwest Research Station.

Raphael, M. G., G. A. Falxa, D. Lynch, S. K. Nelson, S. F. Pearson, A. J. Shirk, and R. D. Young. 2016. Status and trend of nesting habitat for the marbled murrelet under the Northwest Forest Plan. Chap. 2 in *Northwest Forest Plan—the first 20 years (1994–2013): Status and trend of marbled murrelet populations and nesting habitat.* Technical coordination by G. A. Falxa and M. G. Raphael. General Technical Report PNW-GTR-927. Portland, OR: USDA Forest Service, Pacific Northwest Research Station.

Reeves, G. H., B. R. Pickard, and K. N. Johnson. 2016. *An initial evaluation of potential options for managing riparian reserves of the Aquatic Conservation Strategy of the Northwest Forest Plan.* General Technical Report PNW-GTR-937. Portland, OR: USDA Forest Service, Pacific Northwest Research Station.

USDI (US Department of the Interior, Bureau of Land Management). 2016. *Proposed resource management plan/final environmental impact statement: Western Oregon.* Portland, OR: US Department of the Interior, Bureau of Land Management, Oregon/Washington State Office.

US Fish and Wildlife Service. 2011. *Revised recovery plan for the Northern Spotted Owl* (Strix occidentalis caurina). Portland, OR: US Fish and Wildlife Service.

APPENDIX

Common and Scientific Names of Species

Note: Capitalization conventions for common names differ by taxon and publisher; to standardize, we have not capitalized, either here or in the text.

STANDARD ENGLISH NAME	SCIENTIFIC NAME
Trees	
alder, red	*Alnus rubra*
Alaska yellow-cedar	*Callitropsis (Chamaecyparis) nootkatensis*
cedar, incense	*Calocedrus decurrens*
chinquapin	*Chrysolepis* spp.
chinquapin, giant	*Chrysolepis chrysophylla*
cottonwood, black	*Populus balsamifera*
dogwood, Pacific	*Cornus nuttallii*
Douglas-fir	*Pseudotsuga menziesii*
fir, grand	*Abies grandis*
fir, noble	*Abies procera*
fir, Pacific silver	*Abies amabilis*
fir, subalpine	*Abies lasiocarpa*
hemlock, mountain	*Tsuga mertensiana*
hemlock, western	*Tsuga heterophylla*
madrone, Pacific	*Arbutus menziesii*
maple, big-leaf	*Acer macrophyllum*
oak, California black	*Quercus kelloggii*
oak, canyon live	*Quercus chrysolepis*
oak, Oregon white	*Quercus garryana*
pine, lodgepole	*Pinus contorta*
pine, ponderosa	*Pinus ponderosa*
pine, shore (coastal variety of lodgepole)	*Pinus contorta contorta*

STANDARD ENGLISH NAME	SCIENTIFIC NAME
pine, sugar	*Pinus lambertiana*
pine, western white	*Pinus monticola*
redcedar, western	*Thuja plicata*
redwood, coast	*Sequoia sempervirens*
spruce, Engelmann	*Picea engelmannii*
spruce, Sitka	*Picea sitchensis*
tanoak	*Notholithocarpus (Lithocarpus) densiflorus*
yew, Pacific	*Taxus brevifolia*

Shrubs

ceanothus	*Ceanothus* spp.
hazelnut	*Corylus cornuta*
huckleberry	*Vaccinium* spp.
maple, vine	*Acer circinatum*
salmonberry	*Rubus spectabilis*
thimbleberry	*Rubus parviflorus*

Herbs

beargrass, common	*Xerophyllum tenax*
camas	*Camassia* spp.
cattail	*Typha* spp.
dwarf mistletoe	*Arceuthobium* spp.
fern, maidenhair	*Adiantum* spp.
fern, deer	*Blechnum spicant*
lupines	*Lupinus* spp.
nettle	*Urtica dioica*
trillium, western white	*Trillium ovatum*
wheatgrass	*Elymus* spp.

Fungi

American matsutake/ Tanoak mushroom	*Tricholoma magnivelare*
chestnut blight	*Cryphonectria parasitica*
fungus, laminated root rot (cedar)	*Phellinus weirii*

Standard English Name	Scientific Name
fungus, laminated root rot (Douglas-fir and hemlock)	*Phellinus sulfurascens*
fungus, velvet-top	*Phaeolus schweinitzii*
snake fungal disesase	*Ophidiomyces ophiodiicola*
sudden oak death	*Phytophthora ramorum*
Swiss needle cast	*Phaeocryptopus gaeumannii*
white pine blister rust	*Cronartium ribicola*
white-nose syndrome	*Pseudogymnoascus destructans*
whirling disease	*Myxobolus cerebralis*
Insects	
beetle, Douglas-fir bark	*Dendroctonus pseudotsugae*
beetle, mountain pine	*Dendroctonus ponderosae*
beetles, bark	Subfamily Scolytinae
bumblebees	*Bombus* spp.
caddisflies	Order Trichoptera
hemlock wooly adelgid	*Adelges tsugae*
Mollusks	
slugs, jumping	*Hemphillia* spp.
slugs, taildropper	*Prophysaon* spp.
Fish	
salmon, Chinook	*Oncorhynchus tshawytscha*
salmon, coho	*Oncorhynchus kisutch*
salmon, sockeye	*Oncorhynchus nerka*
trout, coastal cutthroat	*Oncorhynchus clarkii clarkii*
trout, rainbow/steelhead	*Oncorhynchus mykiss*
Amphibians	
frog, coastal tailed	*Ascaphus truei*
salamander, coastal giant	*Dicamptodon tenebrosus*
salamander, Dunn's	*Plethdon dunni*

Standard English Name	Scientific Name
salamander, Oregon slender	*Batrachoseps wrighti*
salamander, Scott Bar	*Plethodon asupak*
salamander, Van Dyke's	*Plethodon vandykei*
salamanders, torrent	*Rhyacotriton* spp.
salamanders, woodland	Plethodontidae

Birds

brown creeper	*Certhia americana*
finch, purple	*Haemorhous (formerly Carpodacus) purpureus*
flycatcher, olive-sided	*Contopus cooperi*
flycatcher, Pacific-slope	*Empidonax difficilis*
flycatcher, willow	*Empidonax traillii*
hummingbird, rufous	*Selasphorus rufus*
jay, scrub	*Aphelocoma coerulescens*
lark, streaked horned	*Eremophila alpestris strigata*
meadowlark, western	*Sturnella neglecta*
murrelet, marbled	*Brachyramphus marmoratus*
nuthatch, white-breasted	*Sitta carolinensis*
owl, barred	*Strix varia*
owl, northern spotted	*Strix occidentalis caurina*
plover, western snowy	*Charadrius nivosus nivosus*
sparrow, chipping	*Spizella passerina*
sparrow, Oregon vesper	*Pooecetes gramineus affinis*
sparrow, savannah	*Passerculus sandwichensis*
thrush, Swainson's	*Catharus ustulatus*
thrush, varied	*Ixoreus naevius*
vireo, warbling	*Vireo gilvus*
warbler, hermit	*Setophaga occidentalis*
warbler, orange-crowned	*Leiothlypis celata*
warbler, Townsend's	*Setophaga townsendi*
warbler, Wilson's	*Cardellina pusilla*
warbler, yellow	*Setophaga petechia*
woodpecker, downy	*Picoides pubescens*
woodpecker, hairy	*Picoides villosus*

Standard English Name	Scientific Name
wood-pewee, western	*Contopus sordidulus*
wren, house	*Troglodytes aedon*
wren, winter	*Troglodytes troglodytes*

Mammals

Standard English Name	Scientific Name
bats	Order Chiroptera
bear, grizzly	*Ursus arctos*
beaver, American	*Castor canadensis*
deer	*Odocoileus* spp.
elk	*Cervus elaphus*
fisher	*Pekania pennanti*
fox, montane red	*Vulpes vulpes necator*
lynx, Canada	*Lynx canadensis*
marten, American	*Martes americana*
marten, Pacific	*Martes caurina*
squirrel, northern flying	*Glaucomys sabrinus*
vole, red tree	*Arborimus longicaudus*
wolf, gray	*Canis lupus*
wolverine	*Gulo gulo*

ABOUT THE EDITORS

DEANNA H. OLSON

Dede's work as an ecologist is devoted to sustainability of our natural heritage. Her work has encompassed every vertebrate class (fish, amphibians, reptiles, birds, mammals), with a focus on amphibians. Her bachelor's degree at the University of California, San Diego, intersected with the first Conservation Biology Conference there in 1978, helping to build the foundation for her passion for biodiversity conservation. In 1981, her PhD degree from the Department of Zoology at Oregon State University brought her to the Pacific Northwest, with its natural grandeur from the sea to the forests, mountains, and high deserts. Currently, she is a research ecologist with the USDA Forest Service Pacific Northwest Research Station in Corvallis, Oregon. She also serves as courtesy faculty at Oregon State University and associate editor for *Herpetological Review* and is past president of the Society for Northwestern Vertebrate Biology and past co-chair of Partners in Amphibian and Reptile Conservation.

BEATRICE VAN HORNE

Bea has an interest in the processes within and among species that manifest as visible ecological communities. Beginning with PhD research on the effects of clearcut logging on small mammal populations in Southeast Alaska, her work has sought to untangle the factors driving small-mammal, ground squirrel, and bird populations. After seventeen years as a professor of biology at Colorado State University, she spent ten years in the Washington, DC, area serving in research program leadership with the US Forest Service and the US Geological Survey. She then returned to the western United States, where she has been a research manager for the USDA Forest Service Pacific Northwest Research Station in Corvallis, Oregon, for the past five years. She currently serves as the director of the USDA Pacific Northwest Climate Hub.

ABOUT THE CONTRIBUTORS

SUSAN J. ALEXANDER is an agricultural economist with the US Forest Service Pacific Northwest Research Station in Juneau, AK. She specializes in forest economics, community forestry, and policy analysis.

PAUL D. ANDERSON is a research program manager with the US Forest Service Pacific Northwest Research Station in Corvallis, OR. He specializes in ecophysiology and silviculture of Pacific Northwest forests.

STANLEY T. ASAH is an associate professor in the School of Environmental and Forest Sciences, University of Washington, Seattle, WA. He specializes in the human dimensions of natural resource management.

KEITH B. AUBRY is a research wildlife biologist (emeritus) with the US Forest Service Pacific Northwest Research Station in Olympia, WA. He specializes in the ecology and conservation of rare and elusive forest carnivores.

CHELSEA BATAVIA is a graduate student in the Department of Forest Ecosystems and Society, Oregon State University, Corvallis, OR. Her research focus is the ethics of forest management and conservation.

MATTHEW G. BETTS is an associate professor in the Department of Forest Ecosystems and Society at Oregon State University, Corvallis, OR. He studies the influence of landscape composition and pattern on animal behavior, species distributions, and ecosystem function.

DALE J. BLAHNA is a research social scientist with the US Forest Service Pacific Northwest Research Station in Seattle, WA. He conducts research on cultural ecosystem services and the use of social science in environmental and natural resource planning and ecosystem management.

BERNARD T. BORMANN is a professor and the director of the Olympic Natural Resources Center, School of Environmental and Forest Sciences, University of Washington, Forks, WA. His work focuses on developing and evaluating new ways to manage forests over the long run, considering people as integral to the ecosystem, with the goal of achieving and sustaining high levels of both ecological and community well-being.

JULIA I. BURTON is a research assistant professor in the Department of Wildland Resources at Utah State University, Logan, UT. She specializes in the ecology and conservation of forest understory plant communities and associated ecosystem services in moist coniferous forest ecosystems.

JOHN L. CAMPBELL is an assistant professor and senior researcher in the Department of Forest Ecosystems and Society, Oregon State University, Corvallis, OR. He investigates the effects of disturbance on forest structure, function, and development.

MICHAEL J. CASE is a postdoctoral researcher in the School of Environmental and Forest Sciences, University of Washington, Seattle, WA. He specializes in assessing climate change vulnerability and integrating mechanistic and empirical models.

LEE K. CERVENY is a research social scientist with the US Forest Service Pacific Northwest Research Station in Seattle, WA. She is an anthropologist who studies forest governance, public engagement, and resource planning in the context of public lands management.

WARREN B. COHEN is a research forester with the US Forest Service Pacific Northwest Research Station in Corvallis, OR. He specializes in translation of remotely sensed data into ecological information.

DAVID V. D'AMORE is a research soil scientist with the US Forest Service Pacific Northwest Research Station in Juneau, AK. He studies soil and watershed biogeochemistry related to watershed management and mixed-species forested ecosystems.

ROBYN L. DARBYSHIRE is the regional silviculturalist with the US Forest Service Pacific Northwest Regional Office, Portland, OR. She worked on the Long-Term Ecosystem Productivity project for seventeen years.

EMILY JANE DAVIS is an assistant professor and extension specialist in the Department of Forest Ecosystems and Society, Oregon State University, Corvallis, OR. Her research and extension work focuses on collaborative natural resource management.

RAYMOND J. DAVIS is the monitoring leader for old forests and northern spotted owls for the Northwest Forest Plan Interagency Monitoring Program, USDA Forest Service, Pacific Northwest Region, Corvallis, OR.

ROBERT L. DEAL is a research forester with the US Forest Service Pacific Northwest Research Station in Portland, OR. His research focuses on ecosystem services and applied silviculture, including developing new directions for incorporating ecosystem services into a framework for managing public lands.

THOMAS H. DELUCA is a professor and director of the School of Environmental and Forest Sciences, University of Washington, Seattle, WA. His research includes a diverse program in environmental and forest sciences with specific interests in forest soils and natural resource sustainability.

JASON B. DUNHAM is a supervisory aquatic ecologist with the US Geological Survey, Forest and Rangeland Ecosystem Science Center, Corvallis, OR. His work is focused on climate and land-use change, species conservation, biological invasions, and development of tools to engage managers and stakeholders in science applications.

A. PAIGE FISCHER is an assistant professor in the School of Natural Resources and Environment, University of Michigan, Ann Arbor, MI. The goal of her research is to increase scientific understanding of human behavior as it relates to the sustainability of socioecological systems.

REBECCA L. FLITCROFT is a research fish biologist with the US Forest Service Pacific Northwest Research Station, Corvallis, OR. She specializes in broad-scale analysis of aquatic systems, with special emphasis on disturbance processes and native fishes.

JERRY F. FRANKLIN is professor of Environmental and Forest Sciences, University of Washington, Seattle, WA. He is a leading authority on sustainable forest management and the maintenance of healthy forest ecosystems, integrating ecological and economic values into harvest strategies.

CHERYL ANN FRIESEN is the USFS science liaison for the US Forest Service, Willamette National Forest, McKenzie Bridge, OR. She facilitates linkages between scientists and managers involved with national forest management issues.

HANNAH GOSNELL is an associate professor of Geography, College of Earth, Ocean, and Atmospheric Sciences, Oregon State University, Corvallis, OR. Her research interests are in environmental governance in the context of rural working landscapes and the ways in which laws and institutions evolve to reflect changing geographies and facilitate socioecological transformation when necessary.

ANDREW N. GRAY is a research ecologist with the US Forest Service Pacific Northwest Research Station, Corvallis, OR. He specializes in regional assessments of change in forest composition, structure, and function.

MICHAEL S. HAND is a research economist with the US Forest Service Rocky Mountain Research Station, Missoula, MT. His current research focuses on how people and communities derive benefits and value from publically managed natural resources and how public agencies allocate resources to manage the flow of ecosystem goods and services provided by public lands.

MARK E. HARMON is a professor and Richardson Chair in Forest Science, Department of Forest Ecosystems and Society, Oregon State University, Corvallis, OR. He is a forest ecologist who examines the effects of disturbance and management on carbon in forests and the forest sector.

JEFF HATTEN is an assistant professor, Department of Forest Engineering, Resources and Management, Oregon State University, Corvallis, OR. He conducts research in soil carbon, soil productivity, effects of fire on soil, and effects of silviculture on sediment transport and water quality.

RICHARD W. HAYNES retired as a research forester from the US Forest Service Pacific Northwest Research Station, Portland, OR, where he was a program manager and economist specializing in the market impacts of broad-scale forest management strategies.

PAUL E. HENNON is a research plant pathologist, retired from the US Forest Service Pacific Northwest Research Station, Juneau, AK. He studies the roles of forest diseases in the functioning of coastal temperate rain forests.

SUSAN STEVENS HUMMEL is a research forester with the US Forest Service Pacific Northwest Research Station, Portland, OR. Her expertise is in transdisciplinary research that expands the knowledge and practice of silviculture.

SHERRI L. JOHNSON is a supervisory research ecologist with the US Forest Service Pacific Northwest Research Station, Corvallis, OR. She is an aquatic ecologist specializing in forested stream and reservoir food-web dynamics, biogeochemistry, water quality, and ecosystem responses to natural and anthropogenic disturbances.

JOHN B. KIM is a research biological scientist with the US Forest Service Pacific Northwest Research Station, Corvallis, OR. He studies climate change impacts on vegetation using dynamic vegetation models.

EINI C. LOWELL is a research forest products technologist with the US Forest Service Pacific Northwest Research Station, Portland, OR. Her work identifies ways to best use our forest resources, including by-products of forest restoration and fuel-reduction treatments, with an emphasis on value-added products that benefit communities.

THOMAS MANESS is the Cheryl Ramberg-Ford and Allyn C. Ford Dean and director of the Oregon Forest Research Laboratory in the College of Forestry, Oregon State University, Corvallis, OR. He has conducted research in forest policy, land-use planning, and sustainable forest management and has also explored extensively in the desert canyons of the Colorado Plateau to study the ecology, geology, and archeological evidence of former inhabitation with respect to climate change and sustainable living.

BRUCE G. MARCOT is a research wildlife biologist with the US Forest Service Pacific Northwest Research Station, Portland, OR. He works in assessment and conservation planning of rare species and habitats.

TEODORA MINKOVA is an ecologist with the Washington Department of Natural Resources, Olympia, WA. She leads the research, monitoring, and adaptive-management program at the Olympic Experimental State Forest in Washington state.

CLAIRE A. MONTGOMERY is a professor in the Department of Forest Engineering, Resources and Management, Oregon State University, Corvallis OR. She is a forest economist, specializing in bioeconomic modeling of forested landscapes.

MICHAEL PAUL NELSON is the Ruth H. Spaniol Chair of Renewable Resources and a professor in the Department of Forest Ecosystems and Society, Oregon State University, Corvallis, OR. He is an environmental philosopher and ethicist whose work focuses on the various philosophical dimensions of conservation as well as the confluence between ecology, social science, and ethics.

BROOKE E. PENALUNA is a research fish biologist with the US Forest Service Pacific Northwest Research Station, Corvallis, OR. She is an aquatic ecologist who specializes in the ecology of Pacific salmon and trout and their links to riparian ecosystems.

STEVEN S. PERAKIS is a supervisory research ecologist with the US Geological Survey, Forest and Rangeland Ecosystem Science Center, Corvallis, OR. He studies biogeochemical cycles of carbon and nutrients in forests.

KLAUS J. PUETTMANN is the Edmund Hayes Professor in Silviculture Alternatives in the Department of Forest Ecosystems and Society, Oregon State University, Corvallis, OR. He is a silviculturist specializing in developing an ecological understanding of managing diverse structured forests; more recently, he has become interested in understanding how forests adapt to changing conditions and how we can manage forest ecosystems to facilitate such adaptations.

MARTIN G. RAPHAEL is an emeritus scientist with the US Forest Service Pacific Northwest Research Station, Olympia, WA. He specializes in the ecology, conservation, and management of threatened and endangered wildlife in forests of the Pacific Northwest.

GORDON H. REEVES is a research fish ecologist with the US Forest Service Pacific Northwest Research Station, Corvallis, OR. His work focuses on the ecology and conservation of Pacific salmon and other native aquatic biota of the Pacific Northwest and the freshwater ecosystems on which these organisms depend.

LEO A. SALAS is a quantitative ecologist with the Climate Change and Informatics Group, Point Blue Conservation Science, Petaluma, CA. He develops analyses, models, and data visualizations for many conservation projects at Point Blue, including models assessing the impact of climate change on birds and their habitats, and is the developer of the statistical analysis server for Avian Knowledge Network.

LAURENCE R. SCHIMLECK is department head and a professor in the Department of Wood Science and Engineering, Oregon State University, Corvallis, OR. He is a wood scientist.

MARK SCHULZE is the forest director for Oregon State University's H. J. Andrews Experimental Forest, Blue River, OR. He is a forest ecologist, with an interest in issues related to sustainable forest management.

CATALINA SEGURA is an assistant professor in the Department of Forest Engineering, Resources and Management, Oregon State University, Corvallis, OR. Her interests are focused on the interactions between fluvial geomorphic, hydrologic, and ecological processes, in order to understand hydrologic processes and to formulate predictions of natural and anthropogenic disturbances that contribute to the sustainable management of forests and river systems.

KENNETH E. SKOG is a supervisory research forester, retired from the US Forest Service Forest Products Laboratory, Madison, WI. He is a forest economist who specialized in forest sector outlook studies and evaluation of the carbon-emission effects of wood products production, use, and forest regrowth.

JANE E. SMITH is a research botanist with the US Forest Service Pacific Northwest Research Station, Corvallis, OR. Her expertise is in fungal ecology, below-ground ecosystem dynamics, and soil recovery in response to wildfire, restoration thinning, and prescribed fire.

THOMAS A. SPIES is a research forester with the US Forest Service Pacific Northwest Research Station, Corvallis, OR. He is a forest landscape ecologist who specializes in the ecology, conservation, and restoration of old-growth forests and uses landscape models to explore interactions of social and ecological systems.

FREDERICK J. SWANSON is a research geologist (emeritus) with the US Forest Service Pacific Northwest Research Station, Corvallis, OR. He specializes in the study of physical disturbances in forest ecosystems and watersheds—floods, wildfire, landslides, volcanic eruptions, forestry operations, and roads.

JULIE A. VANO is a project scientist for the National Center for Atmospheric Research, Boulder, CO. She is a hydrologist with expertise in climate science and water resource applications.

DANA R. WARREN is an assistant professor in the Department of Forest Ecosystems and Society and the Department of Fisheries and Wildlife, Oregon State University, Corvallis, OR. He studies aquatic ecology with a focus on biota and ecosystem processes in headwater streams.

BYRON K. (KEN) WILLIAMS is the executive director and CEO of The Wildlife Society, Bethesda MD, an international organization of wildlife professionals. He is responsible for strategic leadership and planning, organization management, government affairs, publications and communications, professional development, outreach to the conservation community, and scientific engagement.

VIKRAM YADAMA is an associate professor in the Department of Civil and Environmental Engineering and Composite Materials and Engineering Center, Washington State University, Pullman, WA. He specializes in conversion of underutilized wood resources, such as small-diameter timber from hazardous fuel treatments, into wood composites for structural and nonstructural applications.

INDEX

Page numbers followed by "b", "f", and "t" indicate boxes, figures, and tables. "PL" indicates plates in insert.

Island Press | Board of Directors